Cold Rolling Precision Forming of Shaft Parts

Jianli Song · Zhiqi Liu · Yongtang Li

Cold Rolling Precision Forming of Shaft Parts

Theory and Technologies

国防工業出版社
National Defense Industry Press

Springer

Jianli Song
Beijing Information Science & Technology University
Beijing
China

Yongtang Li
Taiyuan University of Science and Technology
Taiyuan
China

Zhiqi Liu
Taiyuan University of Science and Technology
Taiyuan
China

ISBN 978-3-662-54046-6   ISBN 978-3-662-54048-0 (eBook)
DOI 10.1007/978-3-662-54048-0

Jointly published with National Defense Industry Press, Beijing, China, Original Chinese edition published by National Defense Industry Press, China, 2013

The print edition is not for sale in China Mainland. Customers from China Mainland please order the print book from: National Defense Industry Press.

Library of Congress Control Number: 2017930801

Printed on acid-free paper

This Springer imprint is published by Springer Nature
The registered company is Springer-Verlag GmbH Germany
The registered company address is: Heidelberger Platz 3, 14197 Berlin, Germany

Preface

With the rapid development of mechanical manufacturing industry, the demands for high-strength and high-precision shaft parts are increasing gradually in the fields of aeronautics and astronautics, automobile, wind power, petrochemical, national defense industry, and major equipment manufacturing. Such components as thread, hollow thread, leadscrew, spline, and screw rod are very important for power transmission and play a key role for the safety operation of equipments. Traditional machining method of shaft parts as spline and screw is mainly metal cutting, which has the characteristics such as lower efficiency, and higher material and energy wasting. Meanwhile, because the metal fibers are cut off in the machining, poor mechanical property and surface quality of the components will be resulted in, and the requirements of various industrial fields cannot be satisfied.

The principle of cold rolling precision forming technology is plastic forming. It is a less and non-cutting technology, in which the metal billets are continuously rolling at room temperature to produce shaft parts with complex features. Compared with traditional metal cutting processes, the production rate of this process can be improved for about 30 ~ 40 times, the forming microstructure and comprehensive mechanical property of the components can be greatly improved, the materials can be saved for 15% ~ 30%, and the production cost can also be decreased for about 30%. This technology has been more and more focused in the production of higher strength and precision complex components, which will have an extensive application prospect, and good social economic benefit can be obtained.

Cold rolling forming of shaft parts is a continuous local deforming process, the deforming mechanism is very complex, and relatively fewer researches on the process theory and mechanical analysis of spline and thread components have been carried out. In present production, experiences of technicians and workers are mainly depended on, which has limited the popularization and application of this kind of high efficient and material saving chipless technology in conditions lacking of systematical and advanced theoretical guiding.

The research team of the authors has established a cold rolling precision forming joint laboratory with the Cold Rolling Technology Research and Development Center of Qingdao Shengjian Machinery Factory, Shandong, China. The cold rolling forming process of shaft parts, theory, and equipments has been studied systematically, and parts of the contents of this book are based on the research results collaborated with cold rolling research and development center and the theory and engineering practices of the research team. The research has also been supported by the National Natural Science Foundation of China (No. 50675145, No. 51275331), the Major Special Science and Technology Projects of Shanxi Province (No. 20111101034), and the Research Foundation for the Doctoral Program of Higher Education of China (No. 20131415110001).

In this book, spline and thread components have been selected as the main research objects. The theory, process, and equipments of cold rolling precision forming of shaft parts have been introduced. This book is divided into 10 Chapters. In Chap. 1, the current situation and development trends of cold rolling precision forming technology have been introduced. In Chap. 2, the principle of the cold rolling process and the mechanism of the forming process of spline have been analyzed. Chapter 3 introduces the design of the cold rolling precision forming process parameters of spline, which includes the contact area of the spline shaft, rolling force and rolling moment, the calculation of the cold rolling workpiece diameter, and so on. In Chap. 4, finite element modeling and numerical simulation analysis of the spline cold rolling precision forming process have been carried out. And the metal flow and deforming rules of the cold rolling spline tooth during the forming process, the accuracy, defects, and the forming quality control of the cold rolling forming process have been described in Chap. 5. Chapter 6 analyzes the cold rolling precision forming process and principle of threads. In Chap. 7, the cold rolling forming process parameters of threads have been narrated, including calculation of the workpiece billet diameter, selection of the rolling force, rolling speed, feed rate, and the minimum wall thickness criteria in cold rolling of hollow threads. In Chap. 8, the rolling force, velocity field, and stress and strain field in the cold rolling forming of threads are analyzed with numerical simulation methods. The failure process of threads has also been analyzed. In Chap. 9, metal flow rules and parameter optimization of thread cold rolling precision forming are analyzed. In Chap. 10, design of the cold rolling forming machine and dies and measurement of the mechanical parameters of the cold rolling precision forming equipment at home and abroad have been described. Cold rolling forming equipment production enterprises, product properties, characteristics, and technique parameters have also been briefly introduced. Hope this book would provide some theoretical references for scholars and technicians who engaged in the researches and production of cold rolling precision forming of shaft parts.

The authors would be grateful for the assistance of the research team colleagues and graduate students, and this book could not have been written without their help, especially doctor Zhang Dawei in Xi'an Jiaotong University and vice professor Qi Huiping and director Zheng Quangang in Cold Rolling Technology Research and Development Center of Qingdao Shengjian Machinery Factory. Many thanks for their contribution and supports.

During the edition, many research achievements and essences have been adopted from the literatures in the Internet and the Web site of the cold rolling equipment manufacturers. The authors may not be listed distinctly, but the contributions should be specifically acknowledged. The authors would be very grateful for any comments from the readers regarding any suggestions about the contents presented or any errors that may escaped our attention during the preparation of this book.

Many thanks to director Yingheng Tang and the executive editor Shu Xiao of the National Defence Industry Press for their beneficial suggestions and kind help.

Beijing, People's Republic of China
Taiyuan, People's Republic of China

Jianli Song
Zhiqi Liu
Yongtang Li

Contents

Introduction

Cold rolling precision forming technology is one of the less or non-cutting advanced manufacturing technologies, which has the advantages of high productivity, good microstructure and properties of components, energy and material saving, etc., and has been widely applied in the production of high-strength and precision shaft parts in the fields of aeronautics and astronautics, wind power, petrochemical industry, automobile, heavy equipment manufacturing, and so on. This book introduces the theory, process, and equipment of cold rolling precision forming technology in detail, taking spline and thread shaft parts as research objects. The main contents include research status at home and abroad of cold rolling precision forming technology, design and calculation of the process parameters, numerical simulation of the cold rolling forming process, and the cold rolling forming equipments. This book should be most suitable to serve as a reference book for researchers and engineers who are engaged in the study and production of cold rolling precision forming technology, as well as a teaching reference book for undergraduates and postgraduates majored in materials forming and mechanical manufacturing.

Abstract

Cold rolling precision forming technology is a kind of less or non-cutting processing technology. Metal billet is locally rolled continuously in room temperature, complex shaft parts such as threads, spline, and worm gears can be manufactured by plastic deformation. Spline and thread components have been selected as the main research objects. The theory, process, and equipments of cold rolling precision forming of shaft parts have been introduced.

Chapter 1
General Introduction

Abstract Characteristics of cold rolling precision forming technology, present situation of shaft parts manufacturing technology, research progress on the cold rolling precision forming technology, and development trends on cold rolling precision forming technology of shaft parts are introduced.

1.1 Introduction of the Cold Rolling Precision Forming Technology

With the rapid development of aeronautics and astronautics, automobile, wind power, petrochemical, national defense, and major equipment manufacturing industries, the requirement for higher strength and higher precision shaft parts such as threads, hollow threads, lead screw, spline, and screw rod is increasing constantly. These power transfer components play an important role for the safe operation of equipments [1]. Traditional manufacturing method of spline and thread shaft parts is mainly metal cutting, which has the characteristics of low efficient, and waste of materials and energy. At the same time, because the metal fiber has been cut off during the machining, the mechanical properties and surface quality have been impacted, some high-strength and high-precision screw and spline even cannot be produced with this method, and the requirements of different fields cannot be satisfied.

Cold rolling precision forming technology is a kind of less or non-cutting processing technology. In this technology, the metal billet is locally rolled continuously in room temperature utilizing the principle of plastic forming. Shaft parts with complex features such as threads, spline, and worm gear and so on can be produced. Additionally, knurling, finish rolling, groove, and other special shape processing can also be carried out with this technology. According to the number of the roller, cold rolling precision forming can be classified as three types: single-roller cold rolling, double-roller cold rolling, and three-roller cold rolling, as shown in Fig. 1.1 [2]. Single-roller cold rolling forming process is commonly used to study the rolling forming of gear and spline with a rolling machine modified from universal machine

J. Song et al., *Cold Rolling Precision Forming of Shaft Parts*,
DOI 10.1007/978-3-662-54048-0_1

tools as lathe and milling machine, and this kind of forming experimental method is also called as single-direction rolling, in which the wheel rolls toward the workpiece from a single direction, under the radial force of the roller, flexural deformation has been induced on the workpiece, and the forming accuracy of the spline shaft is not perfect, as shown in Fig. 1.1a. In double-roller cold rolling forming process, a couple of rollers with the same parameters and forming dies are installed in the two transfer spindles, which are rotating at the same direction synchronously, and one spindle or double of them should be feed radially with a constant rolling velocity, till the end of the forming. While the two wheels rolled relatively at the same time, the force that results in the flexural deformation of the billet has been counteracted; therefore, the stability of the processing system can be enhanced, as shown in Fig. 1.1b. The three-wheel rolling machine has three rollers which are installed at an average angle of 120° around each workpiece. The purpose to use three rollers is mainly to balance the rolling force acted on the workpiece and to guarantee the accuracy of the rolled components. Three-wheel rolling forming method is suitable for rolling hollow shaft spline, hollow threads, and components with the same contour diameters along the axial direction, as shown in Fig. 1.1c.

The internal and surface quality of the components can be effectively increased adopting cold rolling precision forming technology. The radial stress induced during the forming process can not only increase the hardness and decrease the roughness of the component surface, but also increase the fatigue and torsion strength of the workpiece greatly. Compared with traditional metal cutting technology, in cold rolling forming technology, the production rate will increase 30–40 times, the materials can be saved for 15–30%, the cost can be decreased about 30%, and the microstructure and comprehensive mechanical property of the formed components can also be greatly improved [3, 4]. Cold rolling forming is a kind of ideal high-efficient, energy saving, and lower consumption technology, which has been increasingly focused in the production of high-strength and high-precision complex shaft parts, and will have a wide application prospect and good social and economical benefits.

Threaded components (lead screw, worm gear, screw rod, and so on) are key transmission and connecting parts in mechanical manufacturing, and the level of the manufacturing process and production quality has a direct influence on various of mechanical assemblies. Spline connection has a higher loading capability, higher

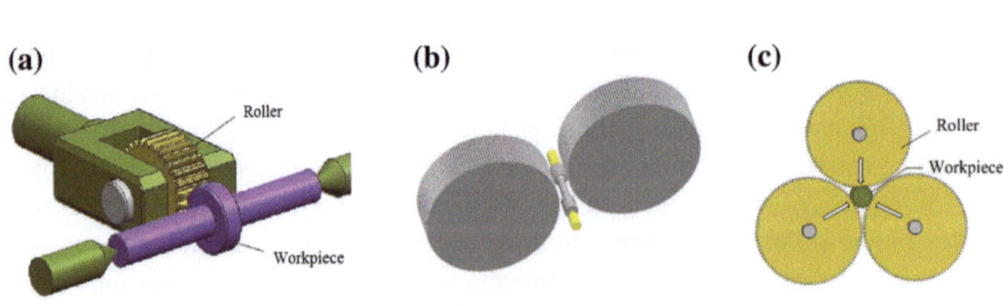

Fig. 1.1 Classification of cold rolling precision forming process. **a** Single roller; **b** double-roller; **c** three-roller

alignment accuracy, better guiding characteristics, shallow tooth root, lower stress concentration, and less weakening between the shaft and the wheel hub. As the key components transforming force and torsion moments, spline connection has been widely utilized in the field of automobile, various vehicles, and equipment manufacturing. In China, the annual output of spline shaft has exceeded 100 million, the production batch is very large, and the requirement for accuracy is higher. In volume production of threads and spline shaft parts, less or non-cutting cold rolling precision forming technology is an inevitable development trend.

In this book, spline and thread shafts are taken as research objects, and the theory and technology of cold rolling precision forming technology are discussed systematically.

1.2 Present Situation of Shaft Part Manufacturing Technology

In present, the commonly used manufacturing methods of spline and threads can be classified into two types: The first type is metal cutting, such as slotting and hobbing, and the other one is less and non-cutting processing technology.

The manufacturing methods of spline shafts include three types: metal cutting, electrical discharge method, and plastic processing, as shown in Fig. 1.2. In metal cutting methods, the shape of the components is obtained by removing of metals, which include multiple types of methods, such as milling, slotting, broaching, planning, and grinding [5]. In plastic processing, the raw materials are processed to the required shape by plastic forming method, and there are no chips produced during the production process. Because the tooth height of the spline is not higher and the modulus is smaller, it can usually be formed in room temperature, and the processes mainly include cold extrusion, thread rolling, cold striking rolling, and cold rolling. According to different application conditions and requirements of accuracy, the above-mentioned methods have all been used in different situations [6].

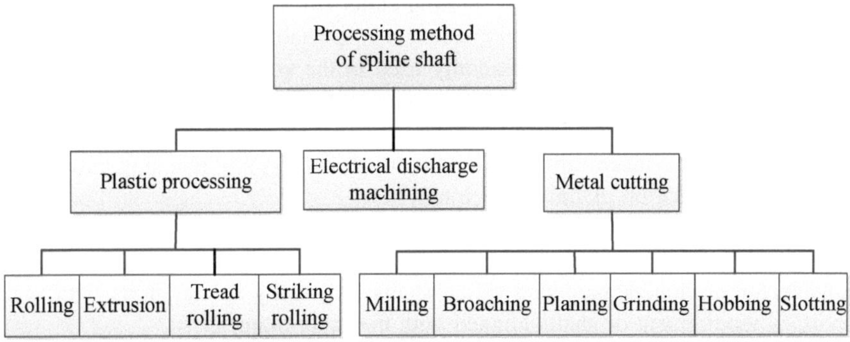

Fig. 1.2 Manufacturing methods of spline shafts

1.2.1 *Cutting of the Spline Shafts*

Traditional manufacturing method of spline shafts is material-removing metal cutting method, which can be classified into two types according to different moving methods of the cutting tools on the workpiece: forming cutting and generating cutting [7].

1.2.1.1 Cutting with Formed Cutter

The tooth shape of the spline can be formed on a machine by directly cutting the metals into the gear tooth space with a formed cutter of the same tooth profile. The commonly used forming cutting method of the spline shafts is milling, which is usually used in the machining of rectangular external spline.

The milling process is carried out on a milling machine using a forming milling cutter with an axial section the same as the shape of the cutting spline tooth space. While a tooth space has been machined, the gear billet is rotated for 360°/z by the indexing mechanism; then, the next tooth will be cut; and then, every tooth space in order. Several commonly used milling patterns are shown in Fig. 1.2. If a forming milling cutter is not used in the milling process, two disk milling cutter can be used to mill both of the tooth flanks at the same time, while every tooth has been milled in turn, and the dedendum diameter can be dressed with one of the disk cutter. The production rate and the accuracy of milling method are relatively lower; therefore, it is mainly used in the small batch production of spline shaft aligning with the external diameter and in the rough machining before hardening.

In the forming milling of involute spline shaft with different spline parameters (modulus, tooth number, and pressure angles), the tooth profile is different, and different forming milling cutters should be adopted to ensure the precision of the tooth profile. Therefore, specialized forming cutter should be designed according to different spline parameters, which will dramatically increase the cost of the practice production. In production, an approximate method has often been used to mill the tooth of the spline, which is to mill the splines with the same modulus and different tooth numbers by the same type of milling cutter [8] (Fig. 1.3).

In order to improve the machining efficiency, specialized spline milling machine and forming milling cutter are generally used in the volume machining of spline shafts. The operation can also be realized on general horizontal milling machine in the enterprises without specialized milling machines. In addition, spline shafts can also be machined with forming planning method and forming planning cutter with specialized automation planning machine [7].

In grinding method, the tooth flank and dedendum diameter of the spline can be ground with a forming grinding wheel with spline shaft grinding machine, which is suitable for machining hardened spline shafts, especially spline shafts with high accuracy requirement or shafts aligned with the inner diameter.

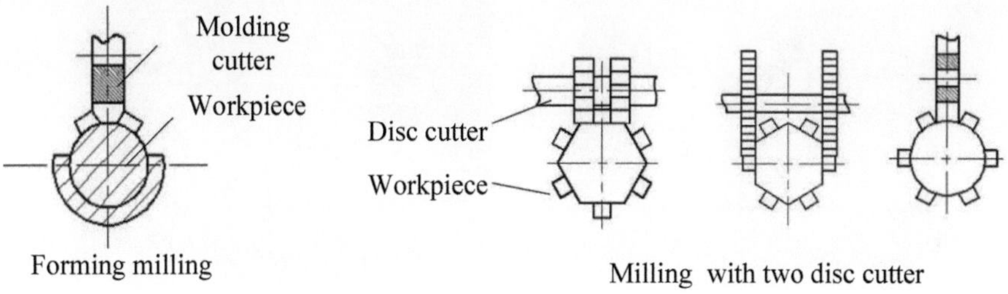

Fig. 1.3 Milling of the spline shafts [8]

Internal spline can be slotted tooth by tooth with forming slotting cutter on slotting machines, the production rate and accuracy is lower, and this method is mainly used in single or small batch production. Broaching of spline with broaching machining has the characteristics of higher production rate and accuracy, and has been widely used. However, because the broaching cutter must be manufactured customized, the cost is higher and it is only suitable for the large volume production of spline shafts. Grinding of the internal spline with a small diameter forming grinding wheel can be used to machining hardened internal spline with large diameters or with requirement of higher accuracy.

1.2.1.2 Generating Cutting Method

The method for machining involute spline according to the principle of non-backlash meshing is called as generating cutting method, in which the tooth profile of the two wheels acting as the envelope curve of each other [9]. In generating cutting of spline shaft, the cutter and the workpiece keep moving with a constant angular velocity ratio. The cutting motion of the cutter is along the axial direction of the billet, and the envelope curve of the cutting edge at every position of the billet is actually the tooth profile of the involute spline [10].

The same with the gear hobbing method, external spline can be machined with generating cutting method on hobbing machines. Figure 1.4 shows the machining principle of spline shafts on a vertical gear hobbing machine. With this method, continuous machining of spline shafts can be realized with higher production rate and accuracy. While cutting involute spline shaft on a gear hobbing machine, the same hobbing cutter can be used to manufacturing spline shafts with the same modulus and pressure angles.

In the process of spline shaft production, the methods of material removing have been widely applied due to their strong adaptability and technical maturity. In China, more than 80% of spline shafts are still produced with various cutting methods. However, traditional cutting method has the defects such as lower production rate and higher time and labor consumption, large quantity of materials have to be removed

Fig. 1.4 Hobbing of the spline shaft on a hobbing machine

during the machining process, and the utilization ratio of materials is only about 60%. What is more, the metal fiber of the workpiece has been cut off during the machining, which results in a poor mechanical property and surface quality of the components. In conditions, a higher surface precision is required; tooth grinding procedure is often being arranged to improve the surface quality [11].

In recent years, aeronautics and astronautics, and automobile and mechanical industry developed rapidly in China and all over the world, the production and demands for involute spline shafts increased gradually, and higher precision and mechanical properties are required. Therefore, it is of great urgent to develop new types of spline shaft production processes with higher efficiency, better quality, higher precision, lower energy, and materials consumption.

1.2.2 Plastic Forming of Spline Shafts

Nowadays, with the intensifying competition of the world's manufacturing industry, massive (bulk) metal cold plastic precision forming technology has already become one of the most important research and development directions in the field of mechanical manufacturing. Bulk metal cold forming refers to the plastic forming process, in which the deformation occurs under the crystallization temperature, companying with work hardening but without recovery and recrystallization (vice versa, it can be called as metal bulk hot plastic forming). Using cold plastic precision bulk forming process, the tooth shape of the spline can be directly formed without the need of other cutting operations, and the formed components can satisfy the requirement of applications. Bulk metal cold forming technology has the characteristics of higher precision, higher efficiency, green production, and energy saving, and it is suitable for high-level fields especially aerospace, national defense, high-speed railway, and wind power, in which a large number of high-strength and high-precision spline shafts must be manufactured with cold plastic precision forming methods to improve the strength of the components. The integration and control of "forming/property" has become one of the hot research points in the field of metal forming manufacturing.

In the bulk cold precision forming process of involute spline shafts, cold extrusion, cold rolling forming, cold striking rolling, and thread rolling processes and equipments have already been developed [12–17]. With the progress of the equipment, process, lubrication, and other matching techniques, various precision forming technologies have been improved gradually. The bulk cold precision forming process manufacturing complex tooth profile components such as spline shafts has the following typical characteristics:

(1) Forming process is carried out at room temperature, it is energy saving, and near-net shape high-precision component can be obtained.
(2) Continuous local plastic forming, small resistant deformation force exists during the process, and equipments of small capability can be used to form large components.
(3) For higher material utilization ratio of near-net shape technology, the metal fiber of the formed components is complete, the surface quality is better, and the requirements for higher strength and higher precision equipment manufacturing can be satisfied.
(4) It is liable to realize high-speed automation production with a higher production rate.

Because the tooth of the spline is relatively short, the spline shaft can be formed by plastic deformation from a billet at room temperature utilizing the characteristics of the plasticity of metals. The plastic forming method of spline is chipless, without the need of removing metal from the raw materials, and the commonly used methods include cold extrusion, rolling (striking), thread rolling, and cold rolling forming.

1.2.2.1 Cold Extrusion Forming of the Spline Shafts

In the cold extrusion process, a metal billet is put into the cavity of the die installed on the press, the metal is forced to extrude from the die cavity with a certain speed under the pressure, and then, the spline shaft with required shape and size can be obtained. The process can be classified into axial extrusion and radial extrusion methods. Forward cold extrusion method is the main method processing involute spline [18]. Figure 1.5 shows the radial extrusion forming principle of spline and shaft parts formed with radial and axial extrusion forming methods by Kinefac Corporation of USA [19]. During the cold extrusion forming process, the deforming resistance force is rather large, the billet should be softening annealing treated, and the hardness should be lower than HB160 to improve the formability.

Cold extrusion forming of spline is a process without material waste, with little tolerance and higher repeatable accuracy in batch production, and near-net shape components can be obtained. The tooth shape precision of cold extrusion forming spline is determined by the dies, which is liable to wear. Compared with cutting method, in the cold extrusion forming of spline, the axial extrusion production rate will improve 6 times, and that of the radial extrusion will improve more than 2 times, the surface hardness of the components can improve 30–50%, and the bending and

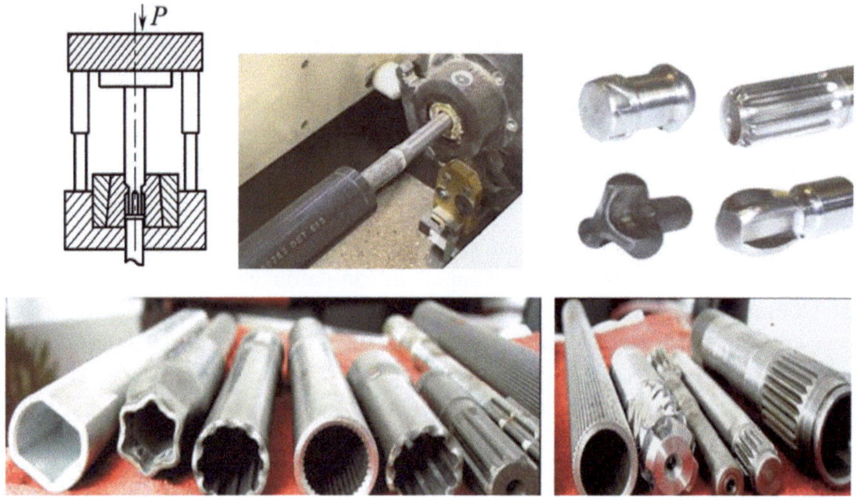

Fig. 1.5 Cold extrusion forming process and the formed shaft parts of Kinefac Corporation USA [19]

torsional strength can be improved simultaneously. The dimension accuracy of the cold extrusion forming spline shaft can reach IT6–IT7, and the surface roughness can reach Ra0.4–Ra0.8. Because the length-to-diameter ratio of long spline shafts is large, bar instability is liable to occur during the forming process, which will result in the bending defection of the forming components. Therefore, cold extrusion forming is not suitable to process long shaft-type spline. The defection of cold extrusion forming spline shafts is that in the extrusion process part of the metal flow along the axial direction, and local upsetting at the end is liable to form. Measures can be taken to reduce the metal flow along the axial direction during the period of punch design to avoid local upsetting defects [20]. Due to the limit of dies, cold extrusion forming technology is suitable for manufacturing in large volume production of small modulus spline [21].

1.2.2.2 Cold Rolling Forming of Spline Shafts

In the cold rolling forming process of spline, the roller with the same tooth shape as the spline shaft is installed on the rolling head of the machine; that is, the section shape of the roller is the same as the cross section of the spline, the roller then rolled freely on the surface of the workpiece, and all of the teeth can be formed in a working stroke of the cold rolling process, as shown in Fig. 1.6. In the cold rolling process, there is no driving needed for the roller. The workpiece is pushed by the press, and the teeth of the spline are formed through the extruding of the rolling head. The shearing stress existing on the extruded surface of the workpiece is very small, and only frictional force of the roller should be overcome, which is very favorable for the cold plastic forming [8].

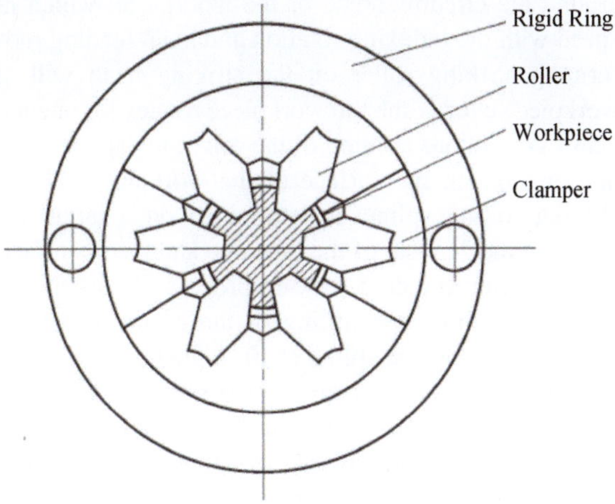

Fig. 1.6 Cold rolling forming principle of spline shafts

The cold roll forming process can be carried out on general press, the equipment is simple. However, the structure of the rolling head is relatively complex and is difficult for adjusting. The forming precision mainly depends on the install precision of the roller. Under too large axial force in the extrusion, upsetting and bending of the workpiece is easy to form and the rigidity is insufficient. Therefore, cold rolling is not suitable for processing slender spline shafts. Limited by the structure of the roller head, the process is also not suitable for processing spline with too many teeth, and the teeth number is commonly less than 20.

1.2.2.3 Striking Rolling of the Spline Shafts

In the 1950s of the twentieth century, Ernst Grob Inc. in Switzerland invented the cold striking rolling process of spline shafts. The striking rolling forming of spline shaft should be carried out on specialized machine, and the principle is shown in Fig. 1.7 [22]. Two striking rollers are located symmetrically on the lateral side

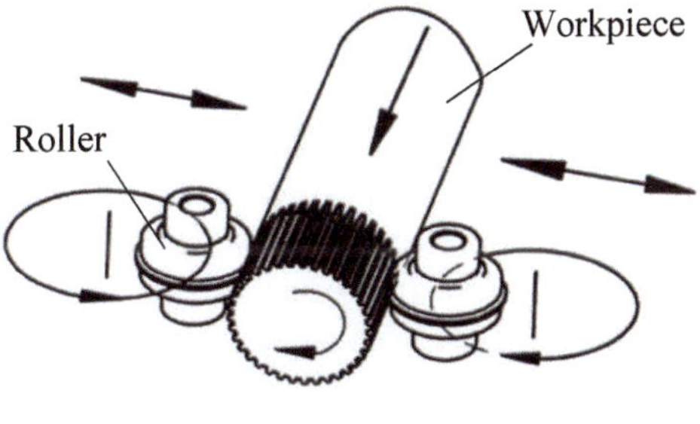

Fig. 1.7 Striking rolling principle of spline shafts [23]

around the circumference of the workpiece, which are rotating at a high constant speed with the indexing rotation and axial feeding movement of the workpiece. The forming striking roller on the striking head will strike the tooth space of the workpiece every time the workpiece rotates for one tooth. Under the high-speed and heavy continuous striking of the roller, the spline will be formed through the plastic deformation on the surface of the workpiece. The forming precision of the cold striking rolling spline is determined on a large degree by the precision of the indexing mechanism of the machine and the manufacturing precision of the striking roller. During the cold striking process, the tooth profile is formed through extrusion and high-speed striking of the roller, and because only local plastic deformation occur on the spline shaft, higher precision and good surface quality of the workpiece can be obtained and the precision of the spline can approach that of the ground spline. Due to plastic deformation and cold work hardening effect, the strength of the tooth surface and the wear resistance of the cold striking rolling spline shaft can also be increased.

During 1960s and 1970s of the twentieth century, the First Automobile Works of China introduced cold striking rolling machine to China from abroad, and some research work had been carried out. Researchers in several institutes such as Shanghai Mechanical Manufacturing and Technique Institute have also studied this technology. In recent year, Cui Fengkui in Henan University of Science and Technology has studied the cold striking rolling or spline systematically, and several research papers have been published [24–26]. According to the meshing principle, he designed the theoretical outline of the roller based on the theory of equal pitch helicoidal surface and modified the theoretical outline of the roller adopting experimental calculation method. He also studied the selection and heat treatment of the roller materials and manufactured the roller according to the modified outline. And cold rolling forming process of involute spline has also been studied by them.

The technological difficult of cold striking rolling spline shafts is the design and manufacture of the striking roller, in which specialized roller and machine are needed for different spline parameters, and the roller is easy to wear. Cold striking rolling technology is suitable for forming larger modulus spline shafts and long shaft spline. Because the roller may interfere with the steps in the process of spline groove forming, this method is not suitable for manufacturing of spline shafts with steps. The surface roughness of the cold striking rolling formed spline shafts can reach Ra0.4–Ra0.8.

1.2.2.4 Cold Rolling Forming of Spline Shafts with Rack Dies

In the 1970s, an inventor in USA invented the method for cold rolling of spline with rack dies, and the principle is shown in Fig. 1.8 [27, 28]. Two parallel cold rolling rack dies distribute at the upper and lower of the workpiece and move relatively and synchronously, and the workpiece is driven to rotate between the rack dies and deformed plastically. With the steady elevating of the cold rolling rack plate, the spline component can be finally formed.

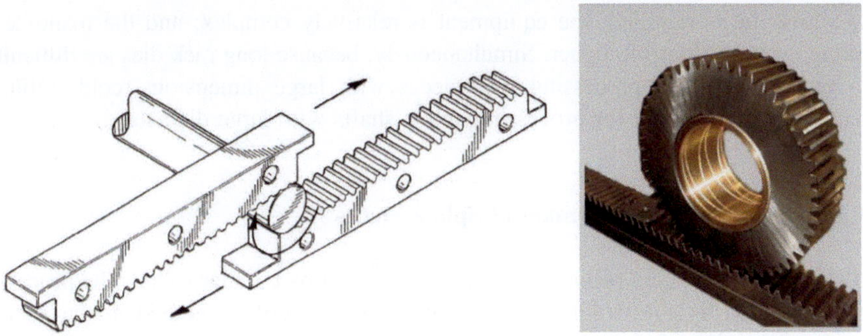

Fig. 1.8 Principle of spline shaft cold rolling with rack dies

During the cold rolling of tooth shaped components, the design and manufacture of dies is the most important. In this field, USA has published many patents of cold rolling rack dies and takes a leading position [29–32]. From the middle of 1980s, China has imported rack die cold rolling machines and dies successively from Japan, Germany, USA, and other countries, and many studies on the precision cold rolling technology began to be carried out by many factories and research institutes [33].

Joseph P. Domblesky of Marquette University in USA and Feng Feng simulated the cold rolling process of external threads with finite element analysis method, and the metal flow characteristics in the deformation process had been obtained [34]. Han Fenglin, Zhang Qing, and Fu Zhuang et al. at Yanshan University of China analyzed the whole working property of the cold rolling machine of spline shafts [35]. In their research work, the frame, sliding and dies, and other related components were taken as the analysis objects. Contact analysis models had been set up to analyze the three-dimensional contact of the cold rolling machine. The positional conditions of the dies after deformation were quantitatively described, and the structural parameters of the machine and the design method were modified according to the simulation results in order to meet the precision requirement of spline [36]. Qiu Ping and Zhang Qing derived the coordinate equation of the cold rolling die tooth profile of involute spline, which provided a basis for the realization of computer-aided design of involute spline dies [37].

Zhang Ziling, Zhang Lianhong, and Wang Haitao et al. at Tianjin University studied the design criteria of involute spline cold rolling dies [38]. Liu Ruiqiu and Peng Xiaonan at Henan University of Science and Technology simulated the cold rolling forming of involute spline with numerical methods and analyzed the elastic recovery of spline cold rolling [39]. The springback value was quantitatively calculated, and it could be taken as a reference in the design and modification of cold rolling cutters.

The surface roughness of cold rolling spline can reach Ra0.4. Rack die cold rolling forming can be used to process small modulus spline nearby the shaft shoulder. In rack die cold rolling of spline, high synchronization precision of the two rack die is required, the equipment is relatively complex, and the manufacturing cost is a litter bit higher. Simultaneously, because long rack dies are difficult to machining while processing workpieces with large dimensions, cold rolling method is not suitable for processing spline shafts with large diameters.

1.2.2.5 Cold Rolling Forming of Spline Shafts

If the rack dies in cold rolling of spline are changed by circular rollers, the process can be called as cold rolling forming, in which the spline teeth with the same sectional profile of the roller can be formed through the meshing and rolling movement between the workpiece and the roller [40]. The principle is shown in Fig. 1.9. A pair of rollers with the same parameters and tooth profile is installed on the two spindles of the cold rolling machines, which is rotating at the same direction synchronously, and one spindle or both of them are feeding in the radial direction at a uniform speed or with a constant rolling force. The workpiece is driven to rotate and deformed plastically under the frictional moment generated at the moment the roller contacting with the workpiece. The rollers are rotating at a constant speed and continuously feeding in the radial direction to exert pressure on the billet, until the involute spline with the same tooth profile with the roller has been finally formed.

In the cold rolling process, two circular rollers are feeding continuously on the billet to form the workpiece. The design, manufacture, and maintenance of the roller are easy, and dies with the same modulus are of long life and can be generally used to form thread and spline shafts on an axle journal with compact structures, while the integrity of the adjacent journal will not be affected, which is difficult to realize in common cutting machining methods. What's more, the manufacture and

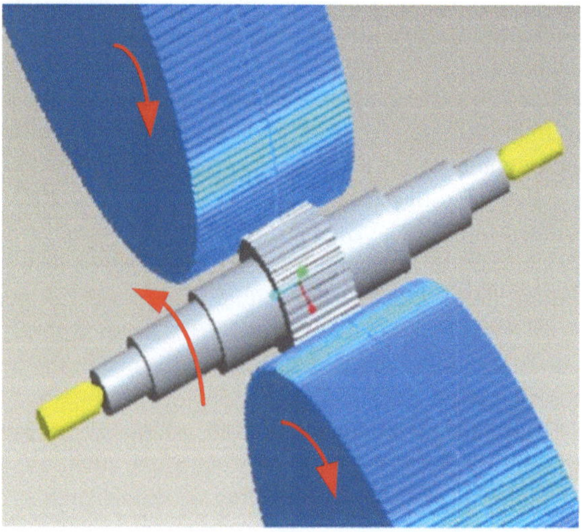

Fig. 1.9 Cold rolling forming principle of the spline shafts

control of the cold rolling equipment are easy to realize. Therefore, this technology is liable for popularization and application [41].

Double-circular roller radial feeding cold rolling precision forming is one of the most representative high-efficient precision forming technologies of spline components. Compared with other chipless cutting methods of spline shafts, it has the characteristics and advantages such as higher processing efficiency, higher precision, simple machine structure, convenient to control, and better generality of the dies. On the other hand, the microstructure of the cold rolling formed components is fine and compact, the property is excellent, and the components can even be used without subsequent heat treatment, while components machining by gear hobbing must be heat treated to improve the service time.

The comparison and analysis of the application scope, characteristics, equipment, and precision of various kinds of spline shaft manufacturing methods are listed in Table 1.1.

With the increasing energy crisis and market competition, high-efficient, high-quality, energy and material, and saving spline processing methods are expected to be developed. Traditional metal cutting methods of spline components have the characteristics such as lower efficiency, higher materials and energy waste. Meanwhile, the metal fibers are cut off in the machining, which will result in a poor mechanical property and surface quality of the components, and the requirements in various industrial fields, for example, aerospace, automobile, wind power, petrochemical, national defense, and major equipment manufacturing, cannot be satisfied.

Compared with traditional metal cutting technology, non-cutting or less cutting spline plastic forming technology will increase the production rate for about 30–40 times, the materials can be saved for 15–30%, and the cost can be decreased for about 30%. Also, the surface hardness and the comprehensive mechanical property of the formed components can be greatly improved. Therefore, cold rolling forming process is an ideal technology for the batch production of spline shafts and will have a great prosperity in the future [42, 43].

1.2.3 Manufacturing Methods of Thread Components

The manufacturing methods of thread components can be classified into five types: cutting, forging, casting, powder metallurgy, and electrodischarge machining, as shown in Fig. 1.10 [44].

In casting method, sand patterns have to be made according to the model of the threads, metal liquid such as steel, iron, aluminum, or copper are cast into the pattern, and threads can be obtained after cooling.

There is no chip in the casting of threads, materials can be saved, and the cost is lower, but the surface of the component is not smooth, and shaping or finishing is needed for most of the thread contours. This technology is suitable for manufacturing nonferrous metals, alloy or plastic thread components with a diameter larger

Table 1.1 Comparison and analysis of various spline manufacturing methods

Manufacturing methods		Equipments	Application scope	Precision	Characteristics
Cutting	Milling	Horizontal milling machine or specialized spline milling machine	Single and small batch production	IT6–IT7	Lower precision of the tooth profile
			Spline of various modulus		
	Planing	Specialized automation machine	Large volume production	IT7–IT8	Lower precision and surfaces quality, suitable for rough machining
	Slotting	Slotting machine	Large volume production	IT6–IT7	Intermittent machining
	Hobbing	Hobbing machine	Large volume production	IT5–IT7	Continuous machining, higher production rate
Chipless processing method	Cold extrusion	Simple machine, the die is liable to wear	Medium and small batch production	IT5–IT7	Higher production rate
			Suitable for small modulus spline, and not suitable for long spline shafts		Strengthening effect of the components
	Cold rolling	Simple machine	Spline of small modulus	IT4–IT6	Higher production rate
		The rolling head structure is complex and difficult to control	Not suitable for long shaft spline, and the tooth number of the spline should be limited in a certain scope		Strengthening effect of the components
	Cold striking rolling	Specialized machine with complex structure	Suitable for spline of large and medium modulus	IT4–IT6	Production rate is higher than the cutting machine but lower than the rack die cold rolling and cold rolling forming
		The roller should be designed specially and is liable to wear	Interfere will occur in the processing of stepping shafts		Strengthening effect of the components
	Rack die cold rolling	Higher cost	Spline nearby the shaft shoulder can be formed, suitable for spline of small modulus,	IT4–IT5	Higher production rate
		Specialized machine with higher cost rack dies			Strengthening effect of the components

(continued)

Table 1.1 (continued)

Manufacturing methods		Equipments	Application scope	Precision	Characteristics
			and not suitable for large diameter spline shafts		
	Cold rolling forming	Design and manufacture of the roller is relatively easy	Suitable for spline of small modulus	IT4 ~ IT6	Higher production rate
			Spline nearby the shaft shoulder can be formed,without limitation of the shaft diameters		Strengthening effect of the components
		Specialized machine is required or modify of the generalized machine			

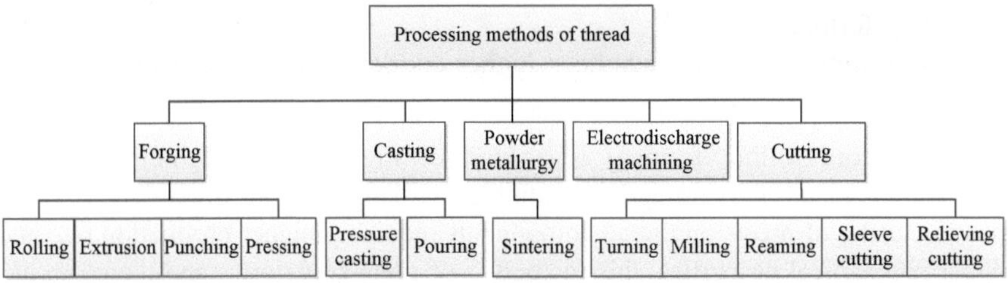

Fig. 1.10 Classification of the manufacturing methods of thread components

than 56 mm, and a pitch above 0.5 mm. Powder metallurgy sintering method is suitable for manufacturing threads with a diameter larger than 6 mm and a pitch above 1.5 mm. The two methods mentioned above are not commonly used.

Forging method can be classified as pressing, punching, extrusion, and rolling. Pressing is used to press plastic and rubber threads with special thread molds under the hot pressing of the mechanical press. Cold punching is suitable for processing of nonferrous metal and alloy thread components with a diameter of more than 5 mm and a pitch of more than 1 mm. And extrusion method is used to extrude internal threads with screw taps on lathe, drilling, and specialized machines.

In present, the most commonly used thread manufacturing methods are cutting and cold rolling. The cutting method has the characteristics that the process is mature and convenient for operation, but the metal fibers are cut off, which decreased the quality of the components, and the demands cannot be satisfied in conditions the number of connecting components is large and higher component quality is required. Cold rolling forming of threads is a kind of chipless forming method, in which the plasticity of certain materials is used to form the components in room temperature. Corresponding threads of the workpiece is formed under the plastic deformation applied by the roller. The principle of this technology is the same with the method of cold rolling forming of spline shafts. Cold rolling forming of threads is based on the theory of metal plastic deformation; therefore, the metal fibers and microstructure of the components are continuous. Compared with metal cutting methods, it has the following advantages:

(1) The surface quality of the thread is much better, and the surface roughness is lower.
(2) The hardness and strength on the surface of the workpiece can be improved, especially the hardness of the dedendum surface.
(3) The production efficient can be improved for several to dozens of times, and automation is liable to be realized.
(4) Cold rolling forming of the threads is a kind of chipless materials saving method.
(5) Slender thread rods not suitable for metal cutting method can be formed by cold rolling forming.
(6) Cold rolling formed threads have higher corrosion resistance than that of the cutting ones.

Thread cold rolling forming is a kind of high-quality, high-efficient, and low-cost advanced non-cutting/less cutting processing method. In particular, in the manufacturing of precision thread components requiring higher production rate and higher mechanical properties, this method is of great superiority and therefore has been widely used [45].

1.3 Research Progress on the Cold Rolling Precision Forming Technology of Shaft Parts

1.3.1 The State of the Art on the Manufacturing of Spline Shafts

Since the 1970s, the cold rolling precision forming technology of spline components has been paid highly attention by the developed countries as Japan, America, Germany, and the former Soviet Union. In particular, in the industry of automation, cold rolling forming is of higher production rate and precision in manufacturing large volume of spline shafts, and better benefit can be obtained [46]. Researchers all over the world have developed variety of specialized spline shaft cold rolling machines, which promoted the continuous improvement in the technology.

Researches in the aspect of design of dies (roller), billet preparation, and quality control have obtained satisfied results. For the development on plastic forming theory, numerical simulation also played an important role for promoting the researches of the spline cold rolling forming technology [47, 48].

In recent years, the automobile industry is developing rapidly. In the enterprises of automobile manufacturing, large numbers of spline shafts are manufactured with cold rolling forming methods. EXCELLO and GROB in Germany, MARBAIX and LAPOINTE in England, ANDERSON-COOK in USA, and NACHI Corporation in Japan have successfully developed cold rolling forming machine products in successively. The double-roller radial feeding cold rolling machine reported by ANDERSON-COOK Corporation can be used to manufacture involute spline shafts with a precision of ISO6–ISO7. At a pressure angle of 30°, the maximum modulus can reach 1.5 mm; and with a pressure angle of 20°, the maximum modulus will reach 1 mm. The specialized spline cold rolling forming machine equipped with automatic loading and unloading material manipulator can obtain a production rate of 12–15 workpieces/min, as shown in Fig. 1.11 [49].

In aspect of the research and development of spline cold rolling equipments, Profiroll Ltd. of Germany [50] and Kinefac Corporation of USA are located at the leading positions of the world. In the three-roller servohydraulic driving technology of the high-precision spline shaft cold rolling forming machine invented by Kinefac Corporation, the workpiece is fed along the axial direction, the central distance of the three rollers is constant, and the rollers are driven by dynamic hydraulic servo, as shown in Fig. 1.12 [19]. The precision of the formed spline shaft can reach ISO5; at a pressure angle of 30°, the maximum modulus is 3 mm; and at a pressure angle of 20°, the maximum modulus is 2.5 mm.

The cold rolling precision forming technology has already been successfully utilized in many developed countries. In recent years, China has imported many

Fig. 1.11 Spline shaft cold rolling forming machines equipped with loading and unloading material manipulator

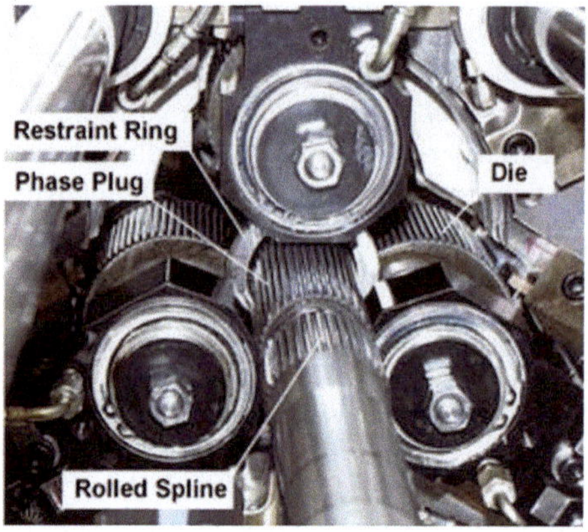

Fig. 1.12 Three-roller dynamic servohydraulic driving cold rolling forming

spline cold rolling forming machines, which are mainly used in the production of the spline in the field of automobile industry. Through the importing of machine and dies or modification of the common lathe or thread rolling machine, qualified involute spline can already been manufactured with domestic machines [51, 52]. Figure 1.13 shows the cold rolling machine modified on the basis of thread rolling machine made in China.

In the fields of specialized spline machine research and manufacturing, Shandong Qingdao Shengjian Machinery Factory listed at the leading position of China, which has a history of more than 40 years in design and manufacturing of thread rolling machines.

During the forming process of spline, the roller and the workpiece disengage from each other transiently, due to the vibration and impact of the system; the requirement for the rigid of the machine and the control system is higher; in

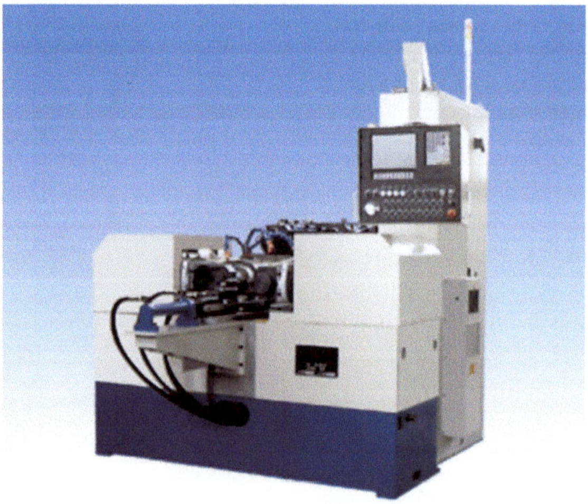

Fig. 1.13 Thread rolling machine made in China

particular, the structure of the main transmission system should be designed specialized. Researches on the force–energy relationship, the forming technique conditions and limited forming capability of the spline cold rolling machine are relatively limited, and the theory and process of the spline cold rolling process should be studied systematically [53].

1.3.2 Theoretical Researches on the Cold Rolling Forming of Spline Shafts

The principle of double-roller radial feeding cold rolling forming of spline shafts are the same as that of the cross rolling. A. Д. Томленов and others of the former Soviet Union used slip line method to solve the unit average pressure on contact surface of simple cross rolling. But the friction factor has not been considered in the process of solving, and it is difficult to be applied in the cold rolling process of spline shafts. The actually used rolling force formulas are still not perfect [40].

German scholar R. Neugebauer studied the methods to decrease the accumulated error of the pitch and improve the manufacturing precision in cold rolling forming of gears with rack dies and two circular rollers, respectively [54–57]. R. Matsunaga and R. Kamashita in Kyushu University and Mitsubishi Heavy Industries LTD of Japan analyzed the rolling process of the spline sleeves used in the automatic clutches of automatic transmission and experimentally verified that the manufacturing precision of the sleeve could meet the demands of applications [58]. Amir A. Kamouneh in the University of Michigan simulated the strengthening process of gears with numerical methods and finite element analysis software DEFORM, and the hardness distribution of the gear tooth after cold rolling can be predicted [59].

Li Yongtang, Zhang Dawei et al. in Taiyuan University of Science and Technology set up the calculation formula of the billet diameter before cold rolling forming of the spline, developed the calculation software system of the billet diameter, set up the model of the contact area of the external spline in the cold rolling precision forming, and analyzed the influence of geometrical parameters on the rolling force [60–64].

Researchers in Qingdao Shengjian Machinery Factory studied the precision and process of threads and splines with experimental methods [41]. Recently, finite element method has been widely used in the analysis of plastic forming [65–67]. On the simulation analysis of spline cold rolling, Zheng Kun, Cai Anjiang and Jia Shuang took small modulus spline as the study object, and FEM software Ansys and Pro/E was adopted to set up the three-dimensional model of the spline shaft, and the modal and harmonic responses were analyzed [68]. They proposed that the vibration of the spline shafts during the forming can be reduced by adopting bending-torsion resistance measures and processes, and improved the dynamic characteristics. Wang Mingfu and Li Yongtang set up the three-dimensional model of the whole spline cold rolling forming process with DEFORM-3D based on the theory of rigid-plastic finite element method, the cold rolling forming process of spline was analyzed with finite element method and the equivalent stress, and the distribution of the equivalent strain and the metal flow rule had been obtained [69]. The researches on the theory of cold rolling precision forming are most qualitative and quantitative analysis; process optimization is very limited; the depth and width of the research should be further improved.

1.3.3 Surface Modification of the Cold Rolling Forming Spline Shafts

When pressure is applied locally on the surface of the metal components, plastic forming and work hardening of the surface layer will happen. The microstructure of the surface layers can be improved. Under the action of the compressive residual stress on the surface layer materials, the abrasive resistance and fatigue resistant strength can be greatly improved [70–72].

(1) Surface cold rolling technology can be used to strengthen the metal components through deformation hardening of the materials, which is especially effective for improving the fatigue resistance and prolonging the lifetime of the components.
(2) The strengthening mechanism is mainly reflected in the change of microstructures and the value and distribution of the compressive residual stress.
(3) The effects of the surface rolling process are mainly reflected in three aspects such as the surface roughness, hardness, and compressive residual stress. The most important rolling process parameter is the optimized rolling force.
(4) Cold rolling surface strengthening is one of the most important developing directions of the surface rolling technology.

In the aspect of surface cold rolling strengthening, Gong Guanjun analyzed the cold rolling strengthening process and proposed that the chipless cold rolling process is of high production efficiency and not only the surface roughness of the components can be decreased, but also the surface strengthening and fatigue resistance improvement can also be realized [73]. Li Fenglei, Xia Wei, and Zhou Zhaoyao proposed the theoretical relationship between the microplastic deformation depth on the surface of the workpiece and the rolling force based on the analysis of the rolling process [74]. Johnson–Cook flow stress model was adopted to verify the cold rolling of two typical steels AISI 1045 and IN718 high-temperature alloy; the correction of the assumed conditions, and the analytical formula have been verified, which provided a theoretical basis for the further study of the cold rolling forming mechanism, especially the forming of residual stress and work hardening.

The strengthening and modification of the spline tooth can be realized by the cold rolling precision forming technology. The microstructure on the tooth surface layer of the spline shaft can be refined, and the compressive residual stress can be generated, and thus increases the surface hardness, the loading capacity, the abrasive resistance, the fatigue resistance, and the corrosion resistance and obtains better comprehensive mechanical properties. The deformation mechanism of the cold rolling precision forming of spline shaft is relatively complex. Therefore, it is necessary to study the cold rolling precision manufacturing of the spline shaft. Through the shape forming and property control of the spline shafts, not only the life span of the spline shafts can be extended, but also the aim of materials saving and lightweight of the components can be realized.

1.3.4 Development Status on Cold Rolling Forming of Threads

The application of cold rolling forming for threads manufacturing has already been lasted for more than 100 years. Scholars all over the world studied and explored the process and principles of cold rolling forming successively, and some achievements have been obtained. However, because the forming process of the threads is complex, there are many uncertain factors existing in the rolling process, especially there are still no parameters or formula that can be utilized directly for the determination of the thread rolling parameters (rolling force, rolling speed, billet diameter and the minimum wall thickness of hollow threads, etc.), experience and trial-and-error methods must be depended in the process, which result in great wastes of materials. The selection of equipments is difficult, and the further development and application of the technology have also been limited.

The rolling of threads at room temperature began from the year of 1831. At the initial stage of the study, screw bolts were rolled by rack die rolling machine to fasten the sleeper of the railways. Before the year 1940, precision rolling forming of threads had not been widely used, because there was no suitable billet material for rolling. During the period of the Second World War, the manufacturing precision of the rolling die was improved with the development of the thread grinding technology, and also because of the demands for weapons, the cold rolling forming technology had been rapidly developed. With the improvement in the billet mechanical properties, the development of the precision rolling tools, and the successful development of higher performance thread rolling machines, thread rolling technology has been greatly popularized and developed. Thread components with variety of materials such as carbon steel, alloys (stainless steel, tool steel), brass, bronze, aluminum, titanium, beryllium, nickel, powder metallurgy material, and plastics can be rolled, for example, screw tap, anchor bolts and power bolts of the engine, spindle, threads of the screw rod, and hollow thin-wall components. In present, the main types of thread rolling equipments include rack die plates, circular rolling machine, and planetary automatic thread rolling machine. In order to guarantee the quality of the rolled threads, control program had been installed on the rolling machine, such as the Helpro and SK system; therefore, the cold rolling process were becoming more mature and the precision of the roller thread were becoming more higher [44].

In volume 16th of the Disc Edition ASM Handbook published in 2001, the cold rolling of thread was introduced, and materials and several commonly used thread rolling methods were analyzed [71].

Joseph P. Domblesky, Ph. Dc of USA, simulated the cold rolling forming process of threads with DEFORM-2D, in which thread rolling was simplified as a plain strain problem, and the influence of elastic deformation and temperature effect was neglected [75]. The billet was considered as a rigid-plastic object, and the forming process and metal flow direction of the tooth profile were obtained. In addition, the three-dimensional numerical simulation was trying to be carried out, but because the forming of the thread belongs to the category of small deformation, the calculation is complex and the performance of the computer cannot meet the requirement, and there were no substantial results obtained.

In the year of 2003, Z. pater, A. Gontarz, and W. Weroñski of Poland provided a new method for the rolling of the railway sleeper bolts and verified its feasibility through FEM simulation and experimental study [76]. In this method, two wedged plate dies with grooves were used for rolling, and the angle of the groove is equal to the helix angle of the thread. Qi huiping and Li Yongtang of Taiyuan University of Science and Technology studied the process parameters, microstructures, and properties of the formed components in thread cold rolling forming with theoretical and numerical simulation methods, and provided a theoretical basis for the design of rollers and rolling equipments [45].

Some scholars studied the thread cold rolling technology from the aspects of practice utilization. Cui Changhua summarized the experience in the cold rolling forming of threads, such as the design and adjustment of the roller, the calculation of the billet diameter, and the main points of typical thread component forming [77]. The manufacturing techniques of slender thread rod, stepped and thin-wall thread components, and the design and manufacturing of the process equipments, design, and manufacture of the advanced relieving roller and the process of thread cold rolling and broaching are introduced in detail by Ji Zhixiao and other scholars. Wang Xiulun discussed several problems in the processing of threads comprehensively [78]. All these researches have provided valuable references for the extension and application of the cold rolling forming process and equipments.

Recently, with the development of modern electrotechnology, the manufacturing level of thread cold rolling equipments has been greatly improved in respect of the loading system, quality control, and automation degree. Advanced technologies such as PLC (programmable logic controller) and CNC (computer numerical control) have been integrated and applied in the thread rolling machines, which greatly improved the automation degree and control ability of the production and enhanced the efficiency and product quality. With the rack die thread rolling

Fig. 1.14 MC-300FI CNC nuclear reactor bolt cold rolling machine

machine GV-3-40#40 manufactured by Saspi S.A of Azcoitia in Spain, the height of the loading material can be controlled through the touch screen and it is readable. USA MC-80 ft/I CNC PowerBox Kine-Roller of Kinefac Corporation has installed Venus250 batch blank feeders, and the loading operation can be uniformly carried out through the inclined end and the weight sensor. The redundant materials can be returned to the feeding end, and the unqualified components can be selected out automatically. The thread cold rolling machines are developing toward the directions of automation, easy to adjust, high quality, and multifunctions [79].

The cold rolling forming equipment of Kinefac Corporation is at the leading position of the world, especially the metal cold extrusion machine and rolling forming machine. Kinefac Corporation is the leading enterprise of USA in the fields of spline, gear, and thread cold extrusion and cold rolling. Figure 1.14 is the thread rolling machine manufactured by Kinefac Corporation, which is one of the machines with the maximum rated rolling force and highest processing precision. The development of the cold rolling machine in China is a little bit backward compared with the top level in the world. The cold rolling machines manufactured by Qingdao Shengjian Machinery Factory can be used to process precision threads and spline, and the property of common thread rolling machine is close to the advanced level of the world with a precision of IT4 [19].

1.4 Development Trends on Cold Rolling Precision Forming Technology of Shaft Parts

Compared with traditional cutting methods, in the spline and thread components cold rolling forming process, the metal fiber is completed, work hardening occurs on the surface, the strength, hardness, and bending-torsion resistance, and the comprehensive mechanical properties of the workpiece have all been improved. The production rate is higher than the common machining method, and energy and materials can be saved. Processing spline and thread shafts with the chipless plastic forming technology have already become an inevitable trend in this field [80].

However, because there still has not a set of normative guiding theory of the cold precision forming, trial-and-error method is usually used in actual production to explore the optimized parameters, such as the rolling force, roller speed, feed rate, billet diameter, and the thickness of the hollow threads. Therefore, the human and material resources are wasted. The mechanical parameters of the forming process cannot be calculated accurately, basic data are deficiency for the design and selection of cold rolling machine, and the design of rolling dies. The lack of deep and comprehensive recognization of the cold rolling forming rules will result in the unsatisfied quality of the cold rolling components, which limits the application of this kind of high efficiency, chipless, and energy saving technology. To sum up, studies on the following respects should be carried out to improve the rapid development of cold rolling forming technology:

(1) The forming theory should be further deepened.

Available cold rolling forming theory is based on the transmission theory of gear meshing, while the forming process is companied with forced metal flow under the rolling of the circular billet and the contact condition is very complex. Therefore, studies on the forming mechanism of the tooth and the precision control technology should be further deepened to perfect the cold rolling forming theory.

(2) Research on the mechanical parameters of cold rolling forming.

With the unremitting efforts of the researchers all over the world, cold rolling precision forming technology has already been applied in the production, and many successful experiences have been accumulated. Because the contact conditions between the roller and the workpiece are very complicated during the forming process, there is no suitable engineering calculation methods for the mechanical parameters, experience must be adopted in the selection and design of equipment, the design of dies (rollers), control of process parameters, and ideal forming precision usually cannot be obtained because of the big error of the components. The exploring of accurate and suitable calculation method of the mechanical methods, and the optimized matching between the design of equipment and dies and process parameters are important problems, and they should be solved to promote the popularity and application of the technology.

(3) The theoretical study on the equipment manufacturing.

On the design and manufacturing of cold rolling precision forming, Profiroll Corporation in Germany and Kinefac Corporation in USA are the most representatives. The research of the fundamental theories on equipment manufacturing and original creation of cold rolling forming should be further strengthened.

(4) Study on the strengthening mechanism and the regulation of shape-property integration.

In cold rolling forming, the material can be strengthened by work hardening, studies on the process of cold rolling forming, and the regulation of shape-property integration is still not very systematical. The strengthening mechanism and shape-property control should be further focused.

References

1. Ming Cuixin. Key and Spline [M]. First Edition, Beijing: China Planning Press, 2004.
2. Zhang Xingwang, Liu Zhiqi, Li Yongtang. Research of the Cold Rolling Technology for Involute Spline [J]. Electromachining & Mould, 2009, S1: 52–54.
3. Xu Chunguo. Theory and Process Research of Flexile Rolling for Shaft Forgings [D]. Beijing: China Academy of Machinery Science & Technology, 2006.
4. Hu Zhenghuan. Rotary Forming (Sixth Chapter, Forging Volume, Forging Handbook) [M]. Beijing: Metallurgical Industry Press, 1996.
5. Wang Xiankui. Gear, Worm Gear and Worm, Spline Machining [M]. Beijing: China Machine Press, 2008: 361–429.
6. Liu Zhongwen, Lai Zhouyi, Zhang Yucheng. Application and Development of the Spline Shaft Cold Forming Technology in China [C]. The 9th Annual Academic Meeting of National Plastic Engineering & the 2nd Global Overseas Chinese Symposium on the Advanced Plastic Processing Technology (II): 2005(10): 52–55.
7. Cui Fengkui, Li Yan, Li Chunmei. Progress in the Processing Method of Involute Spline [J]. Mining & Processing Equipment, 2007, 2: 116–119.
8. Liao Shaohua. Automotive Gear Processing Technology and Typical Equipment [J]. Modern Components, 2009, 1: 83–87.
9. Wu Xutang. Principle of Gearing [M]. Beijing: China Machine Press, 1985.
10. Liu Runai, Zhang Genbao. Development Trend of Gear Hobbing Machine and Gear Hobbing Technology [J]. Machinery Manufacturing Engineer, 2003, 11: 84–86.
11. Bu Yan. Hub Connection (Second Chapter, Twenty-first Volume, Chinese Mechanical Design Collections) [M]. Nanchang: Jiangxi Publication Press of Science and Technology, 2002.
12. Steven P. Underation. State-of-the-Art Process Control for Thread Rolling [J]. Fastener Technology International, 1999, 22 (12): 60–62.
13. Peng Wei, Zhao Yumin, Yang Jine. Influence Factor Analysis of the Extrusion Force in the Spline Tube Cold Extrusion Forming [J]. Machinery, 2006, 44 (504): 35–36.
14. Lv lin. Study of the Stability of Formation of Axle Shaft Spline in the Extrusion [J]. Metal Forming Technology, 2004,(01): 27–29.
15. K. Osakada, X. Wang, S. Hanami. Precision Forging of Spline by Flashless Die Forging with Axially Driven Die [J]. CIRP Annals - Manufacturing Technology, 1997, 46: 209–212.
16. V. V. Klepikov, A. N. Bodrov. Precise Shaping of Splined Shafts in Automobile Manufacturing. Russian Engineering Research. 2003, 23 (12): 37–40.
17. U. S. Dixit, P. S. Robi, D. K. Sarma. A Systematic Procedure for the Design of a Cold Rolling Mill [J]. Journal of Materials Processing Technology, 2002, 121: 69–76.
18. Zhang Jiexin. Cold Extruding Forming Process for Small Modulus Involute Spline Shaft [J]. Construction Machinery and Equipment, 2006, 37: 57–59.
19. http://www.kinefac.com [OL].
20. Zhang Fengkuan, Men Liantong. Study on the Theoretical Model of Open Type Cold Extrusion of Involute Teeth Spline [J]. China Metal Forming Equipment & Manufacturing Technology, 2006, 3: 95–97.

21. Yi Hongzhan, Gao Xin, Duan Xinfeng. Application of Spline Shaft Cold Extrusion [J]. Metalforming Machinery, 2002, 5: 33–36.
22. Cui Fengkui. Study of High-speed Precise Forming with Cold Roll-beating Technique [D]. Xi'an: Xi'an University of Technology, 2007.
23. http://www.ernst-grob.com [OL].
24. Cui Fengkui, Zhao Wei, Xu Yongfu, Yang Jianxi. Research on Fibrous Structure and Surface Work-hardening Test of Cold Rolling Involute Spline [J]. Machine Tool & Hydraulics, 2009, 37 (12): 36–38.
25. Su Zhipeng, Cui Fengkui, Cui Ketian, Du Huiyong, Ge Bianjing. Research and Application of the Mathematical Model of Spline Shaft Cold Rolling Motion [J]. Mining & Processing Equipment, 2006, 36(01): 99–101.
26. Fengkui Cui, Li Yan, Zhou Yanwei, et al. The Design and Simulation of Roller for Machining Involute Spline [C]. 2006 International Conference on Mechatronics and Automation (ICMA 2006), 2006, 4: 1701–1705.
27. Robert L. Miller, Warren Mich. Tooth Generating Tool and Method of Making the Same [P]. United States Patent. 3902349, 1975-09-02.
28. Donald E. Blue, Decatur. Tooth Forming Machine [P]. United States Patent. 3818736, 1974-06-25.
29. Toshio Moro, Akihiro Goto, Tokyo. Rolling Die and Surface Processing Method for Rolling Die [P]. United States Patent. 6314778B1, 2001-11-13.
30. Dale J. Garden,Warren Mich. Method of Conditioning the Surface of a Spline Forming Rack [P]. United States Patent. 5509287, 1996-04-23.
31. Ames T. Killop, Warren Mich. Method and Apparatus for Cold Sizing a Round Workpiece Having Multiple Diameters [P]. United States Patent. 4712410, 1987-12-15.
32. Philippe Monot, Bragny Sur Saone. Gold Forming Tool, Machine and Method [P]. United States Patent. 7040131B2, 2006-05-09.
33. Han Fenglin. Study on Technology and Equipment of Cold Roll Forming of Involute Spline [D]. Qinhuangdao: Yanshan University, 2005.
34. Joseph P. Domblesky, Feng Feng. Finite Element Modeling of External Threading Rolling [J]. Wire Journal International, 2001, 34 (10): 110–115.
35. Zhang Qing, Fu Zhuang, Kong Xiangdong. Error Analysis of Synchro-Mechanism on CRF Machine [J]. China Mechanical Engineering, 2002, 13 (12): 1004–1009.
36. Fu Zhuang. Study on Key Technique of Cold Roll Forming Machine [D]. Qinhuangdao: Yanshan University, 2002.
37. Qiu Ping, Zhang Liling, Zhang Qing. Design and Calculation of the Involute Spline Cold Rolling Dies [J]. Journal of Yanshan University, 2003, 27(03): 244–247.
38. Zhang Ziliang. Cold Rolling Small Module Involute Spline Gear Pitch Cumulative Error Analysis [J]. Automobile Gear, 2006, 2: 34–35.
39. Liu Ruiqiu. The Numerical Simulation of Cold Roll Forming of Involute Spline Based on ANSYS [J]. New Technology & New Process, 2006, 11: 53–55.
40. Zhang Dawei. Theoretical Research on the Process of Spline Cold Rolling [D]. Taiyuan: Taiyuan University of Science and Technology, 2007.
41. Zheng Quangang. Cold Rolling Forming Technology of Cylindrical Spur Gear with Small Module Involute Spline [J]. Automobile Technology & Material. 1997, 7: 16–18.
42. Xu Chunguo. Theory and Process Research of Flexile Rolling for Shaft Forgings [D]. Beijing: China Academy of Machinery Science & Technology, 2006.
43. Wang Minghui. Research on the Process and Key Equipment for Precision Forming of Squr Gears [D]. Changchun: Jilin University, 2010.
44. Wang Xiulun. Cold Rolling Processing Technology of Threads [M]. Beijing: China Railway Publishing House, 1990.
45. Qi Huiping. Research on the Theory and Process Parameters of Cold Thread Rolling [D]. Taiyuan: Taiyuan University of Science and Technology, 2007.
46. R. Boman, L. Papeleux, Q. V. Bui, J. P. Ponthot. Application of the Arbitrary Lagrangian Eulerian Formulation to the Numerical Simulation of Cold Roll Forming Process [J]. Journal of Materials Processing Technology, 2006, 177 (1–3): 621–625.

47. S. Kobayashi, S. I. Oh, Altan. Metal Forming and the Finite-Element Method [M]. New York: Oxford Press, 1989.
48. Zhang Xingwang. The Technological and Experimental Study of Spline Cold Rolling [D]. Taiyuan: Taiyuan University of Science and Technology, 2009.
49. http://www.andersoncook.com/fraser/products.htm [OL].
50. http://www.profiroll. De [OL].
51. Li Jingxin. Process Involute Spline with Screw Rolling Machine [J]. Small & Special Machines, 2003, 2: 29–30.
52. Xue Wenhui. Rolling Machining of Involute Spline on Rolling Machine [J]. Metal Forming Technology, 1999, 17 (3): 46–47.
53. http://www.qdqs.com.cn [OL].
54. R. Neugebauer, M. Putz, U. Hellfritzsch. Improved Process Design and Quality for Gear Manufacturing with Flat and Round Rolling [J]. Annals of the CIRP, 2007, 56 (1): 307–312.
55. R. Neugebauer, U. Hellfritzsch, M. Lahl. Advanced Process Limits by Rolling of Helical Gears [J]. International Journal of Material Forming Supplement, 2008, 1: 1183–1186.
56. Reimund Neugebauer, Dirk Klug, Udo Hellfritzsch. Description of the Interactions during Gear Rolling as a Basis for a Method for the Prognosis of the Attainable Quality Parameters [J]. Production Engineering – Research and Development, 2007, 1: 253–257.
57. R. Neugebauer, D. Klug, M. Hoffmann, et al. Applying Mechatronic Strategies in Forming Technology Using the Example of Retrofitting a Cross Rolling Machine [J]. Recent Advances in Mechatronics, 2007, 3: 345–349.
58. R. Mastunaga, T. Takemasu, T. Ozaki, et al. Spline Rolling of Deep Drawn Cups [J]. Advanced Technology of Plasticity, 1999 (3): 2431–2436.
59. Amir A. Kamouneh, Jun Ni, David Stephenson, et al. Investigation of Work Hardening of Flat-rolled Helical-involute Gears through Grain-flow Analysis, FE-modeling, and Strain Signature [J]. International Journal of Machine Tools & Manufacture, 2007 (47): 1285–1291.
60. Zhang Dawei, Li Yongtang, Fu Jianhua, Zheng Quangang. Rolling Force and Rolling Moment in Spline Cold Rolling Using Slip-line Field Method [J]. Chinese Journal of Mechanical Engineering. 2009, 22 (5): 688–695.
61. Zhang Dawei, Li Yongtang, Fu Jianhua. Tooth Curves and Entire Contact Area in Process of Spline Cold Rolling [J]. Chinese Journal of Mechanical Engineering. 2008, 21 (6): 94–97.
62. Zhang Dawei, Li Yongtang, Fu Jianhua, Zheng Quangang. Mechanics Analysis on Precise Forming Process of External Spline Cold Rolling [J]. Chinese Journal of Mechanical Engineering. 2007, 20 (3): 54–58.
63. Zhang Dawei, Li Yongtang, Fu Jian-hua. Theoretical Calculation and Simulation Analysis on the Rolling Contact Area of External Spline Cold Rolling [J]. Journal of Taiyuan University of Science and Technology, 2007, 1: 64–68.
64. Li Yongtang, Zhang Dawei, Fu Jianhua, Song Jianli. Unit Average Pressure in the External Spline Cold Rolling Process [J]. China Mechanical Engineering, 2007, 24: 2977–2980.
65. J. P. Domblesky, F. Feng. Two-dimensional and Three-dimensional Finite Element Models of External Thread Rolling [C]. Proceedings of the Institution of Mechanical Engineers, Part B. 2002, 216 (134): 507–509.
66. M. Stoica, D.E. Fielden, R. McDaniels, Y. Liu, B. Huang, P.K. Liaw, C. Xu.G. Langdon. An Analysis of the Shear Zone for Metals Deformed by Equal-channel Angular Processing [J]. Materials Science and Engineering, 2005, 41: 239–242.
67. Alves M L, Roderigues J M C, Martins P A F. Cold forging of Gears: Experimental and Theoretical Investigation [J]. Finite Elements in Analysis and Design, 2001, 37: 549–558.
68. Zheng Kun, Cai Anjiang, Jia Shuang. Dynamic Analysis of Small Module Spline Shaft Based on ANSYS [J]. Journal of Mechanical Transmission, 2011, 35 (4): 44–46.
69. Wang Mingfu, Fu Jianhua, Liu Zhiqi, Li Yongtang. Finite Element Analysis of Spline Cold Roll Forming Process [J]. Forging & Stamping Technology, 2010, 35 (4):155–159.
70. Zhou Hang, Zhou Xudong, Zhou Wan. Status and Development Trend in Surface Strengthening Technology for Metal Parts by Surface Rolling [J]. Tool Engineering. 2009, 43 (12): 18–22.

71. Thread Rolling. Dialog on Disc Book [M]. ASM Handbook. 2001.
72. Sun Xitai. Surface Strengthening Technology of Materials [M]. Beijing: Chemical Industry Press, 2006.
73. Gong Guanjun. Cold Rolling Process and the Selection of Tools [J]. Mechanical Technician, 1998, 1: 15–17.
74. Li Fenglei, Xia Wei, Zhou Zhaoyao. Analytic Study and FEA Validation of Micro-plastic Deformation Depth on Component Surface in Burnishing Process [J]. Machinery Design & Manufacture, 2008, 9 (9): 62–64.
75. Joseph P. Domblesky, Feng Feng. Finite Element Modeling of External Thread Rolling [C]. The Wine Association Internation, Inc. Wire & Cable Technical Symposium (WCTS) and 70st Annual convention, Jun.2000, Nashville, Tennessee USA, 2000: 24–31.
76. Z. Pater, A. Gontarz, W. Weroñski. New Method of Thread Rolling [J]. Journal of Materials Processing Technology, 2004, 153–154: 722–728.
77. Cui Changhua. Thread Rolling Process [M]. Beijing: China Machine Press, 1978.
78. Ji Zhixiao, Zhang Rongzhen, Chen Jingtao, Liu Wentian. New Thread Processing Technology [M]. Beijing: The Publishing House of Ordnance Industry, 1990.
79. Round up: Thread Rolling Machine [J]. Fasterner Technology International, 2001, 6: 55–60.
80. Department of Engineering and Materials, National Natural Science Foundation of China. Development Strategy Report of Mechanical Engineering (2011–2015) [M]. Beijing: Science Press, 2010.

Chapter 2
Principle and Mechanical Analysis on the Cold Rolling Precision Forming of Spline

Abstract Principle and process on the cold rolling precision forming of spline shafts have been introduced, mechanical analysis of the process has been introduced. The indexing and biting conditions, the rotating conditions, double-flank non-backlash meshing, the sliding of the contact point, and the slip-line field and the stress field are introduced and analyzed. Metal flow on the tooth surface of the workpiece and average pressure on the contact surface during the forming process have also been analyzed.

2.1 Principle and Process on the Cold Rolling Precision Forming of Spline Shafts

Figure 2.1 shows the cold rolling precision forming of spline shafts. The workpiece is restrained between the two rollers by the front and rear centers, and the two rollers rotate at the same direction synchronously and feed at the radial direction with uniform speed under the driven spindle. The two rollers are with the same parameters, and the workpiece rotates under the driven frictional moment produced by the roller when the feeding roller contacts with the workpiece. The tooth of the roller pressed into the surface of the workpiece results in the plastic deformation of the workpiece. After indentation of the tooth is formed, the roller and the formed spline tooth will mesh each other freely, and finally, the teeth of the spline shaft can be generated by the teeth of the rollers.

In the process of cold rolling precision forming, the main movements between the rollers and the workpiece include the following ones: rotation and radial feeding movement of the rollers, and the rotation movement of the workpiece, in which the movement of the rollers is the main movement and the movement of the workpiece is the passive movement. The teeth of the workpiece are formed through the free generation between rollers and the workpiece, and indexing equipment is not needed.

According to the movement characteristics of the spline shaft cold rolling precision forming, the forming process can be divided into four stages, as shown in Fig. 2.2. In the first stage, the rollers have not contacted with the static workpiece,

J. Song et al., *Cold Rolling Precision Forming of Shaft Parts*,
DOI 10.1007/978-3-662-54048-0_2

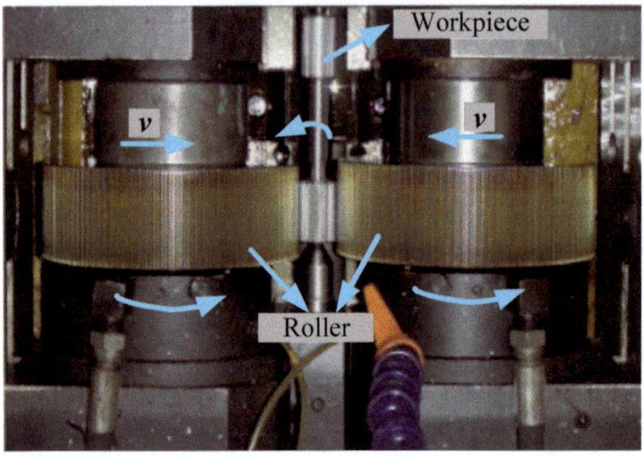

Fig. 2.1 Principle of the cold rolling precision forming process

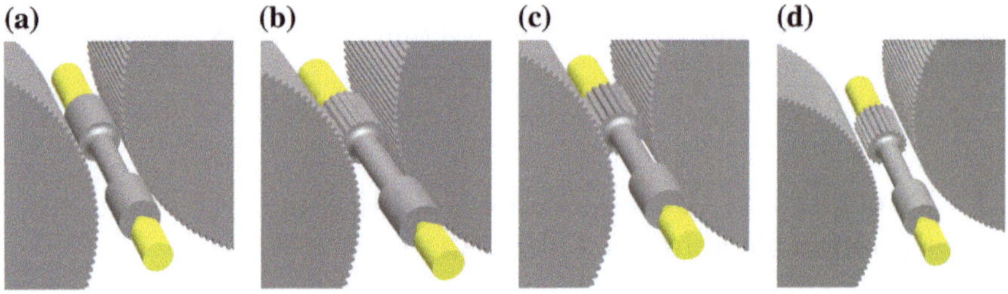

Fig. 2.2 Cold rolling precision forming process of spline shafts **a** stage before contacting; **b** indexing forming stage; **c** finishing stage; **d** exit stage

the two rollers feed rapidly to the workpiece synchronously, and it is called as a stage before contacting, as shown in Fig. 2.2a. In the second stage, the rollers press into and rotate together with the workpiece, and the tooth cogging was extruded in the workpiece by plastic deformation. The two rollers continuously feed at the radial direction to the position of the minimum center space. This stage can be called as the indexing forming period, as shown in Fig. 2.2b. This stage can be subdivided into two stages: The period from the contacting between the roller addendum and the workpiece to half a circle the workpiece rotates is called as the indexing stage. After indexing, the rollers will continuously feed to the minimum central distance, which is called as the pressing stage. Figure 2.2c is the finishing stage. In this period, the rollers will not feed anymore and just rolling freely together with the workpiece to finishing the tooth profile on the workpiece, but the central distance is kept constant. In the stage shown in Fig. 2.2d, the forming process finished, the rollers moved backward and disengaged with the workpiece.

The forming process of the tooth profile is completed in the stages of indexing, pressing, and finishing. In the two stages, the rollers contact with the workpiece, and local deformation occur continuously on the workpiece. In the period of indexing, the circular billet contacts with the addendum circle of the rollers and rotates under the action of the frictional force, and the related movement between the roller and the workpiece is pure rolling of the roller addendum and the external circle of the billet. In the pressing period, the relationship between the billet and the roller is equivalent to the continuously modified gear transmission, and the moving process can be analyzed with the theory of modified gear transmission. In the final finishing stage, the moving relationship between the roller and the workpiece is similar to the double-side non-backlash meshing of a couple of gears. In different forming stages, the shape of the workpiece is different, and the deforming area and the moving characteristics are also different [1, 2].

2.2 Mechanical Analysis

According to the constant volume principle, the cross section of the spline shaft is equal before and after the cold rolling forming; that is, the cold rolling forming of spline can be considered as a plain strain state, there is no deformation along the axial direction during forming process, and the distance and strain along the axial direction are zero [3–5].

While the tooth number of the spline is an even number, the forming principle and the macroforce diagram are shown in Fig. 2.3. Four external forces are exerted on the workpiece by rollers, normal force P and frictional force F_f, which are applied on the workpiece symmetrically. The direction of normal force P will pass through the center of the roller, and the angle between the direction of $\boldsymbol{P}$ and the horizontal axis is defined as θ.

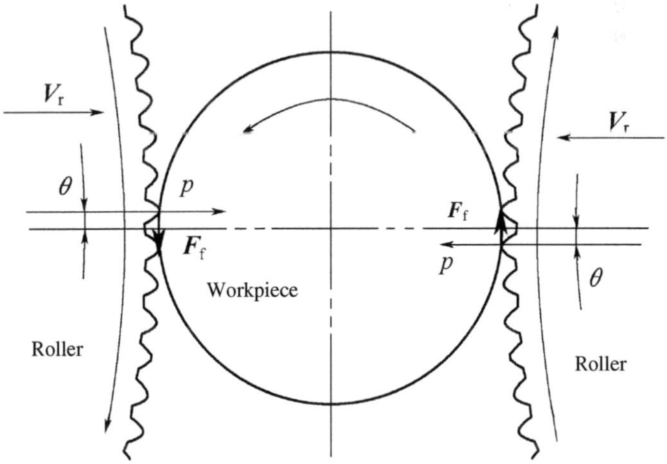

Fig. 2.3 Force diagram of the spline cold rolling forming process

2.3 Analysis on the Cold Rolling Forming Process of the Spline Tooth

2.3.1 *Indexing and Biting Conditions [6]*

In the cold rolling precision forming of spline shafts, the indexing operation is realized by the relative free rolling of the rollers and the workpiece. After contacting with the roller addendum, the circular billet will be driven to rotate along with the rollers, and prescriptive number of teeth will be extruded on the external circle of the billet. The moving parameters should match certain geometrical conditions in the process. In the first stage of the spline cold rolling forming, biting rotating and precision indexing should be guaranteed. The success of the cold rolling precision forming depends on the rotating conditions and precision of the index.

2.3.2 *Rotating Conditions*

Figure 2.3 shows that at the initial stage of contacting indexing, the external circle of the billet contacts with the addendum of the rollers. The movement can be considered as a frictional wheel transmission between the addendum cylinder of the rollers and the external cylinder of the billet, and the contact area can be simplified as a microplane.

According to the principle of friction, the relationship between the macronormal force P and the macrofrictional force $F_{\rm f}$ is as follows:

$$\mu = F_{\rm f}/P \tag{2.1}$$

where μ is the sliding friction coefficient between the roller addendum and the workpiece surface.

The two rollers are mounted symmetrically relative to the workpiece. The forces applied to the workpiece have composed two moments rotating the workpiece. The moment formed by the friction force on the tooth flanks is $M_{\rm f}$, and the moment formed by the bilateral normal force is $M_{\rm p}$. While the friction moment of the contact area $M_{\rm f}$ is larger than the normal force moment $M_{\rm p}$, the workpiece will rotate, and thus, the rotating condition is as follows:

$$M_{\rm f} \geq M_{\rm p} \tag{2.2}$$

That is

$$F_{\rm f} d_{\rm z} \cos\theta > P d_{\rm z} \sin\theta \tag{2.3}$$

where d_z is the diameter of the billet, θ is the angle between the normal force and the connecting line of the two roller centers, $\theta < \pi/Z_2$, and Z_2 is the tooth number of the roller.

From Eqs. (2.1) and (2.3), in order to guarantee the rotation of the workpiece in the preliminary stage of the cold rolling precision forming, the friction coefficient must satisfy the following condition:

$$\mu > \tan\theta \tag{2.4}$$

For the biting rotation of cross rolling, based on the geometrical parameters of the workpiece and the roller, through the control of the relative reduction, the value of θ can meet the biting condition in an appropriate scope.

Define the ratio of the workpiece reduction and the workpiece diameter after the compression $\Delta s'/d$ as the limit relative reduction, and then, the rotation biting condition is as follows [7]:

$$\frac{\Delta s'}{d} \leq \frac{1}{4} \cdot \frac{1 + d/d_{a_2}}{\frac{(1 + d/d_{a_2})^2}{\mu^2} + (d/d_{a_2})^2} \tag{2.5}$$

where d_{a2} is the diameter of the roller addendum.

Taking the limit relative reduction ratio $\Delta s'/d$ and the ratio between the workpiece diameter after rotating for half a circle and the diameter of the roller addendum d/d_{a2} as the parametric variable, the curve obtained from the relation of Eq. (2.5) under different frictional coefficient μ is shown in Fig. 2.4.

The influence of coefficient μ on the rotating condition is more significant than d/d_{a2}. d/d_{a2} decreases with the increase in $\Delta s'/d$; that is, the rotating condition is better with a large ratio of the roller and the workpiece diameter, the relative reduction is larger, and the selection range of the feed rate is more extensive. While d/d_{a2} increased from different values (1, 0.5, 0.4, 0.2, and 0.1) to 0.05, the increased percentage of $\Delta s'/d$ is shown in Table 2.1.

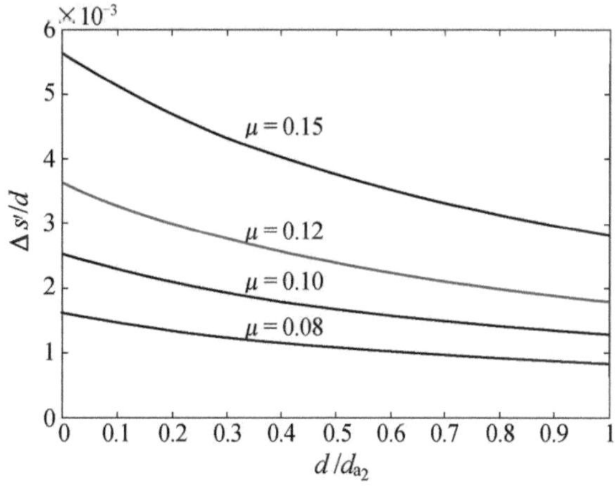

Fig. 2.4 Relationship between the limit reduction ratio $\Delta s'/d$, friction coefficient μ and d/d_{a2}

Table 2.1 Variable percentage of the limit relative reduction

μ	d/d_{a_2}				
	0.1 (%)	0.2 (%)	0.4 (%)	0.5 (%)	1.0 (%)
0.12	4.71	14.33	34.49	43.09	91.16
0.3	4.84	14.57	34.31	44.28	94.76
0.5	4.97	15.07	36.04	46.81	102.37

The table shows that the increased percentage of the limit reduction decreased from the lower right to the top left corner of the table, under conditions $d/d_{a2} < 2$, and the influence becomes weakened, but is still in the controllable scopes. If the diameter of the roller is selected from the perspective of the rotating condition, the diameter of the roller should be more than 5 times of the workpiece diameter.

From Eq. (2.5), the rotating biting condition can be realized by the control of the reduction value from the perspective of process parameter selection. From the aspect of die design, there exist the other two improved rotating biting conditions.

From the geometrical relationship, it is known that the value of θ is not larger than half of the angle between the two teeth of the roller. That is, $\theta < \pi/z_2$.

Therefore, Eq. (2.4) can be modified as follows:

$$\mu > \tan\frac{\pi}{z_2} \tag{2.6}$$

In the rotating frictional condition given by Eq. (2.6), the friction coefficient and the tooth number of the roller can both be guaranteed by the design of the rollers. Two design schemes of the roller are proposed to ensure reliable rotation.

(1) The method to limit the minimum tooth number of the rollers.

The increase in the friction coefficient between the roller and the billet is beneficial to the rotation of the billet. But in order to obtain the required processing precision, the external cylinder surface of the billet is always machined by finishing turning, and the tooth of the rollers should be treated by grinding. Process lubricant should be used in the cold rolling precision forming to prolong the lifetime of the roller, and the value of the friction coefficient μ is about 0.1–0.2 [8]. Therefore, in order to realize the rotation of the workpiece, the minimum tooth number should be restricted, from Eq. (2.5):

$$Z_2 > \frac{\pi}{\arctan\mu} \tag{2.7}$$

In the design of rollers, the minimum tooth number should be verified according to Eq. (2.6). The tooth number of the roller will also influence the accumulated error of the formed spline tooth pitch.

(2) The method of increasing the friction coefficient.

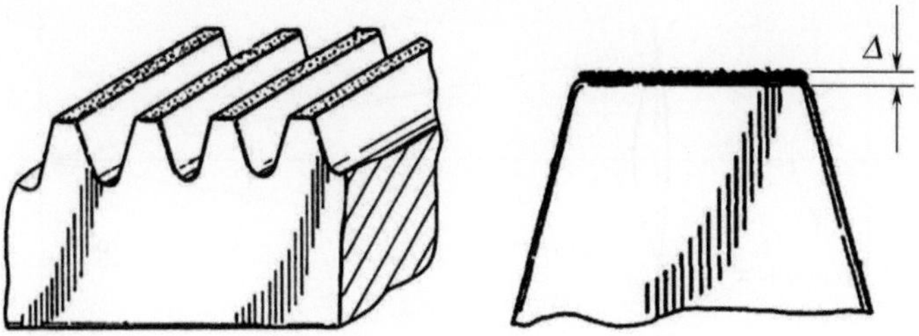

Fig. 2.5 Coating on the addendum of the roller

At the initial indexing stage of the cold rolling precision forming of spline shafts, the external circle contacts with the addendum of the roller and rotates under the driven friction force. Biting rotation can be realized by increasing the friction coefficient between the workpiece and roller addendum. Coating technology has already been widely used long time ago in the field of cutting tool manufacturing. According to an American patent [9, 10], in the cold rolling of spline with rack dies, the plane rollers were coated with WC or TiC for a thickness of $\Delta = 0.035\text{–}0.050$ mm to improve the lifetime of the dies, as shown in Fig. 2.5. The biting rotation condition can be improved by increasing the friction coefficient between the roller and the billet. Because the coating layer has a higher hardness, the wear resistance of the roller can be increased and thus increases the lifetime of the roller.

The forming rotating biting condition should all be considered on the period of cold rolling machine selection, die design, and process parameter planning.

The two schemes mentioned above can be used for the selection of the roller parameters and as the rotating conditions in the manufacturing of the rollers. Equation (2.5) can be used as the rotating conditions realized through the control of the process reduction in the determination of cold rolling machine and dies.

2.3.3 Geometrical Indexing Conditions

At the preliminary stage of the spline shaft cold rolling forming process, the addendums of the rollers contacts with the external circle of the workpiece billet firstly, and the rollers rotate along with the workpiece. To indexing accurately on the workpiece, sliding should not happen on the contact area of the roller addendum and the external circle of the workpiece, and the diameter of the billet should be in a special dimensional relationship with the diameter of the roller addendum. Therefore, the geometrical indexing condition can be deduced.

According to the meshing principle, while the roller rotates for one tooth, the workpiece will rotate for one tooth correspondingly; therefore, the chord length $\overline{AB}$ between the two tooth addendums should be equal to the chord length $\overline{AB}$ on the

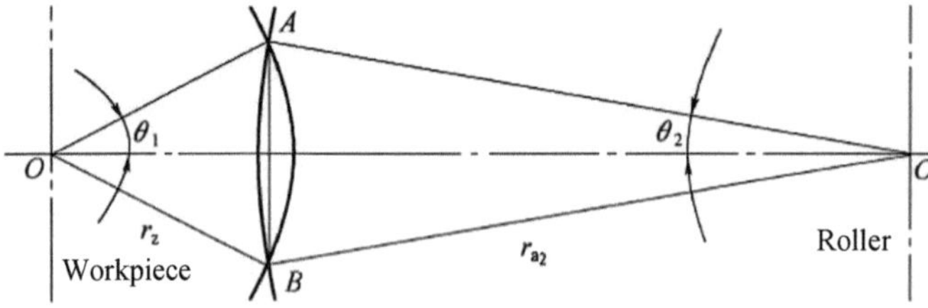

Fig. 2.6 Indexing of the roller on the workpiece billet

external circle of the billet corresponding to the adjacent tooth cogging center of the workpiece billet, and the geometrical relationship is shown in Fig. 2.6.

Thus

$$r_Z \sin \theta_1 = r_{a_2} \sin \theta_2 \tag{2.8}$$

where r_Z is the radius of the billet before rolling forming. θ_1 is half of the angle between the two adjacent teeth on the workpiece, $\theta_1 = \pi/Z$. θ_2 is half of the angle between the adjacent teeth on the roller, $\theta_2 = \pi/Z_2$. And r_{a_2} is the radius of the addendum circle of the roller.

In Eq. (2.8), the relation between the billet radius before rolling and the addendum circle radius of the roller is involved. In practical application, it can usually be expressed with the diameters as follows:

$$d_Z \sin \theta_1 = d_{a_2} \sin \theta_2 \tag{2.9}$$

While determining the cold rolling precision forming parameters, the diameter of the spline shaft billet d_Z and the addendum circle diameter of the roller d_{a_2} should meet the geometrical indexing condition.

Because cold rolling precision forming is a kind of chipless production method, once the formed spline parameters have been determined, the theoretical diameter d_Z of the billet should be calculated according to the volume constant principle before and after cold rolling forming. In the design of rollers, the required indexing condition of Eq. (2.9) can be satisfied through addendum modification to obtain suitable diameter of the addendum circle.

The diameter of the billet d_Z is a very important parameter in the cold rolling forming of spline shaft, which has a great influence on the precision of the forming process.

Because the shape of the spline is complex, there are many parameters describing the shape of the spline, such as modulus, number of the tooth, pressure angle, diameter of the addendum circle, and diameter of the root circle. According to the given parameters of the drawing, calculation of the spline shaft cross-sectional area is rather complex based on the constant volume principle. Presently, empirical formula or trial-and-error method is often used in production to determine the diameter of the billet [11, 12]. Zhang Dawei and Li Yongtang et al.

provided an exact solving method of the billet diameter with numerical simulation calculation and also deduced the calculation formula of the billet diameter of cold rolling precision forming process [13, 14]:

$$d_z = \sqrt{d_f^2 + \frac{4}{\pi} A_{th} Z} = \sqrt{d_f^2 + \frac{Z}{\pi}\left[\frac{d_b^2}{3}(\tan^3\alpha_a - \tan^3\alpha_f) + d_a s_a - d_f s_f\right]} \quad (2.10)$$

where

d_f is the root circle diameter of the spline.
A_{th} is the section area of one of the teeth on the spline.
Z is the tooth number of the spline.
d_b is the base circle diameter of the spline, $d_b = mz\cos\alpha$.
d_a is the addendum circle diameter of the spline.
α_f is the pressure angle of the root circle on the spline, $\alpha_f = \arccos\left(\frac{d_b}{d_f}\right)$.
α_a is the pressure angle of the addendum circle on the spline, $\alpha_a = \arccos\left(\frac{d_b}{d_a}\right)$.
S_f is the tooth thickness of the root circle on the spline, $s_f = d_f\left(\frac{s}{d} + \mathrm{inv}\alpha - \mathrm{inv}\alpha_f\right)$.
S_a is the tooth thickness of the addendum circle on the spline, $s_a = d_a\left(\frac{s}{d} + \mathrm{inv}\alpha - \mathrm{inv}\alpha_a\right)$.

Because the formula is relatively complex and is not convenient for manual calculation, it is programmed in the study, and the billet diameter before cold rolling can be easily obtained after inputting corresponding tooth parameters.

2.4 Double-Flank Non-backlash Meshing

At the initial stage of indexing, after the tooth cogging is pressed on the billet, the cold rolling process can be considered and analyzed as a double-flank non-backlash meshing movement [15]. The state of double-flank non-backlash meshing is different from the single flank meshing of common gear transmission, and researches on this field are very limited. During the cold rolling forming process, the contact area changes instantly, the contacting state of the roller and the workpiece is very complex, and the forces applied on every tooth vary significantly, which has a great influence on the forming quality of the tooth on the workpiece.

2.4.1 *Contact Area in the Forming Process*

If the movements of the roller and workpiece are considered as a meshing movement in the cold rolling precision forming process, in the axial section, the contact of every instant can be considered as point contact, and during the meshing process the roller rotating for a tooth, the number of the contacting points on both sides of the tooth flank is different. The number of the contact points has a close relation with the overlap coefficient, which increases with the increasing tooth number, and is always not an integer number; therefore, the number of the contact points during the meshing process is changing. For example, when the overlap coefficient is in the range of 1–2, single-tooth and double-teeth contact area are all exist on the left and right meshing line. In the cold rolling forming of spline shafts, the contact conditions at the meshing instant of the roller and the workpiece are shown in Fig. 2.7.

In the contact process of the roller and the workpiece shown in Fig. 2.7, a tooth of the workpiece must experience the following period shown in Fig. 2.7a–f from entering into the contact area to disengage from it. In Fig. 2.7a, the tooth ② has just entered into the contact area, the former tooth ① is still not disengaged from the contact, the number of the contact points is 4, and there are two couples of teeth in contact on the left and right meshing lines. At the position of (b), tooth ① on the left meshing line disengaged from contact, there are still two couples of teeth in contact, and the total number of the contact points is 3. After rotating to the position of (c), tooth ① has completely disengaged from contact, only tooth ② are in the contact state on the left and right meshing lines, and the number of the contact points is 2. At the position of (d), the rear tooth ③ enters into the contact area and

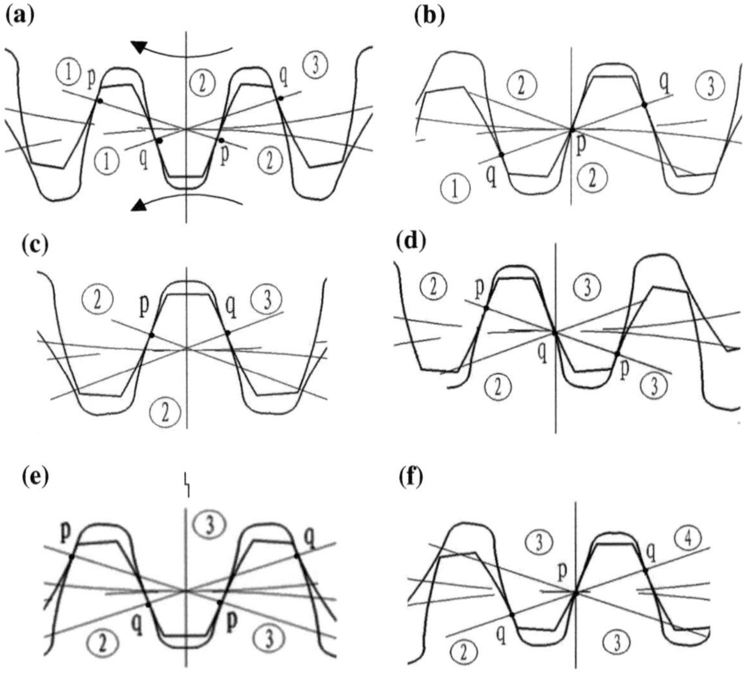

Fig. 2.7 Instant variation of the contact points on the tooth surface

tooth ② and tooth ③ on the left meshing line are in contact, while only tooth ② is in the contact state on the right meshing line, which is just opposite to the contact condition of Fig. 2.7b, and the number of the contact number is 3. At the position of (e), there are both two couples of teeth in contact, which is the same condition as (a) and the contact number is 4, but the contacting teeth are changed from ① and ② to ② and ③. Continuously rotating with this sequence, and at the position of (f), tooth ② will disengage from the contact area and the contact condition is the same as tooth ① which will disengage from the contact area in Fig. 2.7b.

The number of the contact points and the variation in the position is shown in Fig. 2.7, when the workpiece or the roller rotates for one tooth (such as from tooth ② to tooth ③). Therefore, the following conclusions can be drawn:

(1) The number of the contact points and positions is changed periodically, and the period of rotating for one tooth is called as a cycle. The total number of the contact points on the left and right meshing lines is equal to the sum of the contacting teeth on the two lines, and each couple of meshing teeth is corresponded to one contact point.
(2) The contact points p and q on both sides of the tooth flank are moving points at opposite moving directions, and while one of the points moving toward the dedendum of the tooth, the other one will inevitably moving toward the addendum of the tooth, and vice versa. No matter how the roller rotates, the two contact points are always moving on the meshing line with the continuous rotating of the roller and the workpiece.

The above analysis is suitable for the finishing stage of the cold rolling precision forming, the teeth of the workpiece have been formed, and the central distance between the roller and the workpiece will not change any more. During the relative rolling process of the roller and the workpiece, the overlap coefficient is changing, and the number of the contact points is varied periodically. It can be seen from the moving of the contact point on both flanks that the force exerted on the instant contact point is different, and the plastic deformation on both sides of the tooth is also different. With a big modulus of the cold rolling spline, tooth error will be induced in conditions of a large tooth height.

In the process of the radial feeding of the roller, the tooth profile of the workpiece is not perfect. If the overlap coefficient is less than 1, disengagement between the roller and the workpiece will occur, this kind of contact condition is more complex, and this period can be considered as a transitional process, which has not great influence on the final precision of the formed components. In the cold rolling forming process of spline shafts, because of the variation in the number of contact points and the instant meshing and disengagement, the spindle of the machine will be impacted; vibration of the machine, unqualified accuracy of the components, and even failure of the forming process will also be caused. Therefore, damping vibration attenuation structure must be designed in the cold rolling forming machine to reduce the impact applied by the forming force on the spindle transmission system.

2.4.2 *Position of the Contact Points*

In the spline shaft cold rolling forming process, non-backlash meshing between the rollers and the workpiece exists on double flanks; the position of the contact points on the spline tooth flank can be determined on the basis of gear meshing theory, and the relation between the position parameters can be obtained [16]. The rollers are rotating at a constant angular velocity, the position at each moment is fixed, and variation in the contact point position on the tooth surface with time can be expressed by the rotating angle of the roller or the workpiece.

From the above analysis, there are always contact points existing on the lateral surface of the tooth, and complete analytical model of the contact point position can be obtained by expressing the position of contact point on the meshing rollers with the rotating angle of the rollers or workpiece.

2.4.2.1 Position of the Contact Point on the Bilateral Tooth Surfaces of the Spline

The position the symmetrical line of the spline tooth coinciding with the connecting line of the two roller centers can be taken as the reference. With the rotation of the workpiece, the positions of the contact points p and q at both sides of the tooth flanks can be expressed by four different forms as shown in Fig. 2.8. That is, the contact point is on the right meshing line and the left side of the reference position, as shown in Fig. 2.8a; the contact point is on the right meshing line and the right side of the reference position, as shown in Fig. 2.8b; the contact point is on the left meshing line and the left side of the reference position, as shown in Fig. 2.8c; and the contact point is on the left meshing line and the right side of the reference position, as shown in Fig. 2.8d.

While the workpiece is rotating counterclockwise around its center O_1, the deviation angle φ_{qh} with the central reference position is positive; while the workpiece is rotating clockwise around the center, the deviation angle φ_{qh} is negative and $\varphi_{qh} = 0$ at the reference position. The contact condition shown in Fig. 2.8a is taken as an example for the analysis of the relationship between different geometrical parameters. The right side of the formed spline intersects with the right meshing line at the left of the reference position, and the intersection point is q. According to the meshing principle, the contact point of the spline and roller moves along the meshing line; at the reference position, the intersection point q of the right side of the spline tooth and the meshing line is at position 1, and the intersection point will move to position 2 when the spline rotates for a left deviation angle φ_{qh}. That is, the contact point q will move from position 1 to position 2 after the spline rotates for a left deviation angle φ_{qh}.

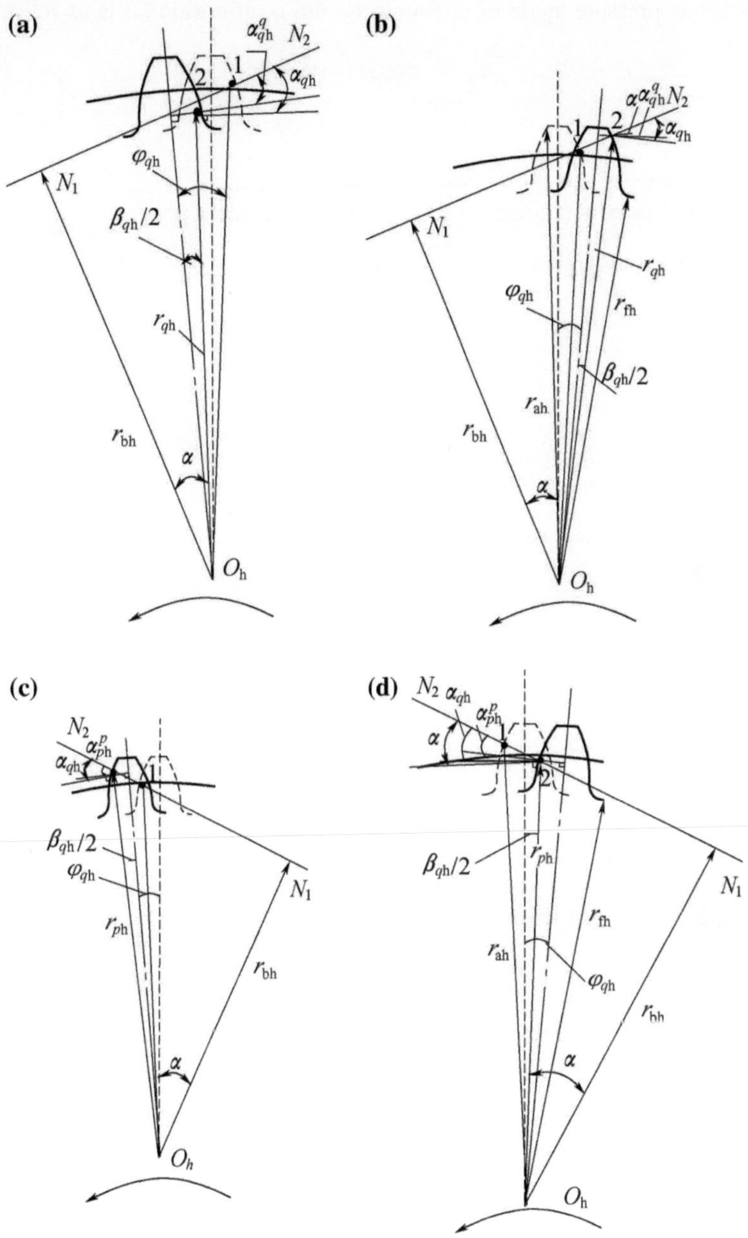

Fig. 2.8 Position of the contact points on the spline tooth **a** *right meshing line* and *left* of the reference; **b** *right meshing line* and *right* of the reference; **c** *left meshing line* and *left* of the reference; **d** *left meshing line* and *right* of the reference

The spline pressure angle of the contact point q on position 2 is as follows:

$$\alpha_{\mathrm{qh}} = \arccos\left(r_{\mathrm{bh}}/r_{\mathrm{qh}}\right) \tag{2.11}$$

where

r_{bh} is the radius of the spline base circle and
r_{qh} is the radius of the circle on the spline at the contact point q.

The tooth width angle on the circle of the contact point is as follows:

$$\beta_{\mathrm{qh}} = \frac{s_h}{r_h} - 2(inv\alpha_{qh} - inv\alpha) \tag{2.12}$$

where

r_h is the radius of the spline pitch circle, $r_h = mz_h/2\,and$
s_h is the tooth width of the spline on the pitch circle, $s_h = m_h/2$.

From the geometrical relation shown in Fig. 2.8a, the deviation angle φ_{qh} is as follows:

$$\varphi_{\mathrm{qh}} = \beta_{qh}/2 + \alpha - \alpha_{qh} \tag{2.13}$$

where α is the meshing angle of the roller and the spline (°), which can be replaced approximately by the pressure of the pitch circle.

Equations (2.10), (2.11), and (2.12) can be simplified as follows:

$$\varphi_{\mathrm{qh}} = \frac{s_h}{2r_h} - \tan\left(\arccos\frac{r_{bh}}{r_{qh}}\right) + \tan\alpha \tag{2.14}$$

Here, the radius of the meshing point is expressed as follows:

$$r_{qh} = \frac{r_{bh}}{\cos\left[\arctan\left(\varphi_{\mathrm{qh}} - \frac{s_h}{2r_h} - \tan a\right]\right.} \tag{2.15}$$

N_1N_2 is the right meshing line of the roller and the workpiece, and further deduced according to Fig. 2.8a, the angle between the meshing line and the normal line of the spline tooth center line on the contact point of the spline positive tooth flank can be expressed as follows:

$$\alpha_{\mathrm{qh}}^{\mathrm{q}} = \alpha_{qh} - \beta_{qh}/2 = \alpha_{qh} - \frac{1}{2}\left[\frac{s_h}{r_h} - 2(inv\alpha_{qh} - inv\alpha)\right] \tag{2.16}$$

If the final meshing point on the tooth dedendum is F, the radius of the circle at F is r_F, the corresponding chord length is t_F, and then, the tooth height at the contact point is as follows:

$$x_{qh} = r_{qh}\cos(\beta_{qh}/2) - \sqrt{r_{Fh}^2 - (t_{fh}/2)^2} \tag{2.17}$$

It can be seen from the equations listed above, α, r_h, and s_h are the parameters of the formed spline, and the position of the contact point can be described by the parameters such as the radius r_{qh} of the meshing circle at the contact point, the angle α_{qh}^q of the meshing line and the normal line of the spline tooth central line at the contact point, and the tooth height x_{qh}. All of these parameters have a unique determined relation with the deviation angle φ_{qh} between the workpiece and the reference position.

At the contact state shown in Fig. 2.8b, φ_{qh} is negative. When contacting at the left tooth flank, φ_{ph}, r_{ph}, x_{ph}, and α_{ph}^p have the same relation as shown in Eqs. (2.10)–(2.16). Also, because $\varphi_{ph} = \varphi_{qh}$, if φ_{qh} is determined, among the scope of the rotation angle of the spline, the position of the contact points p and q on the left and right tooth flanks can be determined.

2.4.2.2 The Position Parameters of the Contact Point on the Rollers

The position parameters of the corresponding contact points on the roller teeth are shown in Fig. 2.9.

The value of r_{qh} on the contact point of the spline tooth can be decided by φ_{qh}, and the geometrical relation is as follows:

$$\rho_{qh} = \sqrt{r_{qh}^2 - r_{bh}^2} \tag{2.18}$$

where ρ_{qh} is corresponded to N_1Q in Fig. 2.9, which is the curvature radius of the contact point on the spline, while QN_2 is the curvature radius at the contact point of roller, and

$$\rho_{qg} = \overline{N_1N_2} - \rho_{qh} \tag{2.19}$$

Calculated with the standard center distance without modification:

$$\rho_{qg} = \left(\frac{r_h + r_g}{2}\right)\sin\alpha - \rho_{qh} \tag{2.20}$$

From $r_{qg} = \sqrt{\rho_{qg}^2 + r_{bg}^2}$, the circle radius r_{qg} at the contact point of the roller tooth can be solved. Similar to the derivation of the contact position on the spline flanks, the position parameters (α_{qg}, β_{qg}, φ_{qg}, α_{qg}^q, and x_{qg}) of the contact point on the left and right tooth flanks of the roller can be obtained. After the mathematical models of the contact point, position parameters have been built, and the boundary conditions can be conveniently determined in the theoretical calculation and numerical simulation process.

Fig. 2.9 Position of the contact point on the roller

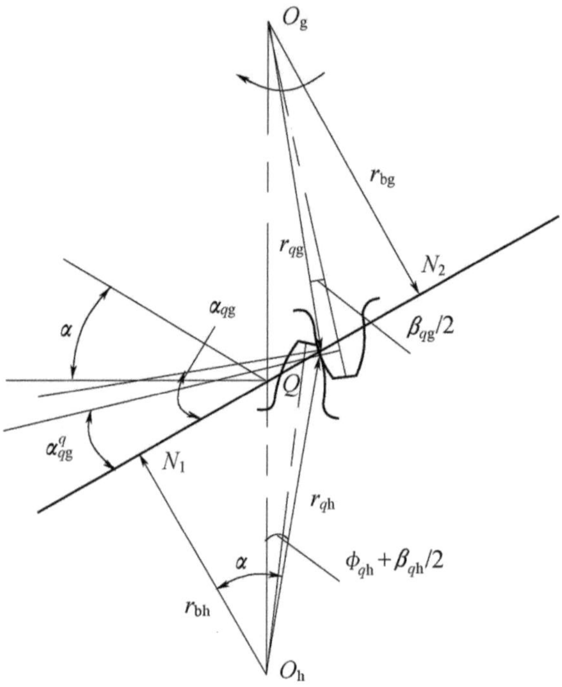

2.5 Sliding of the Contact Point

In the cold rolling precision forming of spline shafts, the main motion of the dies (rollers) is the rotation and radial feeding of the roller. The workpiece is non-backlash meshed with the roller on double flanks. The tooth flank bearing the extrusion force and driving the workpiece to rotate is called as the active flank. Restrained by the teeth of the roller, the other tooth flank contacted by the flowing metal is called as the driven flank. Based on the engagement of the roller and the workpiece, the relative sliding of the contact point is analyzed, and the sliding motion mathematical models of the contact point on the active tooth flank and the driven flank of the workpiece have been set up.

2.5.1 Sliding Motion of the Contact Point on the Active Tooth Flank

The meshing and sliding between the workpiece and the roller on the active tooth flank are shown in Fig. 2.10. In the figure, surface I is the active flank of the forming spline, and II is the tooth flank contacting with it. On the right meshing line, the contact point is M, and the pitch point is P. Let $r = \overline{PM}$; when point M is

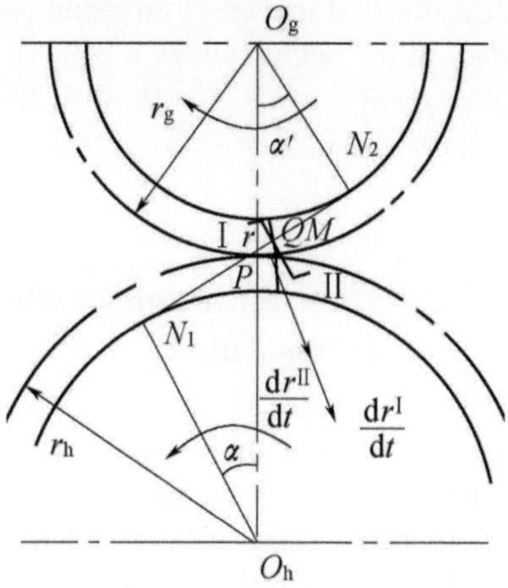

Fig. 2.10 Sliding of the contact point on the active tooth flank

located on the right side of point P, r is positive; when the point M locates on the left side of P, then the value of r is negative.

According to the meshing theory, for the tooth profile of involute spline, the curvature radii at the contact point (meshing point) of flank I and II are as follows:

$$\rho_h = \overline{N_1P} + r = r_h \times \sin\alpha + r \tag{2.21}$$

$$\rho_g = \overline{N_2P} - r = r_g \times \sin\alpha - r \tag{2.22}$$

The sliding coefficient of the contact point on spline tooth I is as follows:

$$\chi_1 = \frac{\frac{d_1 r^{I}}{dt} - \frac{d_2 r^{II}}{dt}}{\frac{d_1 r^{I}}{dt}} = 1 - i_{21} \times \frac{\rho_h}{\rho_g} \tag{2.23}$$

where $\frac{d_1 r^{I}}{dt}$ is the moving speed of contact point M on the spline plane relative to tooth flank I (m/s) and

$\frac{d_1 r^{II}}{dt}$ is the moving speed of contact point M on the spline plane relative to tooth flank II (m/s). Substituting Eqs. (2.21) and (2.22) into (2.23), obtain:

$$\chi_1 = 1 + i_{21} - \left(\frac{r_g \times \sin\alpha + r_h \times \sin\alpha}{r + r_h \times \sin\alpha}\right) \tag{2.24}$$

In the meshing process of the workpiece and the roller, the active tooth flank located on the right meshing line always meshes from the tooth addendum and then enters the tooth dedendum along the contact line. Therefore, the direction of the moving speed of contact point M relative to tooth flank I is always along the tangent line of the workpiece tooth flank and pointed to the tooth dedendum. From the positive and negative of the sliding coefficient, the relationship between the sliding direction and the speed direction can be obtained. While contact point M is on the right of the pitch point P, $r > 0$, $\chi_1 > 0$; while contact point M is on the left of the pitch point P, $r < 0$, $\chi_1 < 0$; while the contact point M is overlap with the pitch point P, then $r = 0$, $\chi_1 = 0$.

2.5.2 *Sliding Motion of the Contact Point on the Driven Tooth Flank*

Similar to the contacting on the active tooth flank, the contact sliding between the driven tooth flank of the workpiece and the roller is shown in Fig. 2.11. In the figure, surface III is the driven tooth flank corresponding to the active flank I on the workpiece and surface IV is the tooth flank of the roller contacted with the driven tooth flank III. On the left meshing line, the contact point is M', and the pitch point is P'. Let $r' = \overline{P'M'}$; while contact point M' is on the right side of the pitch point P', the value of r' is positive; while contact point M' is on the left side of the pitch point P', the value of r' is negative.

The curvature radius of the contact point (meshing point) of flank III and flank IV is as follows:

$$\rho'_h = \overline{N_1P} - r' = r_h \times \sin\alpha - r' \tag{2.25}$$

$$\rho'_g = \overline{N_2P} + r' = r_g \times \sin\alpha + r' \tag{2.26}$$

Accordingly, the sliding coefficient of the contact point on the spline tooth flank is as follows:

$$\chi'_1 = \frac{\frac{d_1 r'^{\mathrm{I}}}{dt} - \frac{d_2 r'^{\mathrm{II}}}{dt}}{\frac{d_1 r'^{\mathrm{I}}}{dt}} = 1 - i'_{21} \times \frac{\rho'_g}{\rho'_h} \tag{2.27}$$

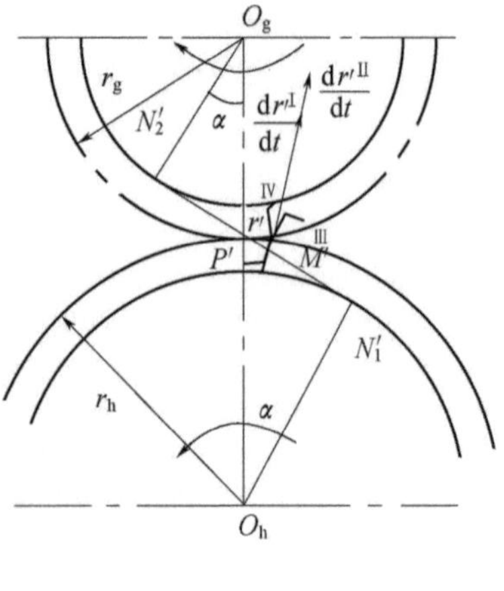

Fig. 2.11 Sliding of the contact point on the driven tooth flank

From Eqs. (2.25), (2.26), and (2.27), obtain:

$$\chi_1' = 1 + i_{21} + \left(\frac{r_g \times \sin\alpha + r_h \times \sin\alpha}{r' - r_h \times \sin\alpha}\right) \tag{2.28}$$

Contrary to the active tooth flank, the driven tooth surface III on the left meshing line is always meshing from the tooth dedendum and then transition to the addendum along the left meshing line; that is, the direction of the relative moving speed $\frac{d_1r'^{\mathrm{I}}}{dt}$ of the contact point M' to tooth flank III is always along the tangent of the tooth surface and pointed to the tooth addendum.

The relationship between the sliding direction and the speed direction can be obtained by analyzing the positive and negative of the sliding coefficient on the contact point χ_1'. While the contact point M' is overlapping with the pitch point P', $r' = 0$, $\chi_1' = 0$; while M' is within the pitch circle, $r' > 0$, $\chi_1' < 0$, the sliding direction of tooth surface III relative to tooth surface IV is the same with $\frac{d_1r'^{\mathrm{I}}}{dt}$; while contact point M' is outside of the pitch circle, $r' < 0$, $\chi_1' > 0$, the sliding direction of tooth surface III relative to tooth surface IV is opposite to $\frac{d_1r'^{\mathrm{I}}}{dt}$.

2.5.3 *Metal Flow on the Tooth Surface of the Workpiece*

Under the action of the rolling force, metals on the tooth surface of the workpiece reach the plastic yield conditions and macroslip of the metals occurs on the surface of the workpiece. Because of the restriction of the roller tooth surface, metals can only flow along the tooth flank of the roller. On the surface of the deforming area, the metal flow direction is consistent with the relative sliding of the contact point. The macro characteristics of the metal flow on the tooth surface can be obtained through the analysis of the sliding conditions of the contact points on the active and driven tooth flank of the workpiece.

According to the above analysis, for the active tooth flank of the workpiece, while $\chi_1 > 0$, that is, when $\frac{d_1r^{\mathrm{I}}}{dt} > \frac{d_2r^{\mathrm{II}}}{dt}$, the sliding direction of surface I relative to surface II is opposite to $\frac{d_1r^{\mathrm{I}}}{dt}$; the direction of the friction shearing force on tooth surface I is the same with $\frac{d_1r^{\mathrm{I}}}{dt}$. Because the metal flow direction coincides with the direction of the shearing force, the direction of $\frac{d_1r^{\mathrm{I}}}{dt}$ is just the metal flow direction on tooth surface I under this situation. While $\chi_1 < 0$, that is, $\frac{d_1r^{\mathrm{I}}}{dt} < \frac{d_2r^{\mathrm{II}}}{dt}$, the relative sliding direction between flank I and II is the same as the direction of $\frac{d_1r^{\mathrm{I}}}{dt}$. Because the friction shear on the tooth surface I of the workpiece is contrary to the direction of $\frac{d_1r^{\mathrm{I}}}{dt}$, the metal flow direction on the tooth surface of the workpiece is opposite to the direction of $\frac{d_1r^{\mathrm{I}}}{dt}$.

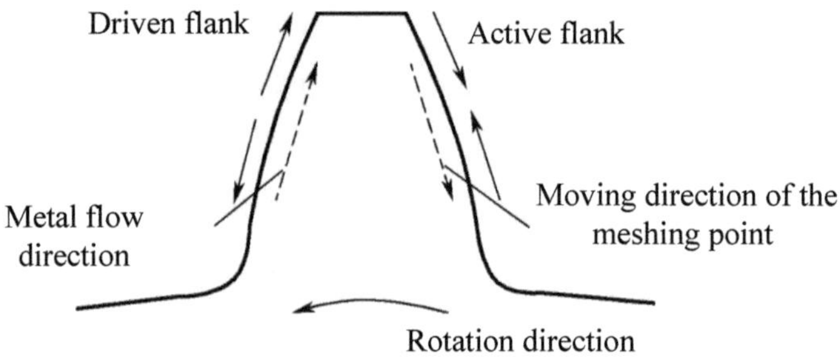

Fig. 2.12 Metal flow conditions on the surface of the workpiece

Accordingly, the analysis method of the metal flow direction on the driven tooth surface is similar to the method of the active surface. While $\chi_1' > 0$, the metal flow direction on surface III is the same as $\frac{d_1 r'^{I}}{dt}$; while $\chi_1' < 0$, the metal flow direction on surface III is contrary to $\frac{d_1 r'^{I}}{dt}$.

During the double-flank non-backlash meshing process, the sliding direction of the contact points on the surface of the workpiece is different from that of the roller contacting surface, which results in the difference of the metal flow direction on the tooth flank of the workpiece, and the metal flow direction is shown in Fig. 2.12.

On the active tooth flank, metals will flow from the addendum and dedendum of the tooth to the surrounding of the pitch point. While on the driven tooth flank, metals will flow from the pitch point to the addendum and dedendum. The metal flow rule will result in the inclination of the tooth profile on the formed cold rolling forming components to a certain side. Under conditions, the modulus of the formed spline is bigger, because the height of the tooth is larger; the forming force is correspondingly very large. The inclination of the tooth will be more serious, which will result in the disqualification of the components due to the large error of the tooth profile. Therefore, cold rolling precision forming process is suitable for the forming of spline shafts with a modulus less than 2.

2.6 Mechanical Analysis on the Cold Rolling Precision Forming of Spline

2.6.1 Basic Assumption of the Stress Analysis

The plastic forming zone of cold rolling precision forming spline varies instantly with the rotation of the workpiece and the roller. In order to analyze the stress of the contacting zone, discrete modeling analysis method according to the contact position is proposed. Based on the built mathematical model of the contact zone, the contacting zone is dispersed according to the angle of the workpiece tooth central line relative to the reference position. At every position, mechanical analysis is carried out using computer-aided calculation method according to the slip-line field theory of plane strain.

On the study of the stress state in the plastic deformation of a point and the solving of the stress components, the following assumptions are proposed:

(1) There is no deformation in the axial direction;
(2) The volume is invariable in the plastic deformation;
(3) The material of the workpiece is ideal rigid-plastic material; and
(4) The normal force and the friction force are evenly distributed in the contact area between addendum of the roller and the workpiece, as well as between the flank of the roller and the workpiece.

The coordinate system of the stress analysis is chosen as follows: taking the center of roller as the zero point; the line connecting the center of roller and the tiptop of roller addendum as y-axis; and the axial direction of the workpiece or the roller as z-axis. xOy plane is the main stress plane, and stress along z-direction is the medium main stress. The stress state of an arbitrary point in the plastic deformation zone of the workpiece is shown in Fig. 2.13.

In the forming process, y-axis is rotating with the roller, while y-axis is perpendicular to the contact surface, and σ_y is the average pressure force on the contacting plane.

The frictional model: The contact between the roller addendum and the workpiece is mainly indentation, the contact between the flank and the tooth of the workpiece is rolling and pressing, and the friction forms on different contact positions are different. According to the contact characteristics, when analyzed with the slip-line theory, constant friction model should be used in the contact area between addendum of the roller and the workpiece, and the coulomb friction model should be used in the contact area between flank of the roller and the workpiece.

The stress-equilibrium differential equation, the strain-displacement relation, and the characteristic of plain strain are all suitable for cold rolling of external spline, that is:

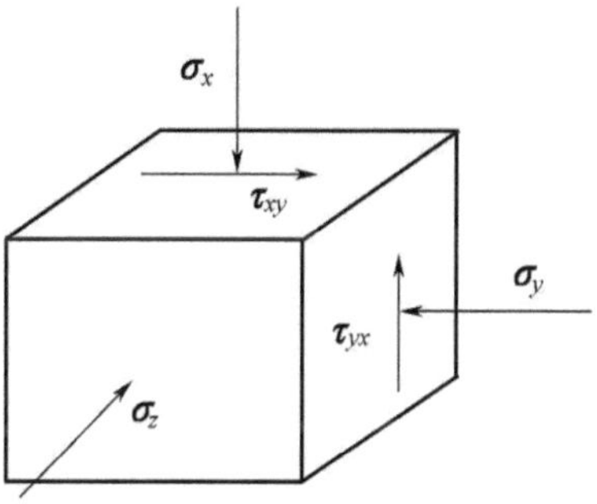

Fig. 2.13 Diagram of the stress state

$$\begin{cases} \sigma_z = \frac{1}{2}(\sigma_x + \sigma_y) = \sigma_m = \sigma_2 \\ \tau_{zx} = \tau_{zy} = 0 \end{cases} \tag{2.29}$$

The other stress components can be expressed as follows:

$$\begin{cases} \sigma_x = \sigma_m - K \sin 2\omega \\ \sigma_y = \sigma_m + K \sin 2\omega \\ \tau_{xy} = K \cos 2\omega \end{cases} \tag{2.30}$$

where

K is the shear yield strength;
ω is the direction angle of the slip line; and
σ_m is the mean stress.

Expressed by the principal stress, the yielding criterion is as follows:

$$\sigma_1 - \sigma_3 = 2K \begin{cases} Tresca : K = \sigma_s/2 \\ Mises : K = \sigma_s/\sqrt{3} \end{cases} \tag{2.31}$$

H. Hencky stress equation that reflects the rule of the mean stress in slip line is as follows:

$$\begin{cases} \sigma_m - 2K\omega = \xi(\beta), & \text{(Along line } \alpha) \\ \sigma_m + 2K\omega = \xi(\alpha), & \text{(Along line } \beta) \end{cases} \tag{2.32}$$

2.6.2 *Fundamental Plastic Forming Theory of Stress Analysis*

In the cold rolling precision forming process, there is no deformation along the axis, and the displacement and strain are zero. In the theory of plastic forming, if all of the particles deform only in the same coordinate plane, there is no deformation at the other normal directions, and thus, it can be called as plane deformation or plane strain. Obviously, the cold rolling precision forming is a kind of plane deformation. Characteristics of the plane strain state and stress-equilibrium differential equation and geometrical equation (relation of the displacement increment and strain increment) are all suitable for this process.

In the cold rolling process, the formed teeth must be rolled for several times and work hardening will occur on the surface of the workpiece, and the average stress in the contact area of the forming process can be obtained by setting up of the slip-line field of each tooth from entering to detaching the contact zone.

2.6.2.1 Fundamental Equation of the Plane Strain

Because plastic forming is only occurred in the coordinate plane of xOy, there is no deformation in the plane perpendicular to the deforming plane, and there is no relation between the variations of displacement in the deforming object and coordinate z. Thus

$$u = u(x, y), v = v(x, y), w = 0 \tag{2.33}$$

According to the geometrical equation, the strain components are as follows:

$$\begin{cases} \varepsilon_x = \dfrac{\partial u}{\partial x}, \varepsilon_y = \dfrac{\partial v}{\partial y}, \varepsilon_z = \dfrac{\partial w}{\partial z} = 0 \\ \gamma_{xy} = \dfrac{1}{2}\left(\dfrac{\partial u}{\partial y} + \dfrac{\partial v}{\partial x}\right), \gamma_{xz} = \dfrac{1}{2}\left(\dfrac{\partial w}{\partial x} + \dfrac{\partial u}{\partial z}\right) = 0, \gamma_{yz} = \dfrac{1}{2}\left(\dfrac{\partial w}{\partial y} + \dfrac{\partial v}{\partial z}\right) = 0 \end{cases} \tag{2.34}$$

From $\varepsilon_z = 0$, it can be deduced that $\dot{\varepsilon}_z = 0$.

According to Saint-Venant's principle, $\dot{\varepsilon}_{ij} = \dot{\lambda}\sigma'_{ij}$, then:

$$\dot{\varepsilon}_z = \dot{\lambda}\sigma'_z = \dot{\lambda}(\sigma_z - \sigma_m) = 0 \tag{2.35}$$

where $\dot{\lambda} = d\lambda/dt$, $d\lambda$ is the positive constant;

σ_m is the average stress, $\sigma_m = \frac{1}{3}(\sigma_x + \sigma_y + \sigma_z)$.

There is $\sigma_z - \sigma_m = 0$

$$\sigma_z = \frac{1}{2}(\sigma_x + \sigma_y) = \sigma_m = \sigma_2 = \frac{1}{2}(\sigma_1 + \sigma_3) \tag{2.36}$$

In the plain deformation, the plane of plastic flow is perpendicular to z-axis, there is no relative movement, inclination, and torsion between surfaces, that is:

$$\tau_{zx} = \tau_{zy} = 0 \tag{2.37}$$

The maximum shear stress can be expressed by the first and third main stresses:

$$\tau_{\max} = \frac{1}{2}(\sigma_1 - \sigma_3) \tag{2.38}$$

If the stress state of an arbitrary point Q is expressed by the mean stress σ_m and the maximum shear stress $\tau_{\max}$:

$$\sigma_1 = \sigma_m + \tau_{\max}, \sigma_2 = \sigma_m, \sigma_3 = \sigma_m - \tau_{\max} \tag{2.39}$$

At a plane strain state, the stress Mohr's circle of an arbitrary point Q in the deforming object is shown in Fig. 2.14.

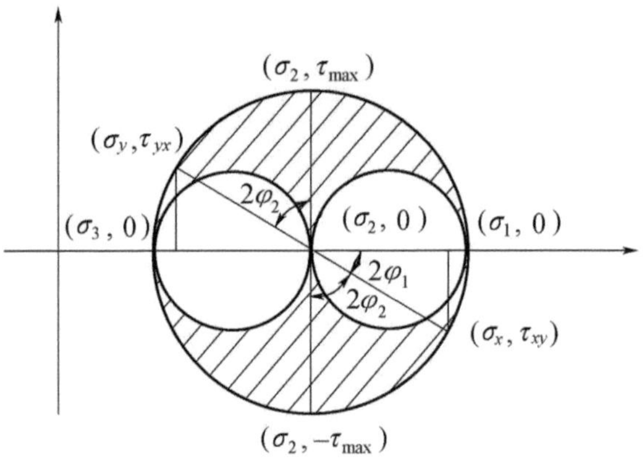

Fig. 2.14 Stress Mohr's circle at plane strain state

From the stress Mohr's circle, the direction angle φ_1 of the main stress and the direction angle φ_2 of the maximum shear stress can be calculated as follows:

$$\tan 2\varphi_1 = -\frac{2\tau_{xy}}{\sigma_x - \sigma_y}, \tan 2\varphi_2 = -\frac{\sigma_x - \sigma_y}{2\tau_{xy}} \tag{2.40}$$

2.6.2.2 Yield Criteria

When the particles of an object are in a multistress state, the stress state of a point can be determined by the six stress components. Under certain deforming conditions (deforming temperature and speed), plastic deforming occur on the particles only when each stress component coincides with a certain relationship.

There are two commonly used yield criteria to determine the plastic deformation state: Tresca criterion and Mises criterion [17]. In this book, Mises criterion is used; that is, under certain deforming conditions, when the elastic potential energy changing the unite volume shape of the material reaches a certain constant, the material will yield. The mathematical description of Mises criterion is as follows:

$$(\sigma_x - \sigma_y)^2 + 4\tau_{xy}^2 = 4K^2 = 4\left(\frac{\sigma_s}{\sqrt{3}}\right)^2 \tag{2.41}$$

where

K is the shear yield strength and
σ_s is the yield stress.

The criterion can be expressed by the main stress as follows:

$$\sigma_1 - \sigma_2 = 2K \tag{2.42}$$

where $K = \frac{\sigma_s}{\sqrt{3}}$.

2.6.2.3 The Strengthening Model

In the cold rolling forming of spine shaft, the final shape of the component is obtained by multipass rolling deformation of the billet applied by rollers. Through the cold plastic deformation and the work hardening of the materials, the mechanical performance of the material can be obtained and thus improved the strength of the spline shafts.

In the modern plastic forming theory, there are many constitutive relation hypothesis and material strengthening models, in which the following three models are most widely utilized in theoretical and engineering analysis: isotropic hardening model, kinematic hardening model, and anisotropic hardening model [18].

(1) Isotropic hardening model: In the plastic deforming process, the subsequent yield surface and the loading surface are enlarged with a similar shape in the stress space and the center is fixed. In the π plane, the loading surface is a curve, which is similar to the initial curve. In the isotropic hardening model, the anisotropy resulted by plastic forming has been neglected. The result coincides with practice under a medium deformation and small variation of the stress deviation ratio.
(2) Kinematic hardening model: It is a simplified model in which Bauschinger effect is considered. The material is assumed to be hardened in the plastic deformation direction and equally softened at the opposite direction. Therefore, in the loading process, with the development of the plastic deformation, the size and shape of the yield surface are invariant and just translation integrally in the stress space.
(3) Anisotropic hardening model: In order to reflect the Bauschinger effect of the materials perfectly, the isotropic hardening model and kinematic hardening model can be combined together. It can be considered that the shape, size, and position of the subsequent yield surface varied with the development of the plastic deformation. Although the result of this model fits well with the experimental result, the application of the model is limited because of its complexity.

Assume that the material is an isotropic strain hardening material, according to the isotropic hardening model, the shape and central position of the subsequent yield locus are unchanged. The loading surface by Tresca criterion is a series of concentric hexagonal prisms surface; while under Mises criterion, the loading surface is a series of concentric cylindrical surface, as shown in Fig. 2.15.

2.6.2.4 The Slip-Line Stress Field Theory

Slip-line theory is a powerful tool for solving ideal rigid-plastic plane strain problems [19–21]. Slip line is the trajectory of the maximum tangent stress direction on the plastic flow plane (plastic zone). Each point Q satisfying the yield

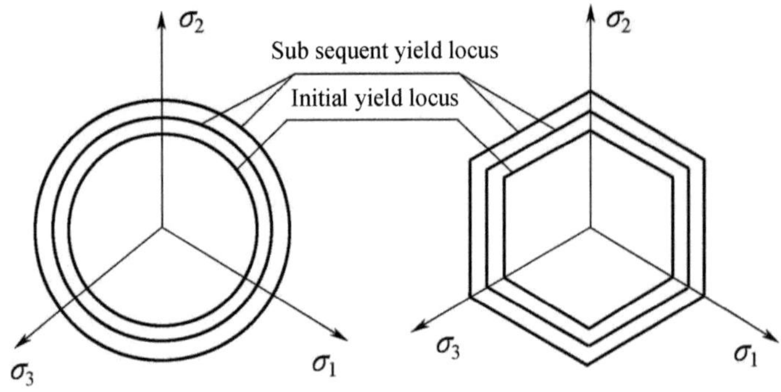

Fig. 2.15 Subsequent yield locus of isotropic strain hardening materials

criterions in the plastic zone has two orthogonal shear directions: the first one and the second one, combining graphic and analysis method, connect each point of the plastic zone along the first and second shear direction, then two group of orthogonal curves α and β can be obtained, which is called as slip line. Mutual orthogonal slip lines consist of the slip-line network and form the slip-line field. Applying the property and stress boundary conditions of the slip line, the distribution of the stress and strain speed, and finally the limited load in the plastic zone can be solved, and thus solve the mechanical problems of plane plastic flow in the cold rolling forming of spline.

The slip line and slip-line field are shown in Fig. 2.16. The direction angle of slip line ω is the angle of the first shear direction, which is the angle formed by the coordinate axis Ox rotating to the direction of the first shear direction clockwisely. The differential equation of α and β group of the slip line is as follows [7]:

Fig. 2.16 Slip line and slip-line field

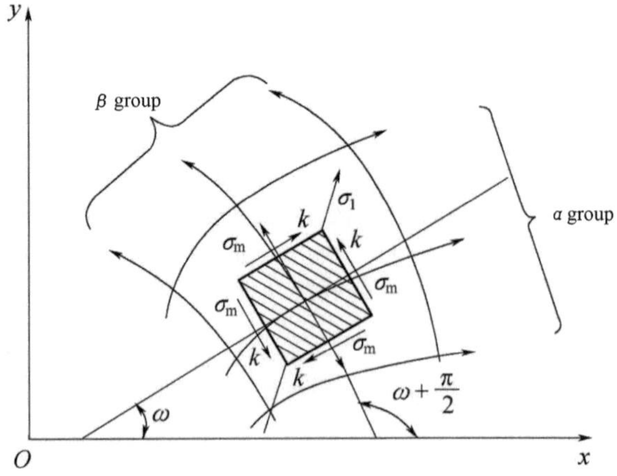

$$\begin{cases} \frac{dy}{dx} = \tan \omega & \text{along } \alpha \\ \frac{dy}{dx} = -\cot \omega & \text{along } \beta \end{cases} \tag{2.43}$$

According to H. Hencky stress equation, the average stress on the slip line can be obtained as follows:

$$\begin{cases} \sigma_m - 2K\omega = \xi(S_\beta) & \text{along } \alpha \\ \sigma_m + 2K\omega = \eta(S_\alpha) & \text{along } \beta \end{cases} \tag{2.44}$$

In the above equation, the value of $\xi(S_\beta)$ (or $\eta(S_\alpha)$) can be determined with the following method: While moving along one of the slip line of group α or β, it is a constant; while moving from one slip line to the other, the constant will change correspondingly. The value of these constants can be obtained by the stress boundary conditions of the slip-line fields.

2.6.3 *Slip-Line Field and Stress Analysis in the Initial Rolling Stage*

In the initial stage, the feed rate is very small; the involute flank of the workpiece has not formed. The plastic deforming area is the contact area between the addendum of roller and workpiece. Taken y-axis as the horizontal direction, the free boundary of the workpiece was an arc, which can be simplified as a straight line. And the arc of the roller addendum circle was divided according to the area of the contact zone. In order to ensure the calculated accuracy, the subsection of arc $\overset{\frown}{AB}$ should satisfy $d\psi < 5°$. According to the stress boundary condition, the slip-line

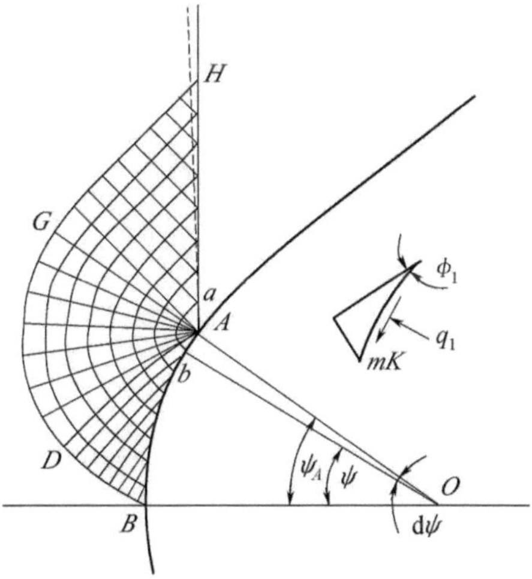

Fig. 2.17 Slip-line field at a small feeding rate

field can be set up, as shown in Fig. 2.17. The boundary shown in dash line is the arc of the external circle on the workpiece before simplified.

In the built slip-line field *ABDGH*, *AH* is the free surface without applied force on the workpiece, and the slip-line field in zone *AGH* is a uniform stress field with a free boundary. The surface *AB* contacted with the addendum of the roller is the non slip-line smooth boundary in the plastic deformation zone. Therefore, in the area *ABD*, the second class of boundary value problem is suitable. The slip-line field can be obtained with the graphic method. The angle between the slip line that contact with the roller addendum and the contact surface is ϕ_1, which is related to friction. Area *ADG* is the central sector field, and area *AGH* is a free boundary uniform stress field.

According to the slip-line field, the stress state on the contact surface can be obtained. Take a point a on the free surface *AH* and a point b on the contact surface *AB*, and points a and b are on the same slip line of ab. The stress states of points a and b are shown in Fig. 2.18a and 2.18b, respectively.

Slip line of ab is the β slip line. Point a is on the free surface, and from the stress boundary types, the slip-line direction angle of point a can be easily obtained as follows:

$$\omega_a = -\frac{3\pi}{4};$$

Additionally, $\sigma_1 = 0$

From the yield criterion, substituting $\sigma_1 = 0$ to Eq. (2.42), it can be obtained as follows:

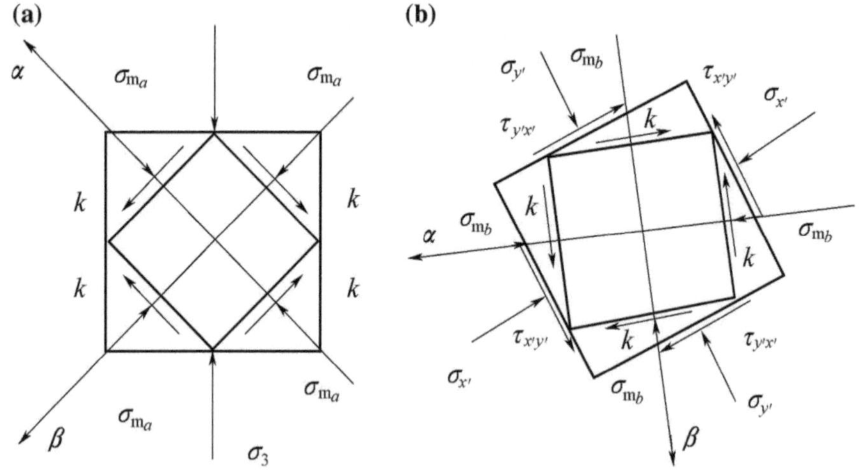

Fig. 2.18 Stress states of points on the free surface and the contact surface **a** stress state of point a; **b** stress state of point b

$$\sigma_3 = -2K$$

The average stress σ_{m_a} of point a on the slip line is as follows:

$$\sigma_{m_a} = -K$$

The direction angle of point b on the contact surface is related to friction:

$$\omega_b = -(\phi_1 + \psi)$$

where ϕ_1 is the angle between the slip line determined by the friction model and the contact surface;

ψ is the angle between Ob and *y-axis*. According to the characteristic that the variation of the average stress along the slip line in the slip-line field is in proportion to the variation of the direction angle of the slip line, the average stress σ_{m_b} on point b on the slip line and the average stress σ_{m_a} of point a has the following relationship:

$$\sigma_{m_b} - \sigma_{m_a} = -2K(\omega_b - \omega_a)$$

The average stress of point b on the contact surface can be obtained as follows:

$$\sigma_{m_b} = -2K\left(\frac{1}{2} + \frac{3\pi}{4} - \phi_1 - \psi\right) \tag{2.45}$$

2.6.4 Slip-Line Field and Stress Analysis in the Stable Rolling Stage

After the preliminary indexing, tooth profile has been formed on the workpiece surface. From the analysis of the contact area, it can be seen that when the workpiece enters the contact zone, the tooth surface contacts with the addendum and the flank of the roller. The contact position of the spline tooth is different at different rotation angles, and only the involute spline on the tooth flank is in contacting. Therefore, the contact condition can be simplified as the contact of two cylinders on the contact point, and the stress can be analyzed with the slip-line field method in simple cross rolling. Here, the slip-line field and stress of the contact zone are mainly analyzed and discussed under the conditions the addendum and the flank contact with the workpiece simultaneously.

y-axis is still chosen as the horizontal direction. The boundary of the contact zone can be simplified by substituting the involute on the roller flank with the tangent line on the intersection point of the pitch circle and the involute of the flank on conditions of rolling meshing. The non-contact free boundary on the workpiece

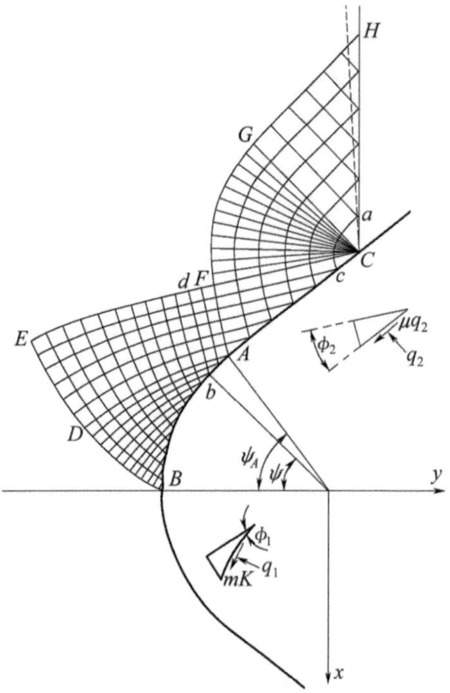

Fig. 2.19 Slip-line field in stable rolling

can also be simplified as a straight line; the arc on the roller addendum can be dealt as the condition at the initial rolling. By analyzing the stress boundary conditions of each contact zone, the slip-line field of the contact zone in stable rolling stage can be set up, as shown in Fig. 2.19.

In slip-line field *ABDEFGHC* established in the stable rolling stage, *CH* is the free surface on the workpiece without external force; therefore, the slip-line field at zone of *CGH* is a free boundary straight-line field. The contact surface *AC* on the roller flank is a non slip-line smooth boundary of the plastic deformation; therefore, the surface slip-line field is the second type of boundary value problem in the area *ABF*. The angle between the slip line contacting with roller flank and the contact surface is ϕ_2 and ϕ_2 is relative with friction. The involute of the flank can be simplified as a straight line, and area *ACF* is a straight-line slip-line field. The area *CFG* located between the contact surface of the roller flank and the free surface of the workpiece is a central sector field. It is the second type of boundary value problem in the area *ABD* at the contact surface *AB* of the roller addendum, and the graphic of the slip-line field in the plastic zone is similar to the slip-line field of the roller addendum contact zone in the initial rolling stage. In the area *ADEF* between the roller addendum and the flank contact surface, the first class boundary value problem is suitable. And the boundary slip-line *AF* of zone *ACF* is a straight line. Therefore, the right upper area of *ADEF* is a straight-line field, other arcs can be substituted by cord lines, and the slip-line field can be obtained with the graphic method.

In order to obtain the stress state of the points on the contact line, point a is taken from the free surface CH and point c is taken from the contact surface AC. Points a and c are in the same group of slip line, and point d is taken from the straight-line slip-line field of area $ADEF$. Points c and d are in the same group slip line. And points d and b are in the same group slip line. The stress states of points a and b are shown in Fig. 2.18, and the stress state of point d is similar to the stress state of point b. According to the stress states, it can be confirmed that lines ac and db are the β slip line, and line cd is the α slip line.

By analyzing of the free surface, it is easy to obtain from the stress boundary type, that, at point a:

$$\omega_a = -\frac{3\pi}{4}, \sigma_{m_a} = -K$$

The direction angle of the slip line at point c is related with friction:

$$\omega_c = -\left(\frac{\pi}{2} - \alpha + \phi_2\right)$$

where ϕ_2 is the angle between the slip line decided by the friction model and the contact surface.

According to the clan property of the slip line, there is:

$$\sigma_{m_c} = -2K\left(\frac{1}{2} + \frac{\pi}{4} + \alpha - \phi_2\right) \tag{2.46}$$

The slip line cd is a straight line, therefore

$$\omega_d = \omega_c, \sigma_{m_d} = \sigma_{m_c}$$

The same as the initial rolling contact, at point b on the contact arc of the roller addendum:

$$\omega_b = -(\phi_1 + \psi)$$

From Hencky stress equation (2.44), obtain:

$$\sigma_{m_b} - \sigma_{m_d} = -2K(\omega_b - \omega_d)$$

It can be obtained as follows:

$$\sigma_{m_b} = -2K\left(\frac{1}{2} + \frac{3\pi}{4} - \phi_1 - \psi\right) \tag{2.47}$$

Equations (2.46) and (2.47) are the formulas of the average stress on contact surface, while the roller addendum and the flank are both contacted with the workpiece. In which Eq. (2.47) has the same format with the average stress expressed in Eq. (2.45) at the initial rolling stage, only the addendum of the roller is contacted with the roller.

2.7 Average Pressure on the Contact Surface During the Forming Process

The theoretical calculation method of the plastic forming force mainly includes main stress method, slip-line field method, upper bound method and finite element method [22].

The unit average pressure on the contact surface of simple cross rolling has been solved by А. Д. Томленов with slip-line method. But friction and work hardening were not considered in the solving process, and it is difficult to be applied in the cold rolling process of the spline. On the basis of this method, modification has been carried out to meet the requirement of the average stress calculation of the spline cold rolling precision forming process.

In the setting up of the slip-line field, different friction conditions are adopted in different contact zones, and in the analysis of the stress, the angle between the slip line and the contact surface should be calculated, respectively, according to the different selected friction models.

Taking stable rolling, for example, in the Fig. 2.19, *AB* zone is the contact zone between the addendum of the roller and the workpiece, where the constant friction condition is more suitable for the contact deformation; therefore, angle ϕ_1 between the slip-line direction on the contact surface and the tangent direction of the contact surface is as follows:

$$\phi_1 = \frac{1}{2}\arccos m \tag{2.48}$$

where m is the constant friction factor, $0 \le m \le 1$.

In Fig. 2.19, zone *AC* is the contact zone of the roller flank and the workpiece, where Coulomb friction condition is more suitable for the contact deforming friction model. Taking the direction perpendicular with the contact surface as axis y', according to the average stress and the maximum shear stress, the normal pressure stress and the tangent stress is as follows:

$$\begin{cases} \sigma_{y'} = \sigma_{mc} + K\sin 2(-\phi_2) = -K\left(1 + \frac{\pi}{2} + 2\alpha - 2\phi_2 + \sin 2\phi_2\right) \\ \tau_{x'y'} = K\cos 2\phi_2 \end{cases} \tag{2.49}$$

According to Coulomb friction condition, Coulomb friction coefficient μ is as follows:

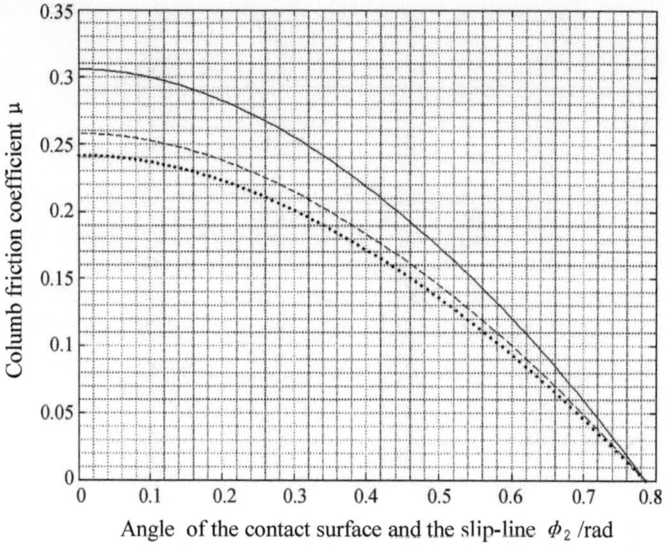

Fig. 2.20 Relationship between Coulomb friction coefficient μ and angle ϕ_2

$$\mu = \frac{|\tau_{x'y'}|}{|\sigma_{y'}|} = \frac{\cos 2\phi_2}{1 + \frac{\pi}{2} + 2\alpha - 2\phi_2 + \sin 2\phi_2} \tag{2.50}$$

It is difficult to solve the angle ϕ_2 from Eq. (2.50) by numerical calculation. Putting a series of ϕ_2 into (2.50), then a series of μ can be obtained. So the diagram can be plotted by the series of ϕ_2 and μ, as shown in Fig. 2.20, and ϕ_2 can thus be obtained from the diagram. In Fig. 2.20, the real line, the broken line, and the point line correspond to $\alpha = 20°$, $\alpha = 37.5°$, and $\alpha = 45°$, respectively. The abscissa represents the change of angle ϕ_2, and the unit of the angle is radian. For example, when $\alpha = 45°$, it can be obtain from Fig. 2.20: When $\mu = 0.1$, $\phi_2 = 0.5855$; and when $\mu = 0.2$, $\phi_2 = 0.3038$.

In the cold rolling precision forming of spline shafts, taking the direction perpendicular to the contact surface as axis y', from the stress analysis, it is known that σ_y is the unit average stress of the contact surface. Homogenizing the unit average stress on the contact zone of the roller addendum, then the normal pressure applied to point b and perpendicular to the contact surface can be expressed as q_1. From the average stress and maximum shear stress of point b, the unit pressure on the contact surface of the addendum contact zone is as follows:

$$q_1 = -\sigma_{y'} = -[\sigma_{m_b} + K \sin 2(-\phi_1)] = 2K\left(\frac{1}{2} + \frac{3\pi}{4} - \phi_1 - \psi + \frac{\sin 2\phi_1}{2}\right) \tag{2.51}$$

Similarly, while contacting on the flank, the unit average pressure is as follows:

$$q_2 = 2K\left(\frac{1}{2} + \frac{\pi}{4} + \alpha - \phi_2 + \frac{\sin 2\phi_2}{2}\right) \tag{2.52}$$

where

ϕ_1 is the function of the constant friction coefficient m;
ϕ_2 is the function of Coulomb friction coefficient μ.

Considering the action of work hardening, put the real stress (subsequent yield stress) at every contacting into the function determined by the yield criterion, for the average pressure of the contact surface with determined spline parameters in cold rolling forming, Eqs. (2.51) and (2.52) can be summed up to:

$$\begin{cases} q_1 = f(\sigma_{zs}) g_1(m, \psi) \\ q_2 = f(\sigma_{zs}) g_2(\mu) \end{cases} \tag{2.53}$$

Function f is determined by the yield criterion. When Mises yield criterion and isotropic hardening model are adopted, the shape of the subsequent yield locus is the same with the initial locus. The size of the yield locus is determined by the properties of materials; during the forming process, the subsequent yield stress s_{zs} will change with the variation of the plastic deformation degree.

Compared with the method of А. Д. Томленов, Eq. (2.53) has considered the position of the contact, different friction models have been used in different contact zones, and the influence of work hardening has been considered. Therefore, it is more close to the real average pressure on the contact surface in the spline shaft cold rolling forming.

Function σ_s is determined by the yield criterion, and once the yield criterion is selected:

$$f(\sigma_s) = C\sigma_s \tag{2.54}$$

where C is a constant.

In the plastic forming, the constant m and Coulomb friction coefficient μ are different under different deforming temperatures, different deforming degrees, and different strain rates. But in practice cold rolling forming, the exact value of the parameters is difficult to measure, and the friction coefficient under different conditions can only be measured with matured methods for the theoretical calculation and numerical simulation.

With a determined constant friction factor m and Coulomb friction coefficient μ:

$$\begin{cases} g_1(m, \psi) = C_1 h(\psi) \\ g_2(\mu) = C_2 \end{cases} \tag{2.55}$$

where C_1 and C_2 are constants.

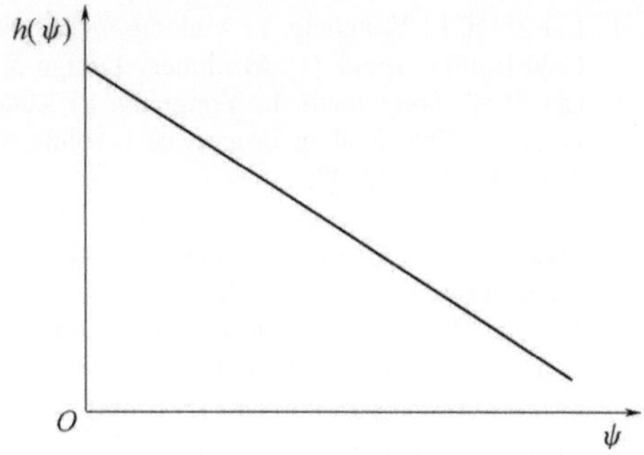

Fig. 2.21 Variation tendency of function $h(\psi)$

Function $h(\psi)$ is the decreasing function of ψ, and the variation trend of it is shown in Fig. 2.21. The maximum and minimum values of g_1 are less than 30%. For the convenience of the calculation, take q_1 on $h(0)$ as the average unit pressure on contact zone (AB) of the contact surface between the roller addendum and the workpiece. Then, Eq. (2.53) can be expressed as follows:

$$\begin{cases} q_{1av} = f(\sigma_s) g_1(m, 0) \\ q_{2av} = f(\sigma_s) g_2(\mu) \end{cases} \tag{2.56}$$

The above deduction is based on the hypothesis of ideal rigid-plastic object, and work hardening of the materials is not considered, what is existed in actual material models. In production, cold plastic forming is always used to improve the mechanical properties of the materials. Therefore, the yield stress before and after plastic forming is different.

For isotropy strain hardening materials, the position of the subsequent yield locus center and the shape after strain hardening is invariable. The shape of the subsequent yield locus is the same with that of the initial one, and the size of the locus is determined by the property of the materials. Therefore, the function f is invariable under determined workpiece materials. However, the stress σ_s should be substituted by the instant flow stress Y (also called as subsequent yield stress), that is:

$$f = f(Y) \tag{2.57}$$

where the real stress Y changes with the variation in the plastic deformation degree.

References

1. Thomas Herlan. Cross Rolling [J]. Metallurgia. 2002, 69(2): FT7.
2. Wu Xutang. Gear Engagement Theory [M]. Beijing: China Machine Press, 1985.
3. Li Yongtang, Zhang Dawei, Song Jianli, Fu Jianhua, Zheng Quangang. Mechanics Analysis and Numerical Simulation on the Precise Cold Roll Forming Process of Spline [J]. China Metal Forming Equipment & Manufacturing Technology, 2007, 6: 79–82.

4. Liu Zhiqi, Li Yongtang, Li Xudong, Qi Huiping. Test and Analysis of the Forming Force in Cold Rolling Spline [J]. Machinery Design & Manufacture, 2011, 1: 188–190.
5. Liu Zhiqi, Song Jianli, Li Yongtang, Li Xudong. Analysis and Experimental Study on the Precision Cold Rolling Process of Involute Spline [J]. Journal of Mechanical Engineering, 2011, 47 (14): 32–38.
6. Liu Zhiqi, Li Xudong, Li Yongtang. Qualification of Tooth Indexing for Cold-rolling Formation of spline [J]. Journal of Lanzhou University of Technology, 2009, 35 (4): 35–38.
7. Zhang Dawei. Theoretical Research on the Process of Spline Cold Rolling [D]. Taiyuan: Taiyuan University of Science and Technology, 2007.
8. Wang Zhongren, Huang Fuhua, Xin Zongren, et al. Special Forming (5th Chapter, First Volume, Forging Handbook) [M]. Beijing: China Machine Press, 1996.
9. Robert L. Miller, Warren Mich. Tooth Generating Tool and Method of Making the Same [P]. United States Patent. 3902349, 1975-09-02.
10. Donald E. Blue, Decatur. Tooth Forming Machine [P]. United States Patent. 3818736, 1974-06-25.
11. Zheng Quangang. Cold Rolling Forming Technology of Cylindrical Spur Gear with Small Module Involute Spline [J]. Automobile Technology & Material. 1997, 7: 16–18.
12. Xue Wenhui. Rolling Machining of Involute Spline on Rolling Machine [J]. Metal Forming Technology, 1999, 17 (3): 46–47.
13. Wu Xiuyi. Rolling Process Technology for Small Module Involute Splines [J]. Machinery, 1998, 1: 16–18.
14. Zhang Dawei, Fu Jianhua, Li Yongtang. Contact Ratio in Process of Spline Cold Roll Forming [J]. China Metal Forming Equipment & Manufacturing Technology, 2008, 4: 80–84.
15. Zhou Guoxiang, Zheng Peng, Liu Chengliang. Analysis and Kinematic Modeling of Ausform-Finishing Process for Manufacturing Hardened Gears [J]. Mechanical Science and Technology, 2005, 24 (5): 562–564.
16. Wu Xutang. Principle of Gearing [M]. Beijing: China Machine Press, 1985.
17. Wang Zhongren. Mechanical Foundation of Plastic Processing [M]. Beijing: National Defence Industry Press, 1989.
18. Jin Junsong. Research on the Key Technology of Near-net Shaped Forming for Gears used in Car [D]. Wuhan: Huazhong University of Science and Technology, 2009.
19. M. H. Parsa, P. H. Matin, M. M. Mashhadi. Improvement of Initial Blank Shape for Intricate Products Using Slip Line Field [J]. Journal of Materials Processing Technology, 2004, 145:21–26.
20. N. S. Das, B. S. Chawla, C. K. Biswas. An Analysis of Strain in Chip Breaking Using Slip-line Field Theory with Adhesion Friction at Chip/tool Interface [J]. Journal of Materials Processing Technology, 2005, 170: 509–515.
21. G. Samolyk, Z. Pater. Application of the Slip-line Field Method to the Analysis of the Die Cavity Filling [J]. Journal of Material Processing Technology, 2004, 153–154:729–735.
22. Yu Hanqing, Chen Jinde. Metal Plastic Forming Principle [M]. Beijing: China Machine Press, 1999.

Chapter 3
Process Parameters in the Cold Rolling Forming of Spline

Abstract Design methods of the spline cold rolling precision forming process parameters, such as the contact area of the spline shaft, rolling force and rolling moment, and the calculation of the cold rolling workpiece billet diameter have been illustrated and analyzed.

3.1 Contact Area in the Cold Rolling Forming of Spline Shafts

In the cold rolling forming process of external spline, the main deformation is the contact deformation between the roller addendum and the workpiece, which is the key factor on the research of the rolling principle and mechanical analysis. The setting up of the mathematical model of the contact surface and the exact solution of the rolling force and moment in the deforming process are of important significance.

In the plastic processing, the external force applied on the workpiece by the movable tools of the machine to produce plastic deformation on the metals is called as forming force. Forming force is one of the necessary mechanical parameters for the selection of machine capability, the proper design of tools, and reasonable planning of the process procedures. The forming force can be calculated through the integration of the stress on the contact areas:

$$F = \iint_{s} \sigma_n ds$$

where

S is the area of the contact zone and
σ_n is the contact stress.

It can be seen from the tooth forming principle of the spline shafts' cold rolling precision forming that the contact zone changes instantly in the rolling forming process, and the stress distribution of the contact zone and the mathematical

J. Song et al., *Cold Rolling Precision Forming of Shaft Parts*,
DOI 10.1007/978-3-662-54048-0_3

description of the contact area are both difficult to determine. Therefore, the solution to the forming force is relatively difficult.

The contact area is varying instantly but periodically in the spline shaft cold rolling forming process. The variation trend of the contact force on a single tooth of the workpiece is similar to that of the former adjacent tooth in the contact zone from engagement to disengagement, only with the increase in the amplitude.

According to the position variation rule of the contact point in the double-flank non-backlash meshing rolling forming, the forming process can be decomposed into the contacting of certain number of single tooth; the contact area in the cold rolling forming process can thus be obtained by the calculation model of the contact area of a single tooth.

3.1.1 Calculation Model

There is no deformation at the axial direction in the cold rolling precision forming of spline shafts; the contact curve perpendicular to the roller and the axial workpiece cross section is also invariable along the axis. At an arbitrary moment of the forming process, the contact area is always the product of the contact curve length on the cross section and the length of the spline. Therefore, by calculating the contact curve length of the spline tooth and the roller cross section in the contact zone, the contact area during the forming process can be obtained according to Eq. (3.1):

$$A = S \cdot L \tag{3.1}$$

where

S is the length of contact curve of the cross section perpendicular to the axis and
L is the axial length of the forming workpiece.

In order to set up the mathematical model of the contact curve on the cross section in the rolling process, the following assumptions are proposed:

(1) The elastic deformation of the workpiece and the roller is ignored;
(2) The center distance of the roller and the workpiece is invariable from entering to leaving of the contact zone of a single tooth; and
(3) The tooth rise per tooth is constant in a rolling process.

The schematic diagram of the contact state on the cross section perpendicular to the workpiece and roller axis direction is shown in Fig. 3.1. Set up the coordinate system Oxy integrated with the workpiece and the coordinate system $O_2x_2y_2$ integrated with the roller, respectively. And θ is the angle of y_2-axis and *x-axis* (counter clockwise is positive and clockwise is negative).

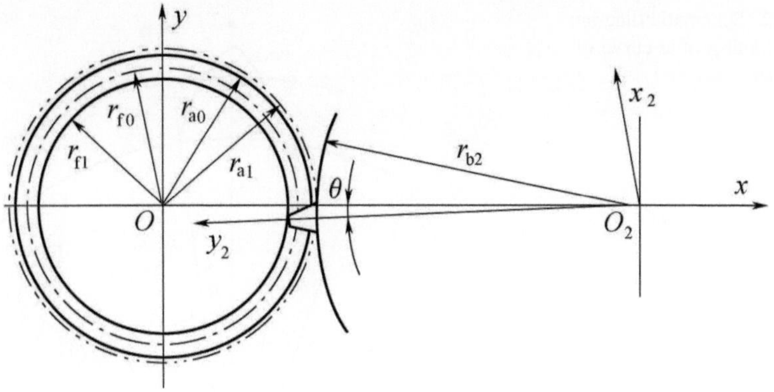

Fig. 3.1 Schematic diagram of the contact state on the cross section

The coordinate transformation from $O_2x_2y_2$ to Oxy is as follows:

$$\begin{pmatrix} x \\ y \end{pmatrix} = \begin{pmatrix} \cos(\frac{\pi}{2}-\theta) & -\sin(\frac{\pi}{2}-\theta) \\ \sin(\frac{\pi}{2}-\theta) & \cos(\frac{\pi}{2}-\theta) \end{pmatrix} \begin{pmatrix} x_2 \\ y_2 \end{pmatrix} + \begin{pmatrix} a \\ 0 \end{pmatrix}$$
$$= \begin{pmatrix} x_2\cos(\frac{\pi}{2}-\theta) - y_2\sin(\frac{\pi}{2}-\theta) + a \\ x_2\sin(\frac{\pi}{2}-\theta) + y_2\cos(\frac{\pi}{2}-\theta) \end{pmatrix} \tag{3.2}$$
$$= \begin{pmatrix} x_2\sin\theta - y_2\cos\theta + a \\ x_2\cos\theta + y_2\sin\theta \end{pmatrix}$$

where a is the relative center distance between the roller and workpiece.

The relative movement between the roller and the workpiece in the spline cold rolling precision forming is called as enveloping motion. According to the meshing principle, considering the feeding reduction of the roller and the tooth rise of the workpiece, the mathematical equation of the roller and workpiece profile can be set up. From the tooth profile equations of the roller and the workpiece in the contact zone, when transformed to the same coordinate system, the contact state can be judged by the coordinate value, and the length of the contact line can be calculated by the program based on MATLAB.

3.1.2 *Tooth Profile Equation of the Roller and the Workpiece in the Forming Process*

3.1.2.1 Tooth Profile of the Roller

Firstly, set up the curve equation of the roller tooth in the coordinate of $O_2x_2y_2$. In the cold rolling forming of spline, a round dedendum of the spline is always

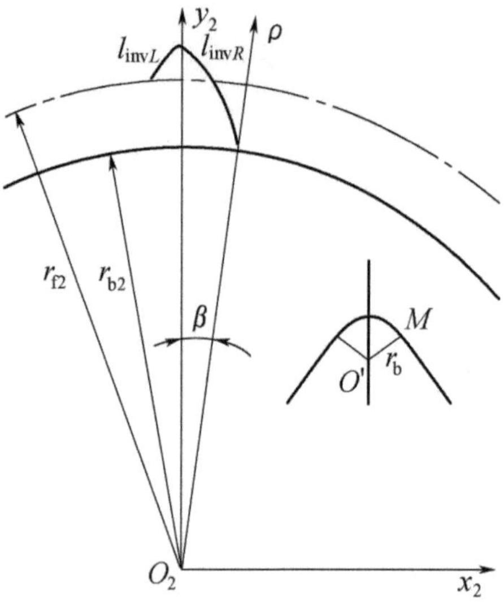

Fig. 3.2 Schematic diagram of the tooth profile curve of the roller

designed. Therefore, in the mathematical model of the roller tooth profile, the corresponding addendum of the roller is a segment of arc tangent with the involute flank on both sides and the addendum circle, as shown in Fig. 3.2. And the profile of the roller can be expressed with the left flank involute l_{invL}, transition arc l_{cirA} of the addendum circle, and the right flank involute l_{invR}.

The polar coordinate equation of the roller flank involute l_{invR} with y_2 as the polar coordinate axis is as follows:

$$\begin{cases} \rho = r_{b_2} \cdot \sec \alpha \\ \theta = \beta - \mathrm{inv}\alpha \end{cases} \tag{3.3}$$

where

β is the angle of the involute polar coordinate axis and y_2

β can be calculated by Eq. (3.4):

$$\beta = \tan \alpha_2 - \alpha_2 + \frac{0.5\pi}{Z_2} \tag{3.4}$$

where

α_2 is the pressure angle of the pitch circle of the roller and

Z_2 is the tooth number of the roller.

The transition arc of the roller addendum and the involute of the flank are tangent at point M. From the geometrical relationship:

$$\angle O_2O'M = \frac{\pi}{2} + \tan\,\alpha_M - \beta$$

In the triangular $O_2O'M$, according to cosine theorem, the polar radius of point M is as follows:

$$\rho_M = \sqrt{r_e^2 + (r_{a_2} - r_e)^2 + 2(r_{a_2} - r_e)r_e \sin(\tan\,\alpha_M - \beta)} \quad (3.5)$$

where

r_e is the radius of the transition arc on the roller addendum.

And,

$$\alpha_M = \arccos\frac{r_{b_2}}{\rho_M} \quad (3.6)$$

Solving the simultaneous equations of (3.5) and (3.6), the mathematical solution of α_M and ρ_M can be obtained by programming calculation method.

In the rolling process, theoretical pressure angle on the nearest point of the workpiece dedendum (the pressure angle on the dedendum of the roller circle with non-top clearance meshing) is as follows:

$$\alpha_{f_2th} = \arccos\frac{r_{b_2}}{r_{a_2} - h_1} \quad (3.7)$$

where h_1 is the tooth height of the spline.

Then, the variation scope of the pressure angle of the roller flank involute is as follows: $[\alpha_M, \alpha_{f_2th}]$.

The equation of the roller flank involute l_{invL} in the coordinate system $O_2x_2y_2$ can be expressed as follows:

$$\begin{cases} x_2 = \rho \cdot \sin(\mathrm{inv}\,\alpha - \beta) = x_2(\alpha) \\ y_2 = \rho \cdot \cos(\mathrm{inv}\,\alpha - \beta) = y_2(\alpha) \\ \alpha \in [\alpha_M, \alpha_{f_2th}] \end{cases} \quad (3.8a)$$

The equation of the transition arc on the roller addendum l_{cirA} in the coordinate system $O_2x_2y_2$ is as follows:

$$\begin{cases} x_2 = r_e \cos\gamma \\ y_2 = r_e \sin\gamma + r_{a_2} - r_e \\ \gamma \in [\tan\,\alpha_M - \beta, \pi - \tan\,\alpha_M + \beta] \end{cases} \quad (3.8b)$$

The involute l_{invR} of the roller flank in the coordinate system $O_2x_2y_2$ is as follows:

$$\begin{cases} x_2 = \rho \cdot \sin(\beta - \operatorname{inv} \alpha) = x_2(\alpha) \\ y_2 = \rho \cdot \cos(\beta - \operatorname{inv} \alpha) = y_2(\alpha) \\ \alpha \in [\alpha_M, \alpha_{f_2 th}] \end{cases} \tag{3.8c}$$

Analysis of the contact conditions should be carried out in the same coordinate system, so the expression of the tooth profile curve of the roller in the coordinate system Oxy can be easily obtained by coordinate transformation in Eq. (3.2).

$$l_{invL}: \begin{cases} x = -r_{b_2} \cdot \sec \alpha \cdot \cos(\operatorname{inv}\alpha - \beta + \theta) + a \\ y = r_{b_2} \cdot \sec \alpha \cdot \sin(\operatorname{inv}\alpha - \beta + \theta) \\ \alpha \in [\alpha_M, \alpha_{f_2 th}] \end{cases} \tag{3.9a}$$

$$l_{cirA}: \begin{cases} x = r_e \cdot \sin(\theta - \gamma) - (r_{a_2} - r_e)\cos\theta + a \\ y = r_e \cdot \sin(\theta - \gamma) + (r_{a_2} - r_e)\sin\theta \\ \gamma \in [\tan \alpha_M - \beta, \pi - \tan \alpha_M + \beta] \end{cases} \tag{3.9b}$$

$$l_{invR}: \begin{cases} x = -r_{b_2} \cdot \sec \alpha \cdot \cos(\beta - \operatorname{inv} \alpha + \theta) + a \\ y = r_{b_2} \cdot \sec \alpha \cdot \sin(\beta - \operatorname{inv} \alpha + \theta) \\ \alpha \in [\alpha_M, \alpha_{f_2 th}] \end{cases} \tag{3.9c}$$

3.1.2.2 Tooth Profile of the Workpiece

The tooth profile curve l_{roller} of the roller consists of three curves: l_{invR}, l_{invL}, and l_{cirA}, which is shown in Eq. (3.9a)–(3.9c).

The addendum circle l_{a_0} of the workpiece in the coordinate system Oxy before rolling is as follows:

$$x^2 + y^2 = r_{a_0}^2 \tag{3.10a}$$

The root circle l_{f_0} of the workpiece in the coordinate system Oxy before rolling is as follows:

$$x^2 + y^2 = r_{f_0}^2 \tag{3.10b}$$

The addendum circle l_{a_1} of the workpiece in the coordinate system Oxy after rolling is as follows:

$$x^2 + y^2 = r_{a_1}^2 \tag{3.10c}$$

The root circle l_{f_1} of the workpiece in the coordinate system Oxy after rolling is as follows:

$$x^2 + y^2 = r_{f1}^2 \tag{3.10d}$$

Before rolling, the billet of workpiece is a cylinder; the equation of the cross section in the coordinate system Oxy is also a circle:

$$x^2 + y^2 = d_z^2/4 \tag{3.11}$$

Calculation of the contact area between the roller addendum and the billet external circle is very simple. The forming process is just the forming process of the workpiece tooth profile. Here, the mathematical model of the workpiece tooth profile in the forming process will be mainly discussed.

In the forming process, the relationship between the roller and the workpiece is presented as an enveloping movement; the tooth profile curve of the workpiece is the conjugate curves of the roller tooth profile curve.

Record the curve group formed by the roller tooth profile curve as C_{roller_θ}, and the enveloping curve of C_{roller_θ} is the tooth profile curve of the workpiece. It can be obtained by solving the enveloping curve of the roller tooth profile.

The tooth profile curve l_{roller} of the roller is composed by three curves. At the position θ, assume that the parameter of the roller tooth profile l_{roller_θ} is t (substitute corresponding α and γ) and the rotating angle of the roller is φ, while the workpiece rotates for an angle of $-i\varphi(i = r_{b_2}/r_b)$, and l_{roller_θ} is at the corresponding position, the coordinate transformation of the conjugate curves l_{enve_θ} is as follows:

$$\begin{cases} x(t,\varphi) = x_2(t)\cdot\cos(i+1)\varphi + y_2(t)\cdot\sin(i+1)\varphi + a\cdot\cos i\,\varphi \\ y(t,\varphi) = -x_2(t)\cdot\cos(i+1)\varphi + y_2(t)\cdot\sin(i+1)\varphi - a\cdot\cos i\,\varphi \end{cases} \tag{3.12}$$

According to the enveloping theory, the equation of l_{enve_θ} is Eq. (3.11) together with the following equation:

$$\frac{\partial y(t,\varphi)}{\partial t}\cdot\frac{\partial x(t,\varphi)}{\partial \varphi} - \frac{\partial y(t,\varphi)}{\partial \varphi}\cdot\frac{\partial x(t,\varphi)}{\partial t} = 0 \tag{3.13}$$

The partial derivative of $x(t,\varphi)$ and $y(t,\varphi)$ is as follows:

$$\begin{cases} \frac{\partial x(t,\varphi)}{\partial t} = \frac{dx_2(t)}{dt}\cos(i+1)\varphi + \frac{dy_2(t)}{dt}\sin(i+1)\varphi \\ \frac{\partial x(t,\varphi)}{\partial \varphi} = -(i+1)x_2(t)\sin(i+1)\varphi + (i+1)y_2(t)\cos(i+1)\varphi - a\cdot\sin i\varphi \\ \frac{\partial y(t,\varphi)}{\partial t} = -\frac{dx_2(t)}{dt}\sin(i+1)\varphi + \frac{dy_2(t)}{dt}cox(i+1)\varphi \\ \frac{\partial x(t,\varphi)}{\partial \varphi} = -(i+1)x_2(t)\cos(i+1)\varphi - (i+1)y_2(t)\sin(i+1)\varphi - a\cdot\cos i\varphi \end{cases} \tag{3.14}$$

Substituting Eq. (3.14) into (3.13), it can be obtained that

$$(i+1)(x_2(t)\frac{dx_2(t)}{dt}+y_2(t)\frac{dy_2(t)}{dt})-i\cdot a(\frac{dy_2(t)}{dt}\sin\varphi-\frac{dx_2(t)}{dt}\cos\varphi)=0 \quad (3.15)$$

Let

$$\begin{cases} \sin\lambda=\dfrac{\frac{dx_2(t)}{dt}}{\sqrt{\left(\frac{dx_2(t)}{dt}\right)^2+\left(\frac{dy_2(t)}{dt}\right)^2}} \\ \cos\lambda=\dfrac{\frac{dy_2(t)}{dt}}{\sqrt{\left(\frac{dx_2(t)}{dt}\right)^2+\left(\frac{dy_2(t)}{dt}\right)^2}} \\ \lambda=\arctan\dfrac{dx_2(t)/dt}{dy_2(t)/dt} \end{cases} \quad (3.16)$$

Then, Eq. (3.14) can be converted into:

$$\sin(\varphi-\lambda)=\frac{(i+1)(x_2(t)\sin\lambda+y_2(t)\cos\lambda)}{i\cdot a} \quad (3.17)$$

Therefore

$$\varphi=\arcsin\frac{(i+1)(x_2(t)\sin\lambda+y_2(t)\cos\lambda)}{i\cdot a}+\lambda \quad (3.18)$$

Equations (3.12) and (3.18) will compose the tooth profile of the workpiece, when l_{enve_θ} is at an angle of θ. For convenience, different tooth profile curves in Eq. (3.12) can be expressed in segments as follows:

$$l_{invL}:\begin{cases} x_2(\alpha)=-r_{b_2}\cdot\sec\alpha\cdot\cos(\mathrm{inv}\alpha-\beta+\theta) \\ y_2(\alpha)=r_{b_2}\cdot\sec\alpha\cdot\sin(\mathrm{inv}\alpha-\beta+\theta) \\ \alpha\in[\alpha_M,\alpha_{f_2th}] \end{cases} \quad (3.19a)$$

$$l_{cirA}:\begin{cases} x_2(\gamma)=r_e\cdot\sin(\theta-\gamma)-(r_{a_2}-r_e)\cos\theta \\ y_2(\gamma)=r_e\cdot\cos(\theta-\gamma)+(r_{a_2}-r_e)\sin\theta \\ \gamma\in[\tan\alpha_M-\beta,\pi-\tan\alpha_M+\beta] \end{cases} \quad (3.19b)$$

$$l_{invR}:\begin{cases} x_2(\alpha)=-r_{b_2}\cdot\sec\alpha\cdot\cos(\beta-\mathrm{inv}\alpha+\theta) \\ y_2(\alpha)=r_{b_2}\cdot\sec\alpha\cdot\sin(\beta-\mathrm{inv}\alpha+\theta) \\ \alpha\in[\alpha_M,\alpha_{f_2th}] \end{cases} \quad (3.19c)$$

The corresponding derivative is as follows:

$$l_{invR}:\begin{cases} \frac{dx_2(\alpha)}{d\alpha}=-r_{b_2}\cdot\sec^2\alpha\cdot\tan\alpha\cdot\cos(\beta-\tan\alpha+\theta) \\ \frac{y_2(\alpha)}{d\alpha}=r_{b_2}\cdot\sec^2\alpha\cdot\tan\alpha\cdot\sin(\beta-\tan\alpha+\theta) \end{cases} \quad (3.20a)$$

$$l_{cirA}: \begin{cases} \frac{dx_2(\gamma)}{d\gamma} = -r_e \cdot \cos(\theta - \gamma) \\ \frac{dy_2(\gamma)}{d\gamma} = r_e \cdot \sin(\theta - \gamma) \end{cases} \tag{3.20b}$$

$$l_{invL}: \begin{cases} \frac{dx_2(\alpha)}{d\alpha} = -r_{b_2} \cdot \sec^2\alpha \cdot \tan\alpha \cdot \cos(\tan\alpha - \beta + \theta) \\ \frac{y_2(\alpha)}{d\alpha} = r_{b_2} \cdot \sec^2\alpha \cdot \tan\alpha \cdot \sin(\tan\alpha - \beta + \theta) \end{cases} \tag{3.20c}$$

3.1.3 Contact Boundary Conditions

As shown in Fig. 3.3, the tooth flank l_{invR} of the roller is tangent with the corresponding tooth flank involute of the workpiece at point K_s at the beginning of contacting. According to the meshing principle, point K_s is the intersection point of the addendum circle on the workpiece and the meshing line N_1N_2. θ_{K_s} is the angle between the coordinate axis y_2 integrated with the roller and the axis of x.

At any time t in the forming process, the radial feeding position of the roller is known. Therefore, the center distance a between the workpiece and the roller can also be determined. The equation of the meshing line N_1N_2 in the coordinate system Oxy is as follows:

$$y = \tan(\frac{\pi}{2} + \alpha') \cdot (x - \frac{a \cdot r_b}{r_b + r_{b_2}})$$

Or

$$y + \cot\alpha' \cdot (x - \frac{a \cdot r_b}{r_b + r_{b_2}}) = 0 \tag{3.21}$$

where α' is the engagement angle under the relative center distance a.

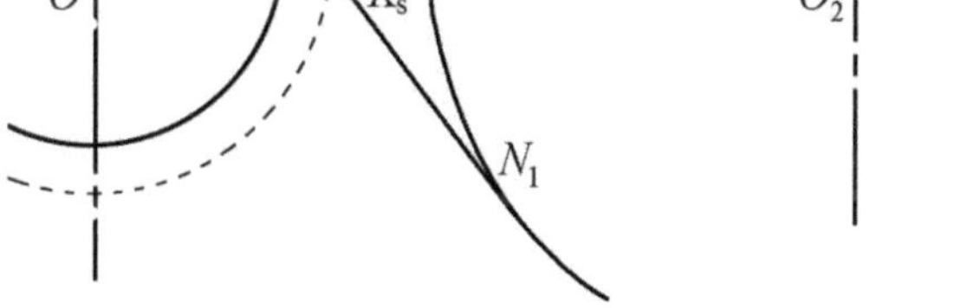

Fig. 3.3 Schematic diagram of the initial contact state

The coordinate (x_{K_s}, y_{K_s}) of point K_s in the coordinate system Oxy can be obtained by solving the simultaneous equations of the meshing line and the tooth addendum circle diameter of the workpiece; and the polar diameter ρ_{K_s} and pressure angle α_{K_s} of point K_s on the roller tooth flank l_{invR} can be further obtained.

$$\begin{cases} \rho_{K_s} = \sqrt{(x_{K_s} - a)^2 + y_{K_s}^2} \\ \alpha_{K_s} = \arccos \frac{r_{b2}}{\rho_{K_s}} \end{cases} \tag{3.22}$$

From the geometrical relationship, angle θ_{K_s} of y_2-axis and x–axis can be obtained while a single tooth entering the contact zone:

$$\theta_{K_s} = -(\arctan\left|\frac{y_{K_s}}{a - x_{K_s}}\right| + \beta - \mathrm{inv}\alpha_{K_s}) \tag{3.23}$$

The angle θ_{K_e} of y_2-axis and x-axis can be obtained similarly at condition the tooth detached from the contact zone. According to the coordinate (x_{K_e}, y_{K_e}) of K_e (the intersection point of the roller addendum circle and the meshing line N_1N_2) in the coordinate system Oxy and the geometrical relationship, it can also be obtained that

$$\theta_{K_e} = \arctan\left|\frac{y_{K_e}}{a - x_{K_e}}\right| - \beta + \mathrm{inv}\alpha_{K_e} \tag{3.24}$$

3.1.4 Calculation of the Contact Area

Discrete the contact region $[\theta_{K_s},\ \theta_{K_e}]$ of the roller and the workpiece, and select the N positions of the roller in the rolling process uniformly, which are expressed with $\theta_1(\theta_{K_s}), \theta_2 \ldots \theta_{N-1}, \theta_N(\theta_{K_e})$, respectively. Setting $\theta_n \in [\theta_{K_s}, \theta_{K_e}]$, at the position of θ_n, the tooth profile curve of the roller l_{roller_n}, and that of the workpiece is the envelope line of the roller tooth profile curve l_{enve_n}. The contact line formed by the roller and the cross-sectional surface perpendicular to the axis on the workpiece is the contact position of the tooth profile curve l_{roller_n} on the roller and the tooth profile curve l_{enve_n} on the workpiece.

Set the contacting criterion as follows: Taking a small positive number δ, when the distance between corresponding points on the roller tooth profile l_{roller_n} and the workpiece tooth profile l_{enve_n} meet the condition of $d < \delta$, it is a contacting state; otherwise, it is a non-contact state.

Numerical method has been used for the calculation of the contact arc. In the zone of $\theta_n \in [\theta_{K_s},\ \theta_{K_e}]$, the distance of corresponding points on the tooth profile

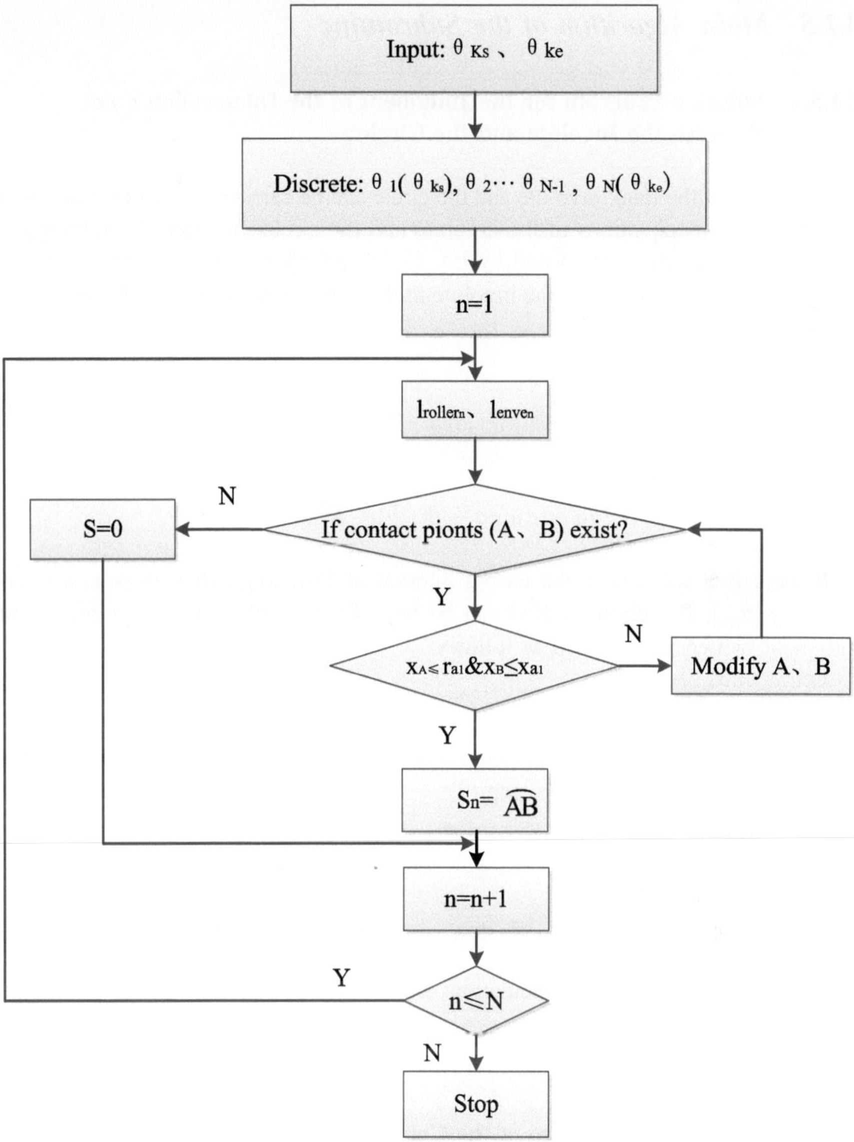

Fig. 3.4 Flowchart of the contact arc calculation program

curve of the roller (l_{invR}, l_{cirA}, l_{invL}) and the tooth profile curve of the workpiece should be calculated successively, and the contact state can be determined through the contacting criterion. In condition of a fixed center distance, the flowchart of the calculation program of the contact arc on the cross section is shown in Fig. 3.4.

3.1.5 Main Algorithm of the Subroutine

3.1.5.1 Solution Program for the Judgment of the Intersection Point Between the Involute and the Circle

The solution of the flank involute and the circle can be carried out by the solution of the simultaneous equations of the involute and the circle. In detail, by solving the simultaneous equation composed by Eq. (3.9a) and (3.9c) and the circle equation $x^2 + y^2 = R^2$, the equation of the involute in the coordinate of Oxy can be expressed as follows:

$$\begin{cases} \rho = \frac{r_{b2}}{\cos \alpha} \\ \rho^2 + a^2 - 2a\,\rho\,\cos(\beta - inv\,\alpha + \theta) - R^2 = f(\alpha) = 0 \end{cases} \tag{3.25a}$$

$$\begin{cases} \rho = \frac{r_{b2}}{\cos \alpha} \\ \rho^2 + a^2 - 2a\rho\,\cos(inv\,\alpha - \beta + \theta) - R^2 = f(\alpha) = 0 \end{cases} \tag{3.25b}$$

If there is a solution in the closed interval of $[\alpha_M, \alpha_{f_2th}]$, then there is a intersection $\rho(\alpha_{cp})$. Bisection method can be used for the solution programming and judgment, which is described as follows:

Let $a_1 = \alpha_M, b_1 = \alpha_{f_2th}$.

For $k = 1, 2, \cdots, M$:

(1) If $f(a_k) \cdot f(b_k) > 0$, then the output intersection is not exist and the program stopped; otherwise, turn to step (2);
(2) Calculate $x_k = \frac{a_k + b_k}{2}$;
(3) If $|a_k - b_k| \le \varepsilon\ |f(x_k)| \le \eta$, stop calculation, and let $\alpha_{cp} \approx x_k$; otherwise, turn to step (4);
(4) If $f(a_k) \cdot f(x_k) < 0$, then, let $a_{k+1} = a_k$, $b_{k+1} = x_k$; otherwise, $a_{k+1} = x_k$; $b_{k+1} = b_k$;
(5) If $k < M$, continue the calculation; otherwise, output information of failure iteration.

3.1.5.2 Calculation Program of the Contact Arc Length on the Roller Tooth Addendum Transition Circle

On the cross section, the contact part of the roller addendum transition arc and the workpiece is actually the arc length. If the two end points $A(x_A, y_A)$ and $B(x_B, y_B)$ of the solved arc and θ_n are known, the corresponding circle center of the roller tooth addendum transition circle is $O'(x_e, y_e)$, as shown in Fig. 3.5. In the triangular

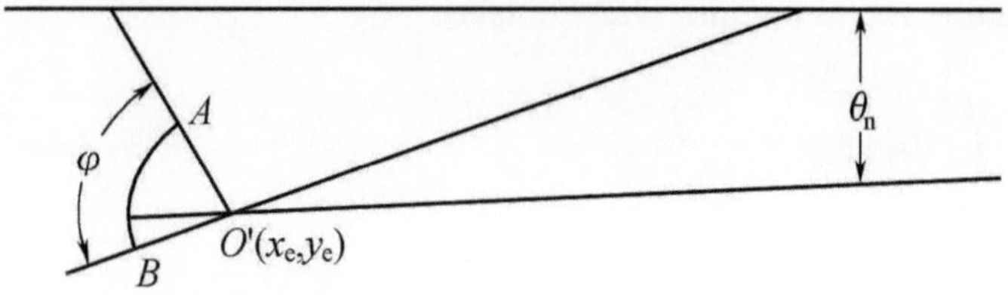

Fig. 3.5 Central angle correlated with arc $\widehat{AB}$

$\Delta AO'B$, the central angle φ correlated with arc $\widehat{AB}$ can be obtained applying cosine theorem:

$$\varphi = \arccos \frac{AO'^2 + BO'^2 - AB^2}{2AO' \cdot BO'} \tag{3.26}$$

Arc $\widehat{AB}$ can be obtained as follows:

$$\widehat{AB} = r_e \cdot \varphi \tag{3.27}$$

3.1.5.3 Calculation Program of the Arc Length on the Involute

On the cross section, the contact arc length of the roller tooth flank and the workpiece is one segment of the involute arc length. The pressure angles α_a and α_b of the two end points A and B on the solved involute arc are given. According to the differential formula of the arc:

$$S = \int_{\alpha_b}^{\alpha_a} \sqrt{[x'(\alpha)]^2 + [y'(\alpha)]^2} d\alpha \tag{3.28}$$

Derivative equation (3.8a)–(3.8b) about α, there is:

$$l_{invR}: \begin{cases} x'(\alpha) = r_{b_2} \sec\alpha \tan\alpha[\sin(\beta - \mathrm{inv}\,\alpha) - \tan\alpha\cos(\beta - \mathrm{inv}\alpha)] \\ y'(\alpha) = r_{b_2} \sec\alpha \tan\alpha[\cos(\beta - \mathrm{inv}\,\alpha) + \tan\alpha\sin(\beta - \mathrm{inv}\alpha)] \end{cases}$$

$$l_{invL}: \begin{cases} x'(\alpha) = r_{b_2} \sec\alpha \tan\alpha[\sin(\mathrm{inv}\alpha - \beta) + \tan\alpha\cos(\mathrm{inv}\alpha - \beta)] \\ y'(\alpha) = r_{b_2} \sec\alpha \tan\alpha[\cos(\mathrm{inv}\alpha - \beta) - \tan\alpha\sin(\mathrm{inv}\alpha - \beta)] \end{cases}$$

Substituting into $[x'(\alpha)]^2 + [y'(\alpha)]^2$ obtains:

$$r_{b_2}^2 \sec^2\alpha \tan^2\alpha(1 + \tan^2\alpha) \tag{3.29}$$

Substituting Eq. (3.29) into (3.28) obtains:

$$S = \int_{\alpha_b}^{\alpha_a} r_{b_2} \tan\alpha \sec^2 \alpha d\alpha = r_{b_2} \frac{\tan^2\alpha}{2}\bigg|_{\alpha_b}^{\alpha_a} = \frac{r_{b2}}{2}(\tan^2\alpha_a - \tan^2\alpha_b) \quad (3.30)$$

3.1.5.4 Important Parameters Necessary for Calculation

(1) Arc radius of the addendum transition circle

The roller tooth profile curve on the cross section is shown in Fig. 3.2. If the tangent line on the intersection point K of the reference circle and the involute is used to substitute the involute (if the roller has been modified, then the reference circle should be substituted by the pitch circle), as shown in Fig. 3.6, the relationship between the angle with *x-axis* and the pressure angle α on the pitch circle is $\frac{\pi}{2} - \alpha$, and the coordinate of K is as follows:

$$\begin{cases} x_k = 0 \\ y_k = m \cdot Z_2/2 \end{cases} \quad (3.31)$$

where

m is the modulus of the spline and
Z_2 is the tooth number of the roller.

The equation of line AB in the rectangular coordinate system is as follows:

$$x + \tan\alpha \cdot y - (x_k + \tan\alpha \cdot y_k) = 0 \quad (3.32)$$

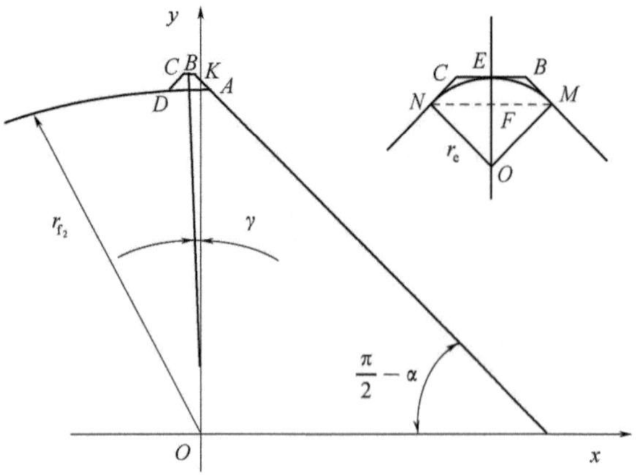

Fig. 3.6 Sketch diagram of the simplified roller tooth profile structure

From the geometric relationship, the coordinate of the transition circle center O' is as follows: $((r_{a2} - r_e)\sin\gamma, (r_{a2} - r_e)\cos\gamma)$; γ is the angle of OE and *y-axis*, $\gamma = 0.5\pi/Z_2$.

From the formula of distance from point to line, the equation of r_e can be set up as follows:

$$r_e = \frac{|(r_{a2} - r_e)(\tan\alpha \cdot \cos\gamma - \sin\gamma) - \tan\alpha \cdot y_k|}{\sqrt{\tan^2\alpha + 1}} \tag{3.33}$$

If the tooth number of the roller $Z_2 > 150$, then $\tan\alpha \cdot \cos\gamma - \sin\gamma \approx \tan\alpha$. Solving Eq. (3.33), it can be obtained that

$$r_e = \sec\alpha|\tan\alpha \times r_{a_2} - \tan\alpha \times y_k| - \tan\alpha \times (\tan\alpha \times r_{a_2} - \tan\alpha \times y_k) \quad (r_e > 0) \tag{3.34}$$

From Figs. 3.3, 3.4, 3.5, and 3.6, the geometrical relationship is obtained as follows:

$$\angle EO'M = \frac{\pi}{2} - \alpha + \gamma$$

Therefore, the chord height of the transition arc is as follows:

$$\overline{EF} = r_e(1 - \sin(\alpha - \gamma)) \tag{3.35}$$

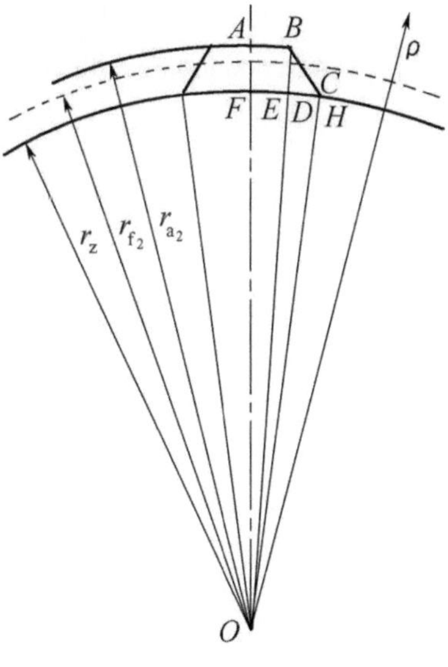

Fig. 3.7 Tooth height of the workpiece under feed rate of *S*

(2) The radius of the workpiece addendum circle under certain feed rate

Assuming that the feed rate at time t is S, the tooth profile of the workpiece (spline) is shown as Fig. 3.7.

The radius of the workpiece root circle at time t is as follows:

$$r_{f_t} = r_Z - f \tag{3.36}$$

where

r_Z is the billet radius before rolling forming.

The pressure angle of the pitch circle at time t is as follows:

$$\alpha_t' = \arccos\frac{m \cdot \cos\alpha \cdot (Z + Z_2)}{2 \cdot (r_{a_2} + r_{f_t})} \tag{3.37}$$

where

m is the modulus of the spline;
α is the pressure angle of the spline pitch circle; and
Z and Z_2 are the tooth numbers of the spline and the roller.

The tooth width of the spline dedendum at time t is as follows:

$$s_{f_t} = 2\pi \cdot r_{f_t}/Z - (2\pi \cdot r_f/Z - s_f) \tag{3.38}$$

where

r_f is the radius of the spline tooth root circle and
s_f is the tooth width of the spline tooth dedendum circle.

The radius and tooth width on the pitch circle of the spline at time t is as follows:

$$\begin{cases} r_t' = \frac{mZ}{2} \cdot \frac{\cos\alpha}{\cos\alpha_t'} \\ s_t' = s_{f_t} \cdot \frac{r_t'}{r_{f_t}} - 2r_t'(inv\alpha_t' - inv\alpha_{f_t}) \end{cases} \tag{3.39}$$

where

α_{f_t} is the pressure angle of the spline tooth root circle at time t.

The pressure angle of points D and B is as follows:

$$\begin{cases} \alpha_D = \arccos\frac{r_b}{r_{f_t}} \\ \alpha_B = \arccos\frac{r_b}{r_{a_t}} \end{cases} \tag{3.40}$$

where

r_{a_t} is the radius of the spline addendum circle.

The polar angle of the key points in the polar coordinate system is as follows:

$$\begin{cases} \theta_D = \tan\alpha_D - \alpha_D \\ \theta_B = \tan\alpha_B - \alpha_B \\ \theta_A = s'_t/2r'_t + \tan\alpha'_t - \alpha'_t \end{cases} \tag{3.41}$$

The area of a single tooth above the root circle is as follows:

$$S_t = r_{a_t}^2(\theta_A - \theta_B) + \frac{r_b^2}{3}(\tan^3\alpha_B - \tan^3\alpha_D) + 2r_e^2(\tan\beta - \beta) - r_{f_t}^2(\theta_A - \theta_D) \tag{3.42}$$

where

2β is the corresponding angle of the dedendum transition circle.

According to the constant volume principle:

$$\pi \cdot r_Z^2 = \pi \cdot r_{f_t}^2 + Z \cdot S_t$$

Therefore

$$r_{a_t}^2(\theta_A - \theta_B) + \frac{r_b^2}{3} \cdot \tan^3\alpha_B = \frac{\pi \cdot f \cdot (2r_Z - f)}{Z} + \frac{r_b^2}{3} \cdot \tan^3\alpha_D + r_{f_t}^2(\theta_A - \theta_D) - 2r_e^2(\tan\beta - \beta) \tag{3.43}$$

In a given feed rate S, the right part of Eq. (3.43) is a constant. Equation (3.43) is a transcendental equation with r_{a_t} as the variable,which can be solved by programming. When the feed rate is smaller, the transition arc of the spline tooth root circle has not formed completely, and the parameter of 2β must be modified and iterated to receive r_{a_t} by programming calculation. 2β has little influence on the single area above the tooth root circle of the spline.

(3) The addendum radius of the workpiece after a rolling process

After a rolling process, the difference Δs of the workpiece (spline) dedendum radius before and after rolling is called as reduction. The variation in the tooth profile is shown in Fig. 3.8.

$$\Delta s = r_{f_0} - r_{f_1} \tag{3.44}$$

where

r_{f_0} is the root circle radius and r_{f_1} of the spline before rolling;
r_{f_1} is the root circle radius of the spline after rolling.

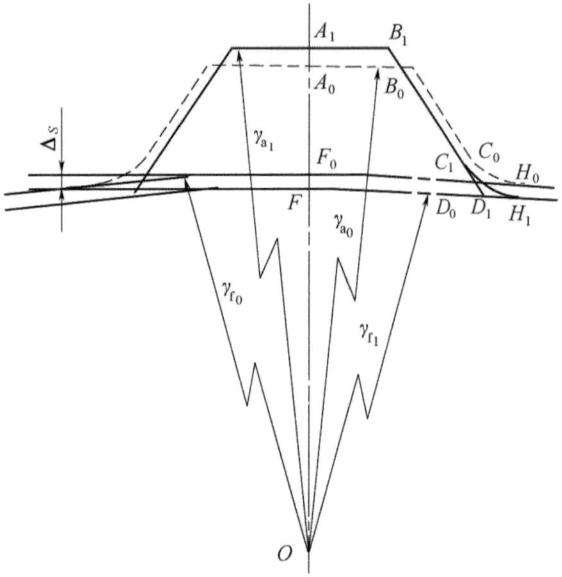

Fig. 3.8 Tooth profile of the workpiece before and after a rolling process

The pressure angle of the pitch circle on the spline before and after rolling is as follows:

$$\begin{cases} \alpha_0' = \arccos\frac{m\cdot\cos\alpha\cdot(Z+Z_2)}{2\cdot(r_{a_2}+r_{f_0})} \\ \alpha_1' = \arccos\frac{m\cdot\cos\alpha\cdot(Z+Z_2)}{2\cdot(r_{a_2}+r_{f_1})} \end{cases} \tag{3.45}$$

Radius of the spline pitch circle before and after rolling is as follows:

$$\begin{cases} r_0' = \frac{mZ}{2}\cdot\frac{\cos\alpha}{\cos\alpha_0'} \\ r_1' = \frac{mZ}{2}\cdot\frac{\cos\alpha}{\cos\alpha_1'} \end{cases} \tag{3.46}$$

Spline tooth width of the root circle before and after rolling is as follows:

$$\begin{cases} s_{f_1} = 2\pi\cdot r_{f_1}/Z - (2\pi\cdot r_f/Z - s_f) \\ s_{f_0} = 2\pi\cdot r_{f_0}/Z - (2\pi\cdot r_f/Z - s_f) \end{cases} \tag{3.47}$$

Spline tooth width of the pitch circle before and after rolling:

$$\begin{cases} s_0' = s_{f_0}\cdot\frac{r_0'}{r_{f_0}} - 2r_0'(inv\,\alpha_0' - inv\,\alpha_{f_0}) \\ s_1' = s_{f_1}\cdot\frac{r_1'}{r_{f_1}} - 2r_1'(inv\,\alpha_1' - inv\,\alpha_{f_1}) \end{cases} \tag{3.48}$$

Pressure angle of points D_0, D_1, B_0, and B_1 is as follows:

$$\begin{cases} \alpha_{D_0} = \alpha_{f_0} = \arc\cos\frac{r_b}{r_{f0}} \\ \alpha_{D_1} = \alpha_{f_1} = \arc\cos\frac{r_b}{r_{f1}} \\ \alpha_{B_0} = \alpha_{a_0} = \arc\cos\frac{r_b}{r_{a0}} \\ \alpha_{B_1} = \alpha_{a_1} = \arc\cos\frac{r_b}{r_{a1}} \end{cases} \tag{3.49}$$

The polar angle of the key points in the polar coordinate system is as follows:

$$\begin{cases} \theta_{D_0} = \tan\alpha_{D_0} - \alpha_{D_0} \\ \theta_{D_1} = \tan\alpha_{D_1} - \alpha_{D_1} \\ \theta_{B_0} = \tan\alpha_{B_0} - \alpha_{B_0} \\ \theta_{B_1} = \tan\alpha_{B_1} - \alpha_{B_1} \\ \theta_{A_0} = s'_0/2r'_0 + \tan\alpha'_0 - \alpha'_0 \\ \theta_{A_1} = s'_1/2r'_1 + \tan\alpha'_1 - \alpha'_1 \end{cases} \tag{3.50}$$

Single-tooth area above the dedendum circle of the workpiece before and after rolling is as follows:

$$\begin{cases} S_0 = r_{a_0}^2(\theta_{A_0} - \theta_{B_0}) + \frac{r_b^2}{3}(\tan^3\alpha_{B_0} - \tan^3\alpha_{D_0}) + 2r_{e_0}^2(\tan\beta_0 - \beta_0) - r_{f_0}^2(\theta_{A_0} - \theta_{D_0}) \\ S_1 = r_{a_1}^2(\theta_{A_1} - \theta_{B_1}) + \frac{r_b^2}{3}(\tan^3\alpha_{B_1} - \tan^3\alpha_{D_1}) + 2r_{e_1}^2(\tan\beta_1 - \beta_1) - r_{f_1}^2(\theta_{A_1} - \theta_{D_1}) \end{cases} \tag{3.51}$$

Because the transition arc of the roller addendum is invariable, the radius of the dedendum transition arc is equal before and after rolling, that is, $r_{e_0} = r_e$. While the involute of the tooth flank has been formed, $2\beta_0 = 2\beta_1$, and $2\beta_0 \approx 2\beta_1$ before rolling. According to the principle of constant volume:

$$\pi \cdot r_{f_0}^2 + Z \cdot S_0 = \pi \cdot r_{f_1}^2 + Z \cdot S_1$$

Therefore

$$\begin{aligned} r_{a_1}^2(\theta_{A_1} - \theta_{B_1}) + \frac{r_b^2}{3}\cdot\tan^3\alpha_{B_1} = {} & \frac{\pi\cdot\Delta s\cdot(2r_{f_0} - \Delta s)}{Z} + r_{a_0}^2(\theta_{A_0} - \theta_{B_0}) \\ & + \frac{r_b^2}{3}\cdot\tan^3\alpha_{D_1} + \frac{r_b^2}{3}(\tan^3\alpha_{B_1} - \tan^3\alpha_{D_1}) \\ & + r_{f_1}^2(\theta_{A_1} - \theta_{D_1}) - r_{f_0}^2(\theta_{A_0} - \theta_{D_0}) \end{aligned} \tag{3.52}$$

With a determined dimension of the spline before rolling, the right part of Eq. (3.52) is a constant. Equation (3.52) is a transcendental equation with r_{a_1} as a variable and can be solved by programming.

3.1.5.5 Analysis of the Factors Influencing the Unit Length Contact Area

According to the solution algorithm shown in Fig. 3.4, corresponding programs and subroutines were written by MATLAB. The variation trend of the unit length contact area of a single tooth in the rolling process can be solved. By converting corresponding impacting factors, the contact area of unit length can be analyzed quantitatively.

In the contact area calculation of the rolling process, the tooth profile curve should be numerically calculated according to the mathematical model of the tooth profile of the roller and workpiece at any position of the rolling process. The contact state between rollers and the workpiece can be presented graphically by MATLAB, as shown in Fig. 3.9. The region of $[\theta_{K_s}, \theta_{K_e}]$ is dispersed into 100 points, and the tooth profile curve of the three typical points θ_5, θ_{50}, θ_{95} are drawn in Fig. 3.9, in which the tooth profile curves of the roller, the workpiece, and the workpiece addendum circle are expressed with full line, dash line, and dash-dot line, respectively. From Fig. 3.9, it is shown that l_{invR} contacts with the workpiece firstly, then line l_{cirA}, and l_{invL}; the contact surface at the medium position is the maximum.

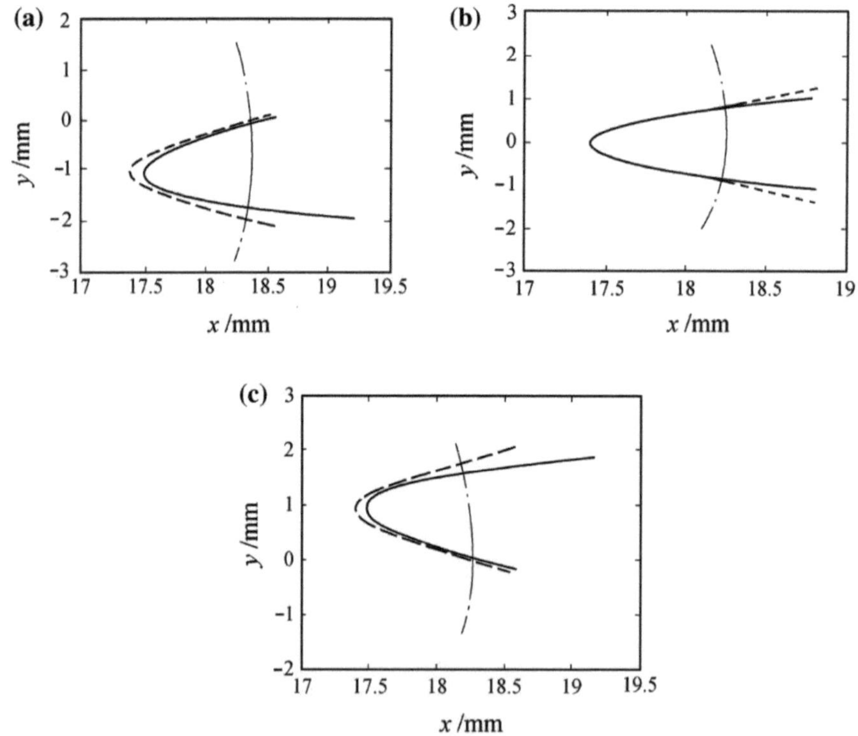

Fig. 3.9 Tooth profile curve of the roller and workpiece at different positions of a rolling process **a** at the beginning of the contact (θ_5); **b** at the medium stage of the contact (θ_{50}); **c** at the finishing stage of the contact (θ_{95})

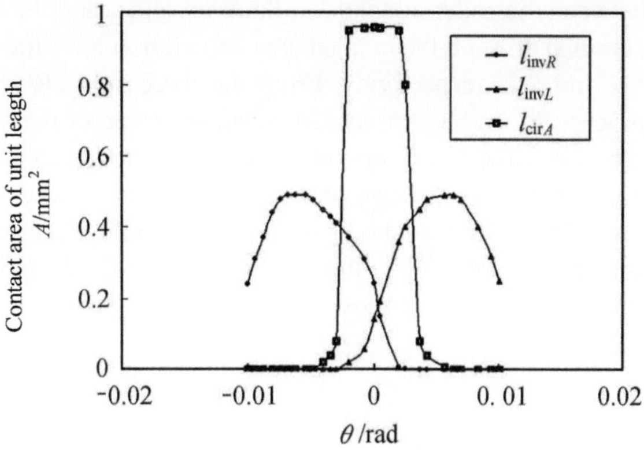

Fig. 3.10 Unit contact area in the cold rolling process

The tooth profile of the roller consists of three segments, and the corresponding calculated contact areas are, respectively, S_{invR}, S_{cirA}, and S_{invL}, as shown in Fig. 3.10. The contact area in the rolling process can be obtained by accumulating of the contact areas between every segment of curves on the roller and the workpiece at corresponding positions. From the figure, it is shown that under different deflection angles, single-side and double-side contacts occur alternatively, which coincides with the double-flank backlash meshing contact analysis described in Sect. 2.4.1. In the calculation, the process is still in the pressing stage, the addendum of the tooth is extruded into the workpiece, and the maximum contact area between the roller addendum and the workpiece is larger than that of the tooth flank. Near $\theta = 0$, the contact area between the roller addendum and the workpiece is the maximum, and the contact curve of the roller addendum is almost completely symmetrical with y_2-axis, which is mainly in the middle part of the variation region of θ.

In the cold rolling forming process, the roller is feeding continuously, and diameter of the workpiece addendum circle under related center distance (certain feed rate) should be considered in the contact arc length calculation. The parameter can be determined according to the volume constant principle and the fact that the tooth profile of the workpiece is constant under corresponding center distance. And the contact area between the roller and workpiece at any time of the rolling process can be obtained by calculation of the contact area under different center distances with special program subroutines.

3.2 Rolling Force and Rolling Moment

3.2.1 Theoretical Calculation of the Rolling Force and Rolling Moment

The process of one tooth of the workpiece entering the contact zone to detaching from it is called as a single rolling process. In the contact zone, the contact area between the roller addendum, flank involute, and the tooth of the workpiece can be obtained through the contact area calculation introduced in Sect. 3.1 and is called as S_{cir} and S_{inv}, respectively. From the force analysis of the workpiece in the rolling process, it can be seen that the resultant force of the normal force q and the friction shearing force τ at directions of x and y is the normal force P and the friction force F_{f}. And the unit average press of the contact surface can be calculated with the unit force calculating method given in this book. Therefore, the macronormal force P and friction force F_{f} in the unit length of a single tooth in a rolling process can be calculated with the following equation:

$$P = q_2 \cdot S_{inv} \cdot [\cos(\frac{\pi}{2} - \alpha) + \mu \cdot \cos\alpha] + q_1 \cdot S_{cir} \tag{3.53}$$

$$F_{\text{f}} = q_2 \cdot S_{inv} \cdot [\sin(\frac{\pi}{2} - \alpha) + \mu \cdot \sin\alpha] \tag{3.54}$$

As shown in Fig. 2.3, if the normal force P and friction force F_{f} is projected along the radial feeding direction of the forming, the rolling force in the cold rolling forming of spline shaft can be obtained as follows:

$$F = P \cdot \cos\theta + F_f \cdot \sin\theta \tag{3.55}$$

In the cold rolling machine, the radial feeding force is provided by the hydraulic system. Equation (3.55) is the forming force for the plastic deformation of the workpiece, and taking the energy loss of the transmission system such as friction force between the roller slider and the guide into consideration, the driving force provided by the hydraulic system should be as follows:

$$F = P \cdot \cos\theta + F_{\text{f}} \cdot \sin\theta + F_{\text{loss}} \tag{3.56}$$

where F_{loss} is the vector sum of the dissipated force of the transmission system of the cold rolling machine.

According to the moment balance principle of the forming process, the moment of the driving roller is as follows:

$$M = F_{\text{f}} \cdot r_{a_2} \tag{3.57}$$

The basic cycle unit of the rolling force and rolling moment periodical variation in the rolling forming process is the single rolling pass of each tooth. The rolling forming process should be discreted by time in the calculation of the rolling force and rolling moment. The contact area and the average force on the contact surface can be calculated with numerical method, and the rolling force and moment in a single rolling process can be calculated. The peak value of the rolling force and rolling moment of each tooth in the rolling forming process can be fitted into the rolling force and moment of the spline in the whole cold rolling forming process.

In the design of the program, it is assumed that the rotation of the workpiece is stable in the rolling forming process; the radial feed rate of the roller is constant, which is called as V.

There are two contact deforming zones in the cold rolling forming of spline shafts. The tooth on the workpiece contacts with the roller once after rotating for half a circle. According to the parity of the tooth number on the spline, the discreted time should be treated differently. If the tooth number of the formed spline is an even number, the time for the workpiece rotating for half a circle is divided into $K = Z/2$ equal segments; if the tooth number is an odd number, the time is divided into $K = (Z + 1)/2$ segments. The value of the $K = (Z + 1)/2$ region is half of the other regions. Thus, the rolling forming process can be discrete by time regions as follows: $[t_0, t_1], [t_1, t_2]\ldots[t_n, t_{n+1}]\ldots$. Assuming that at a certain discrete region, the radius of the workpiece addendum circle and the root circle is, respectively, r_{a_0} and r_{f_0} before rolling, then the radius of the workpiece dedendum is as follows:

$$r_{f_1} = r_{f_0} - \Delta s \tag{3.58}$$

where Δs is the reduction.

According to the volume constant principle before and after rolling, the radius r_{a_1} of the workpiece addendum circle after rolling can be obtained as follows: $r_{a_1} = FUN(r_{a_1})$. Taking the peak values of the rolling force $F_{\max_n}$ and rolling moment $M_{\max_n}$ in every time region, then the peak value variation curve of the rolling force and the rolling moment can be obtained by fitting of the series of $F_{\max_n}$ and $M_{\max_n}$.

The flowchart of the rolling force and rolling moment calculation program in MATLAB is shown in Fig. 3.11.

3.2.2 *Analysis of the Rolling Force and Rolling Moment in the Forming Process*

From calculation, it is obtained that after steady rolling, the rolling force and rolling moment of a single pass rolling process are shown in Fig. 3.12. In both sides of the contact zone, the rolling force and the rolling moment are almost symmetric. The peak value of the rolling force is near the position of $\theta = 0$. At this position, the addendum and both flanks of the roller contact with the workpiece, and the contacting area is the maximum. The maximum rolling moment exists at the positions of $\theta = (1/3 \sim 2/3)\theta_{K_s}$ or $\theta = (1/3 \sim 2/3)\theta_{K_e}$; that is, when entering and exiting from the contact zone, the tooth flank of the roller contacts with the workpiece.

There are many parameters in the spline cold rolling forming process in this book. The influence of material is neglected, and the influences of geometrical parameters, process parameters on the rolling force, and rolling moment are discussed from aspect of process planning.

Fig. 3.11 Flowchart of the rolling force and rolling moment calculation program

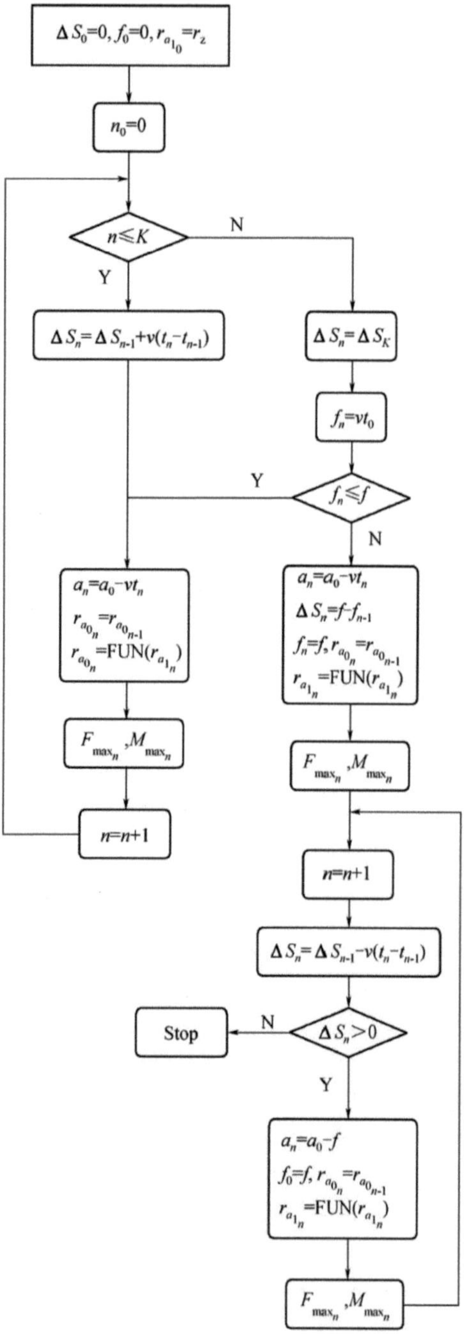

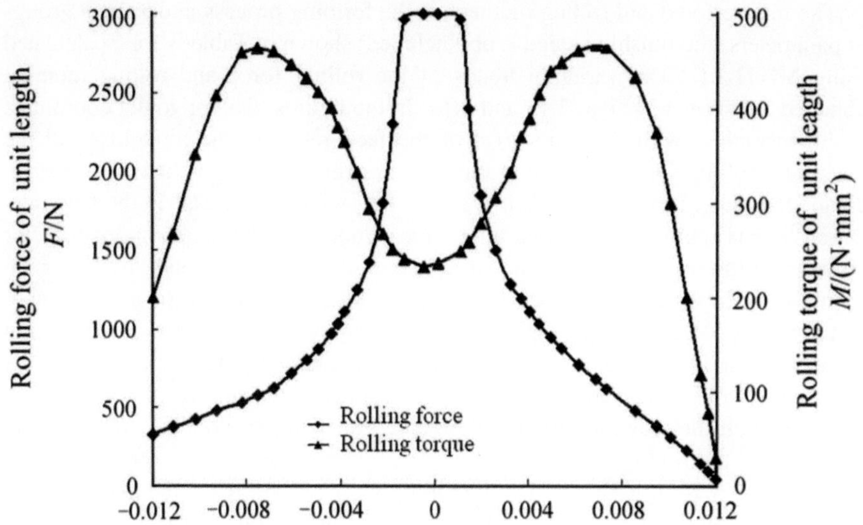

Fig. 3.12 Rolling force and rolling moment in a single rolling process

Selecting the modulus, tooth number, pressure angle of the spline, and the reduction in per rotating, combining with the tooth number and rotation speed of the roller, the relationship between the rolling force and rolling moment is studied. The parameters used in the calculation are listed in Table 3.1, in which the parameters of modulus m, tooth number Z, pressure angle α, and the reduction Δs are designed for 9 group of numbers in the calculation according to orthogonal experimental method. Other parameters are selected combining the actual conditions of the machine according to the principle of equal roller pitch circle, and the selected roller speed is 18r/min.

Table 3.1 Parameters selected in the calculation of the rolling force and rolling moment

Serial number	Modulus m/mm	Tooth number of the spline Z	Pressure angle $\alpha/^\circ$	Reduction Δs/mm	Tooth number of the roller Z_2	Rotation speed of the roller/ (r/min)
1	0.5	19	30.0	0.025	400	18
2	0.5	36	37.5	0.050	400	18
3	0.5	54	45	0.075	400	18
4	1.0	36	30.0	0.075	200	18
5	1.0	54	37.5	0.025	200	18
6	1.0	19	45	0.050	200	18
7	1.5	54	30.0	0.050	140	18
8	1.5	19	37.5	0.075	140	18
9	1.5	36	45	0.025	140	18

The rolling force and rolling moment of the forming process under the 9 groups of parameters (the finishing stage is not included) shown in Table 3.1 are calculated using MATLAB. The variation trends of the rolling force and rolling moment obtained are shown in Figs. 3.13 and 3.14. In the figures, the horizontal coordinate is dimensionless with the ratio vt/m of the feed rate and the modulus, and the maximum rolling force and rolling moment in every time segment is the longitudinal coordinate. From Fig. 3.13, it is shown that with the increase in the feed rate, there is a maximum value of the rolling force. Modulus is the most important factor influencing the rolling process; the larger the modulus of the spline, the larger the rolling force in the forming process is. There is little influence of the tooth number on the rolling force: Pressure angle affects the existing time of the maximum rolling force; with the increase in the pressure angle, the reaching time of the rolling force peak value will be delayed; with a smaller modulus of the spline, the rolling force increases with the reduction per rotation; and with a larger modulus, there is little influence on the rolling force.

From Fig. 3.14, it is shown that the rolling moment will reach the peak value at conditions of the maximum feed rate, and it will last to the finish of the forming

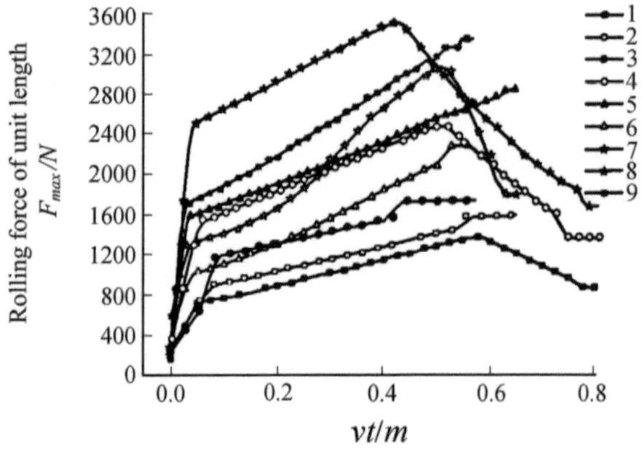

Fig. 3.13 Rolling force under different parameters

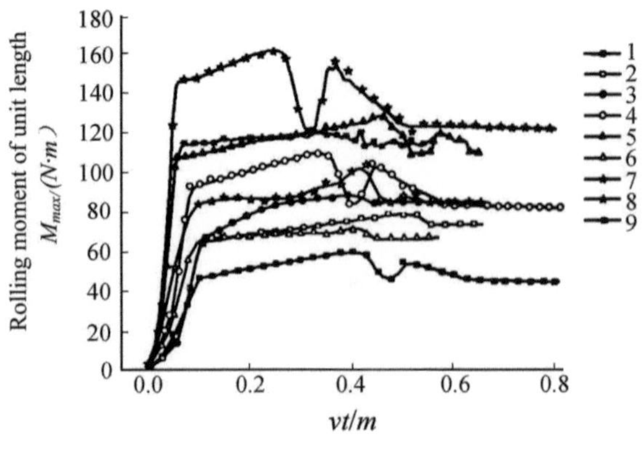

Fig. 3.14 Rolling moment under different parameters

process. The influence of modulus and tooth number has an important influence on the rolling moment, and the influence of the pressure angle on the rolling moment is relatively small. Under a pressure angle of 30°, there will be two obvious peak values of the rolling moment, and the latter is smaller than the previous one. With the increase in the reduction per rotation, the rolling moment increases at small modulus of the spline; and there is little influence on conditions of larger modulus.

3.3 Calculation of the Cold Rolling Workpiece Billet Diameter

3.3.1 Theoretical Calculation Formulas

There is little density variation in the billet before and after rolling forming, the theoretical calculation of the billet diameter d_Z before rolling can be carried out according to the volume constant principle of the spline.

$$\frac{\pi}{4} \cdot d_z^2 \cdot L = \frac{\pi}{4} \cdot d_f^2 \cdot L + S \cdot Z \cdot L + V \tag{3.59}$$

where d_Z is billet diameter of the spline before rolling;

L is the length of the spline;
d_f is the dedendum diameter of the spline tooth;
Z is the tooth number of the spline;
S is the cross-sectional area of a single tooth above the root circle; and
V is the bulging volume at the two ends of the spline.

It is indicated from simulation and experiments that V cannot be compensated by increasing the billet diameter. In actual calculation, V is always neglected. Therefore, calculating according to the constant principle of the cross-sectional area before and after rolling, it can be obtained that

$$d_Z = \sqrt{d_f^2 + \frac{4}{\pi} S \cdot Z} \tag{3.60}$$

3.3.2 Cross-Sectional Area of a Single Tooth Above the Dedendum Circle

The schematic diagram of the spline single-tooth cross section is shown in Fig. 3.15, and the cross-sectional area of a single tooth above the dedendum circle is as follows:

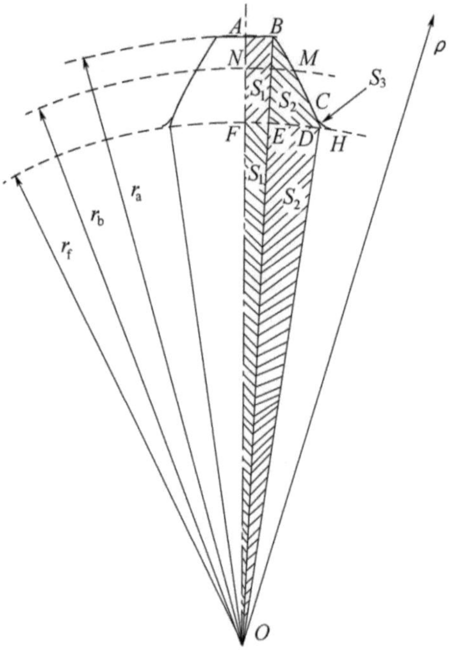

Fig. 3.15 Schematic diagram of a single-tooth cross section of the spline

$$S = 2(S_1 + S_2 + S_3) \tag{3.61}$$

According to the characteristic of involute and differential knowledge, the value of S_1, S_2, and S_3 can be solved, respectively.

According to the polar coordinate equation of the involute:

$$\begin{cases} r_k = r_b / \cos \alpha_k \\ \theta_k = inv\,\alpha_k = \tan \alpha_k - \alpha_k \end{cases} \tag{3.62}$$

where r_b is the radius of the involute base circle.

And, it can be solved that

$$\begin{cases} \theta_D = inv\,\alpha_f = \tan\,\alpha_f - \alpha_f \\ \theta_B = inv\,\alpha_a = \tan\,\alpha_a - \alpha_a \\ \theta_M = inv\,\alpha = \tan\,\alpha - \alpha \end{cases} \tag{3.63}$$

where

α_f is the pressure angle of the spline root circle;
α_a is the pressure angle of the spline addendum circle; and
α is the pressure angle of the spline pitch circle.

According to the differential formula of arc, the length of arc *MN* is as follows:

$$MN = \int_{\theta_M}^{\theta_N} \sqrt{\rho^2(\theta) + [\rho'(\theta)]^2} = r(\theta_N - \theta_M) \tag{3.64}$$

where r is the radius of the spline pitch circle.

And

$$MN = \frac{1}{2}s \tag{3.65}$$

where

S is the tooth width of the spline pitch circle.

Solving the simultaneous equations (3.64) and (3.65) obtains

$$\theta_N = \frac{s}{2r} + \theta_M \tag{3.66}$$

Each part of the area is as follows:

(1) Arc area of the tooth addendum

$$S_1 + S_1' = \int_{\theta_B}^{\theta_N} \frac{1}{2}[\rho(\theta)]^2 d\theta = \frac{r_a^2}{2}(\theta_N - \theta_B) \tag{3.67}$$

where r_a is the radius of the spline tooth addendum.

(2) Area of the involute

From the equation of the involute $d\theta = (\sec^2 \alpha_k - 1)d\alpha_k$.

$$S_2 + S_2' = \int_{\theta_D}^{\theta_B} \frac{1}{2}[\frac{r_b}{\cos\alpha_k}]^2 = \frac{r_b^2}{6}(\tan^3 \alpha_a - \tan^3 \alpha_f) \tag{3.68}$$

(3) Solution of S_3

Arc *CD* and *DH* can be approximately considered as straight lines, which is, respectively, the tangent of the involute and the root circle. Arc *CH* is tangent with the involute and the root circle at points *C* and *H*. As shown in Fig. 3.16,

$$S_3 = R_e^2 \cdot (\tan\beta - \pi \cdot \frac{2\beta}{360}) \tag{3.69}$$

where

2β is the angle of the transition arc and

R_e is the radius of the dedendum transition circle.

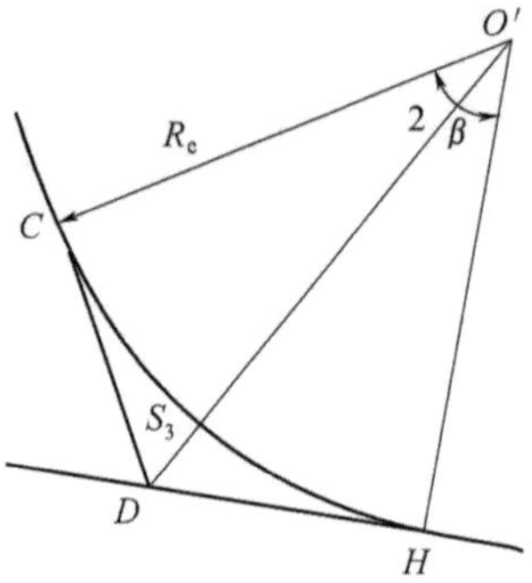

Fig. 3.16 Transition arc of the spline root circle

(4) Area below the tooth dedendum

$$S_1' + S_2' = \int_{\theta_D}^{\theta_F} \frac{1}{2}[\rho(\theta)]^2 d\theta = \frac{r_f^2}{2}(\theta_N - \theta_D) \tag{3.70}$$

where

r_f is the radius of the spline tooth root circle.

Substituting Eqs. (3.67)–(3.70) into Eq. (3.61), then the area of a single-tooth cross section above the tooth root circle can be obtained.

3.3.3 *Radius of the Dedendum Transition Arc*

In the cold rolling forming process, the dedendum of the spline is always designed as a round shape. With a determined round root circle and involute flank, radius R_e of the dedendum transition circle is a certain value, as shown in Fig. 3.17.

The angle δ between the polar axis and *y-axis* is as follows:

$$\delta = inv\alpha - \frac{\pi}{2 \cdot Z} \tag{3.71}$$

The parameter equation of the dedendum transition arc is as follows:

$$\begin{cases} x = R_e \cdot \cos\gamma \\ y = R_e \cdot \sin\gamma + r_e + r_f \end{cases} \tag{3.72}$$

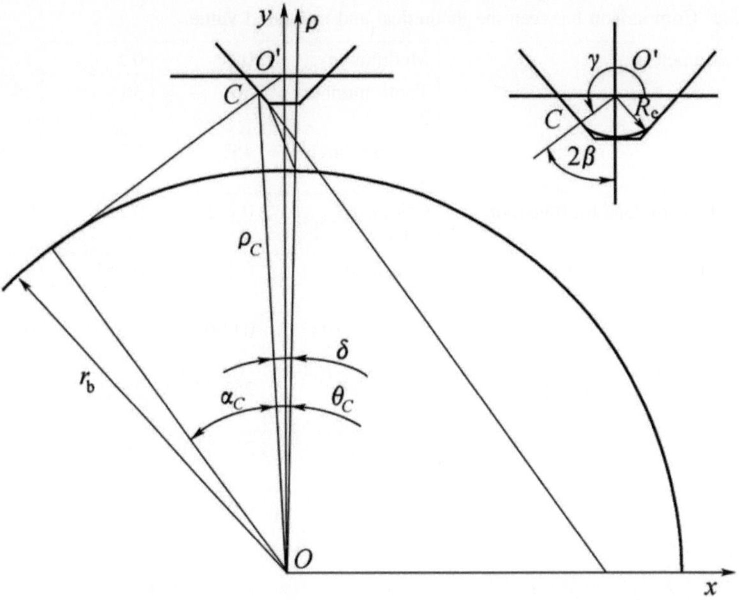

Fig. 3.17 Geometrical sketch diagram of the dedendum part of round dedendum spline

The dedendum transition arc is tangent with the involute at point C, and $O'C$ is the common tangent of the two curves. From the geometrical relationship, it is known that

$$\begin{cases} 2\beta = \frac{\pi}{2} - (\alpha_C + \theta_C - \delta) = \frac{\pi}{2} - \tan\,\alpha_C + \delta \\ \gamma_C = \pi + \tan\,\alpha_C - \delta \end{cases} \tag{3.73}$$

Solving the simultaneous equations (3.62) and (3.72) obtains:

$$\begin{cases} \rho_C = r_b/\cos\,\alpha_C \\ \rho_C \cdot \sin(inv\,\alpha_C - \delta) = R_e \cdot \cos(\tan\,\alpha_C - \delta) \\ \rho_C \cdot \cos(inv\,\alpha_C - \delta) = R_e[1 - \sin(\tan\,\alpha_C - \delta)] + r_f \end{cases} \tag{3.74}$$

Equation (3.74) is a transcendental equation, and the numerical solution of α_C and R_e can be obtained by programming.

Cross section of the spline in actual production is measured to verify the calculation results. According to the standard and dimension of the spline in industrial production and standard handbook, the cross section of the spline is modeling and plotting with CAD/CAM software CAXA. The radius R_e of the dedendum arc and

Table 3.2 Comparison between the theoretical and measured value

Spline parameters	Modulus *m*	0.5	0.5	1
	Tooth number Z	36	36	36
	Pressure angle α	45°	37.5°	37.5°
Radius of the dedendum transition arc/mm	Calculated value	0.225	0.253	0.506
	Measured value	0.225	0.254	0.508
	Relative error (%)	0.000	0.390	0.390
Cross section area of the spline/mm^2	Calculated value	253.575	253.157	1012.592
	Measured value	253.520	253.087	1012.350
	Relative error (%)	0.022	0.027	0.024

the cross section of the spline can be obtained by query, as shown in Table 3.2. The relative error of the dedendum transition arc radius is less than 0.5%, and the pressure angle of the pitch circle has more influence on the error. With a calculation error of the spline cross section area less than 0.05%, the influence of the modulus and the pressure angle is not obvious.

Chapter 4
Numerical Simulation on the Cold Rolling Forming of Spline

Abstract Finite element modeling and numerical simulation analysis of the spline cold rolling precision forming process have been carried out with software DEFORM-3D. The simulated results are compared with those obtained from the theoretical analysis.

4.1 Establishment of the Finite Element Model and Boundary Conditions

4.1.1 Finite Element Model

The cold rolling forming process of spline is numerically simulated and analyzed with finite element analysis software DEFORM-3D. In order to improve the operation speed, the FEM model and initial condition are appropriately simplified. Under conditions, the relative moving pattern between the roller and the workpiece is invariant, and the moving mode of them is equivalently transformed, respectively. Fasten the workpiece, and then, the roller will revolve around its own axis and rotate around the workpiece. Considering the symmetry of the spline tooth, the periodical property of the rolling force, and other related parameters, the model is dealt periodically. Neglecting the angle speed variation of the workpiece during the former half circle, the first and second rotation coordinate axes of the roller and the radial feed rate are designated, and cylinder coordinate system is adopted in DEFORM-3D.

Three-dimensional solid models of the spline billet (single-tooth volume) and the roller are set up, and the STL file is output and imported into the pretreated part of DEFORM-3D to set up the finite element model [1–4].

The surface of the workpiece is divided into fine and dense meshes. When the divided meshes are not appropriate, the simulation of the stress, strain, and force will distorted and the error with the practice results is larger. Under conditions, the meshes of the workpiece surface are not fine enough or the contact structure between the roller and the workpiece is small, the contact notes are rare, and the sliding trend of the workpiece is dominating compared with the rotation movement. Therefore, the sliding phenomenon is expanded.

J. Song et al., *Cold Rolling Precision Forming of Shaft Parts*,
DOI 10.1007/978-3-662-54048-0_4

4.1.2 Simulation Parameters and Constrain Conditions

Single tooth is adopted in the numerical simulation, and corresponding boundary constrain condition is applied. In order to compare the simulation results, the process parameters are constant, and three groups of materials and friction coefficients are alternated in the numerical simulation.

The basic parameters of the spline used in the calculation of the spline billet diameter before rolling and the dimensional parameters of the roller are as follows: modulus $m = 1$ mm, tooth number $Z = 36$, and pressure angle of the pitch circle $\alpha = 45°$. Geometrical models of the workpiece and the roller are set up with 3D solid modeling software Unigraphics (UG). The feed rate and angular speed are invariable in the several groups of simulations.

Two kinds of materials AISI-1015 [70–2000 F (20–1100 °C)] and AISI-1045, and COLD [70 F (20 °C)] are used in the simulation; the stress–strain relationship is $\bar{\sigma} = \bar{\sigma}(\bar{\varepsilon}, \dot{\bar{\varepsilon}}, T)$, Poisson ratio is $\nu = 0.3$, and the modulus of elasticity is $E = 206754$ MPa. The temperature of the material and the environment is 20 °C, and heat transfer is ignored. The workpiece is assumed to be a plastic object, the roller is a rigid one, and elastic deformation is neglected. Mises yield criterion is adopted.

Because the grid degradation speed is very fast, mesh generation occurred frequently. At the same time, in the cold rolling forming process of the spline, the contact time of the tooth flank with the workpiece is longer than that of the tooth addendum. With the increasing feed rate, the contact area of the tooth flank is larger than that of the addendum, and coulomb friction condition is adopted.

4.2 Numerical Simulation of the Cold Rolling Forming Process

4.2.1 Plastic Deformation Zone

Mesh generation is carried out on a certain cross section of the workpiece; the variation of the meshes under different feed rate on the same section is shown in Fig. 4.1.

From Fig. 4.1, it can be obviously seen that the plastic forming area in the cold rolling process is concentrated on the surface of the workpiece. The mesh distortion of the tooth dedendum is the most severe, and the distortion of the addendum is a little mild, but the radial influencing region is larger than that in the dedendum region of the workpiece. The magnified plastic zone in Fig. 4.1b–d is shown in Fig. 4.2.

The radially affected depth of the plastic deformation on the dedendum of the workpiece is smaller than that of the addendum part, because the metal of the workpiece dedendum is flowing to the surface of the addendum and still has a trend to other opposite directions simultaneously, which results in the participating of the central part for deformation, as shown in Fig. 4.3.

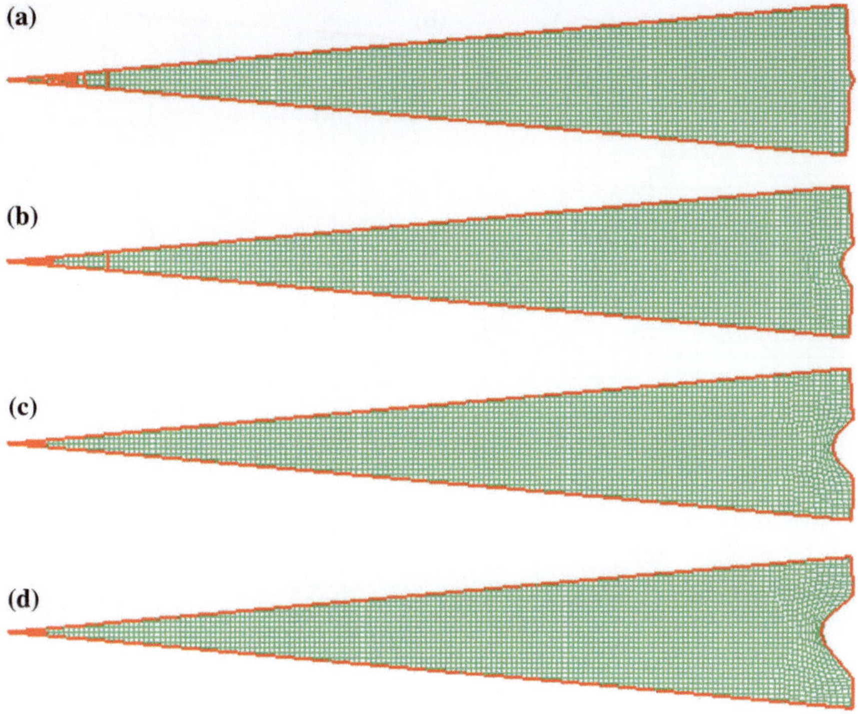

Fig. 4.1 Mesh variation on the cross section during the cold rolling forming process, **a** cross section before cold rolling forming, **b** cross section of the first stage in the cold rolling forming process, **c** cross section of the second stage in the cold rolling forming process, **d** cross section of the third stage in the cold rolling forming process

The plastic deformation zone is expanding with the increase in the feed rate, but the main distortion of the mesh is on the contact surface of the roller and the workpiece; the farer the workpiece surface, the smaller the mesh distortion degree is. Assuming that at time t, the feed rate is f_t and the radius of the workpiece root circle is r_{f_t}, then the metals in the region of $r < r_{f_t} - (1.5 \sim 2)f$ is not participant for the deformation completely.

4.2.2 *Bulge at the End of the Shafts*

In the numerical simulation of the spline cold rolling forming process, if the two ends of the shaft are not restrained, axial bulge will emerged at the dedendum of the end surface. And the dedendum and flank of the spline tooth at the end cannot be fully deformed; therefore, the addendum circle is unqualified, which is not coinciding with the practical production, as shown in Fig. 4.4a.

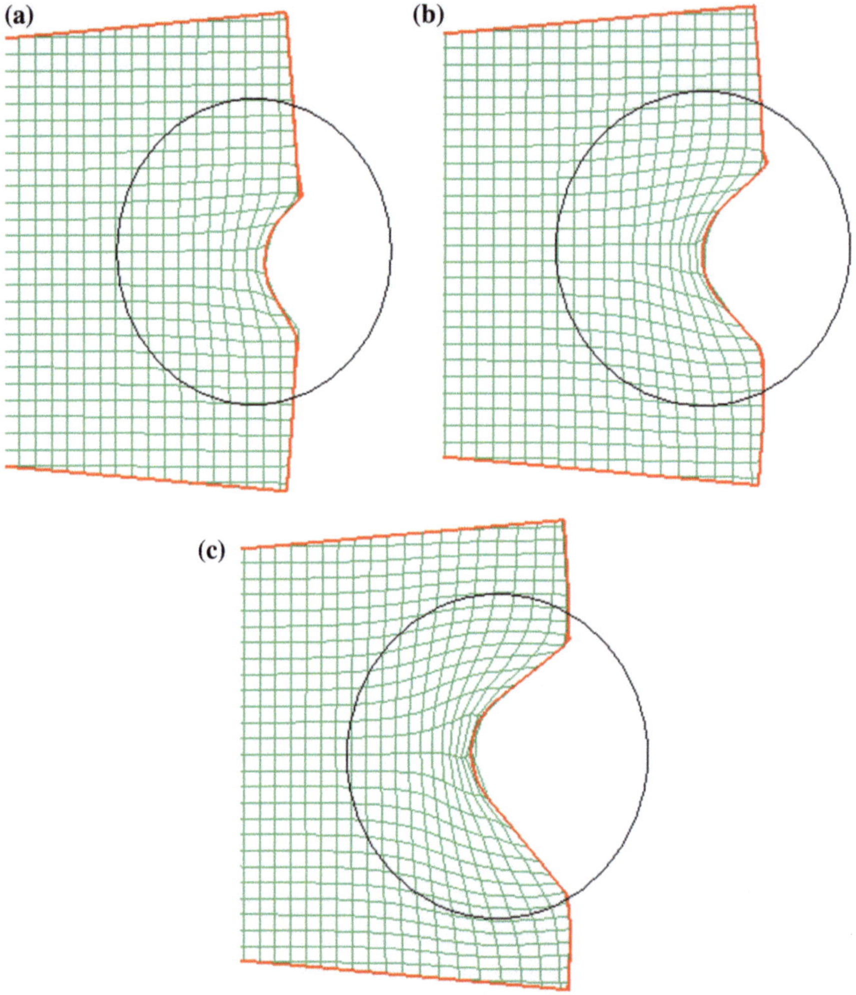

Fig. 4.2 Mesh distortion of the plastic deformation zone, **a** plastic deformation zone of the first stage, **b** plastic deformation zone of the second stage, **c** plastic deformation zone of the third stage

Bulging of the end will result in the insufficient volume and tooth height at a small area of the end. Simulation shows that the defect cannot be compensated by increasing the diameter of the workpiece billet before rolling. The phenomena can be prevented by adding a rod at the end of the workpiece. In the cold rolling production, bulging can also be observed on the end of the spline shaft, as shown in Fig. 4.4b.

It can be seen from the simulation that bulging exists in the plastic deformation area of the workpiece dedendum and will diminish from the dedendum to the center of the workpiece. Therefore, the radial region of the bulging is $r_{f_t} < r < r_{f_t} -$

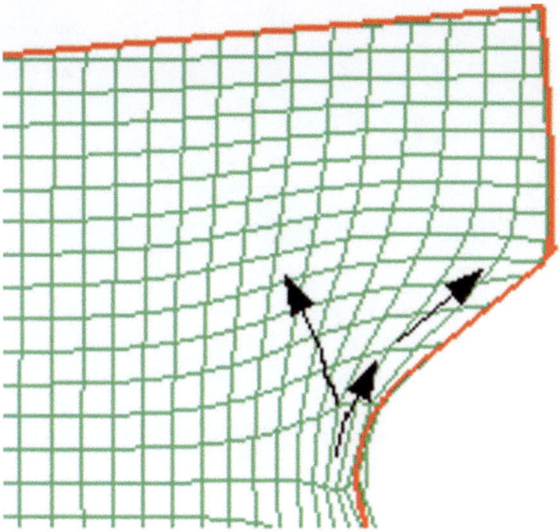

Fig. 4.3 Sketch diagram of the metal flow trends

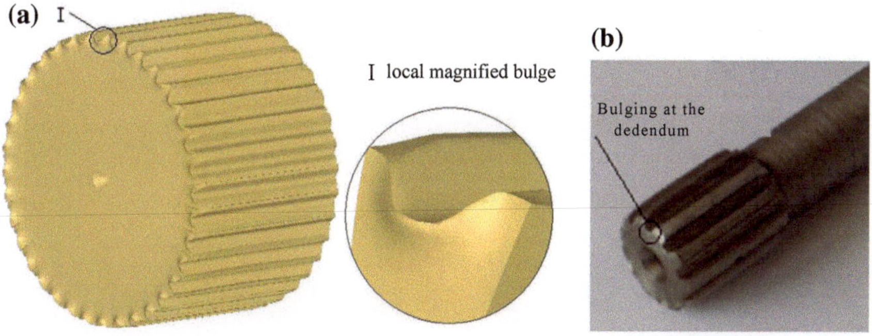

Fig. 4.4 Bulging at the end of the workpiece, **a** numerical simulation result, **b** cold rolling forming spline

$(1.0{-}1.5)f$. The axial constrain of the end in the simulation is appropriate and without distortion.

4.2.3 Stress and Strain Fields

The equivalent stress and strain of different stages in the spline clod rolling process are shown in Figs. 4.5 and 4.6 (at time of the peak rolling force in a single contact process). The fields of equivalent stress and strain are enlarged gradually, and the value of the maximum equivalent strain varies obviously.

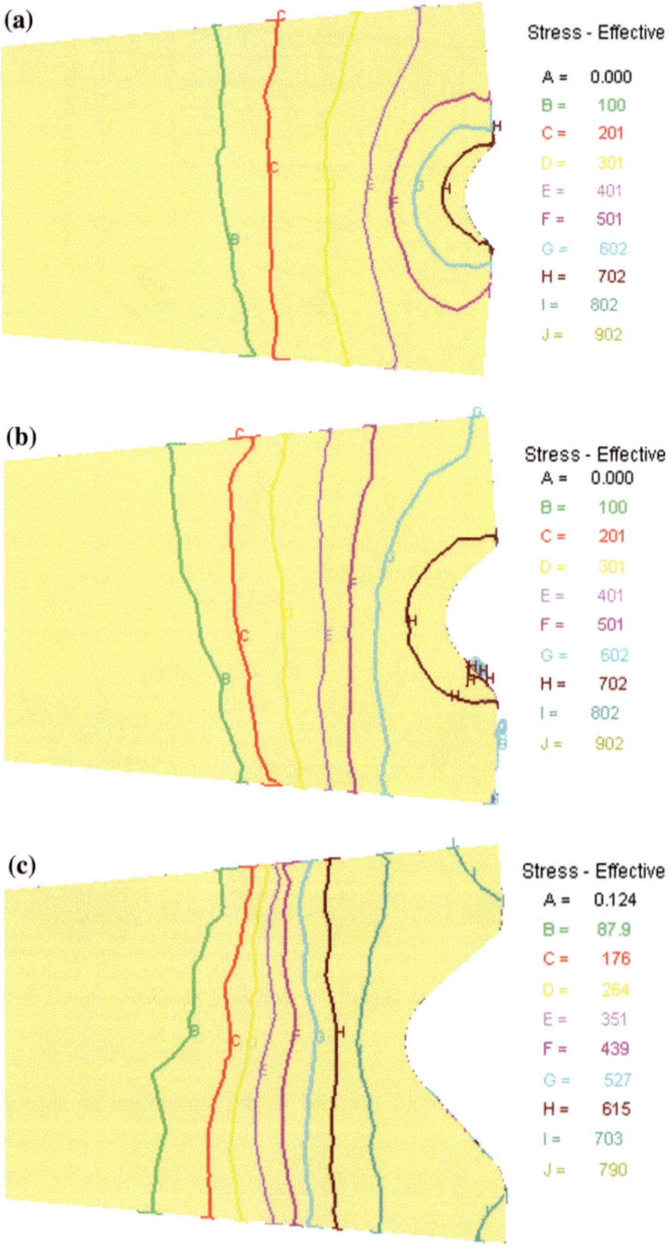

Fig. 4.5 Equivalent stress of the cold rolling forming process **a** the first stage, **b** the second stage, **c** the third stage

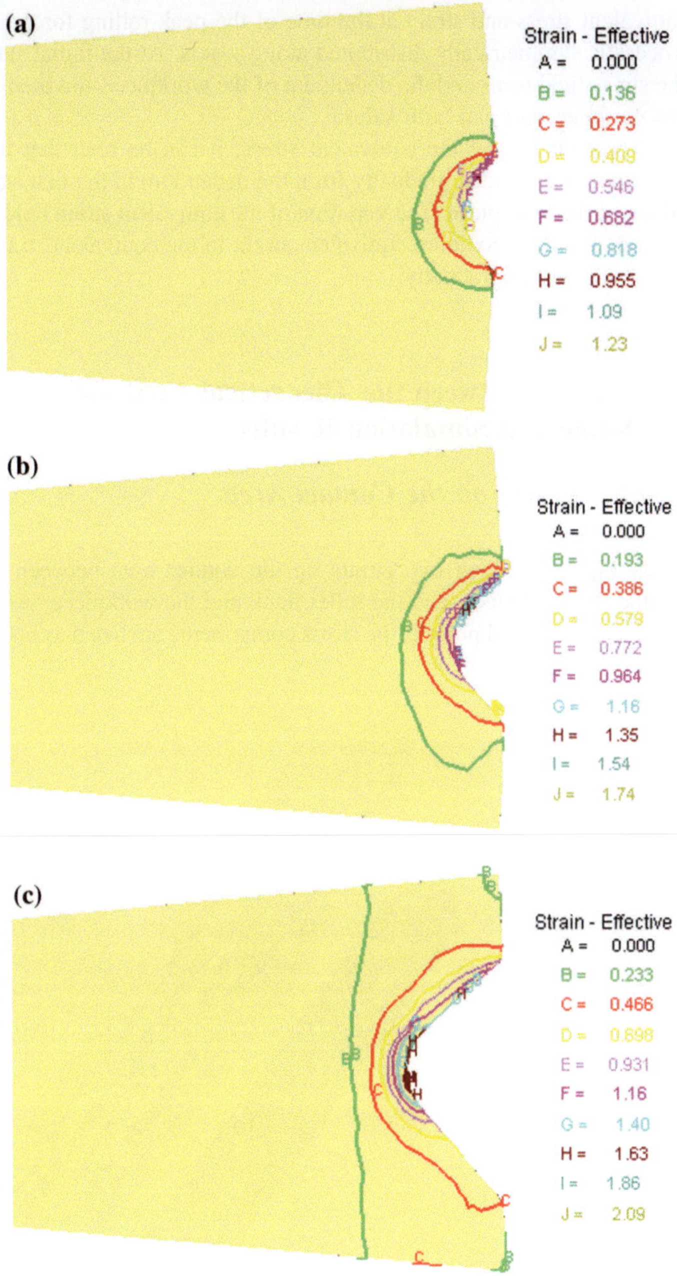

Fig. 4.6 Equivalent strain of the cold rolling forming process, **a** the first stage, b the second stage, **c** the third stage

The equivalent stress and strain at the time of the peak rolling force in a single rolling process is symmetrically distributed along y-axis. At the initial stage of the rolling, the stress field is around the dedendum of the workpiece and then gradually expands to the inner part and both sides.

From the contour line of the equivalent stress, it can be seen that the plastic deformation zone is expanded gradually from the dedendum in the first stage to the whole surface of the workpiece. The variation of the equivalent strain field is also of the same trend, and the maximum equivalent strain in the equivalent strain field of every stage is increasing gradually.

4.3 Comparison Between the Theoretical Analysis and Numerical Simulation Results

4.3.1 Unit Pressure on the Contact Area

As shown in Fig. 4.7, taking two points on the contact area between the roller addendum and the workpiece, and the roller flank and the workpiece, respectively, and recorded as point 1 and point 2, the stress components are listed as in Table 4.1.

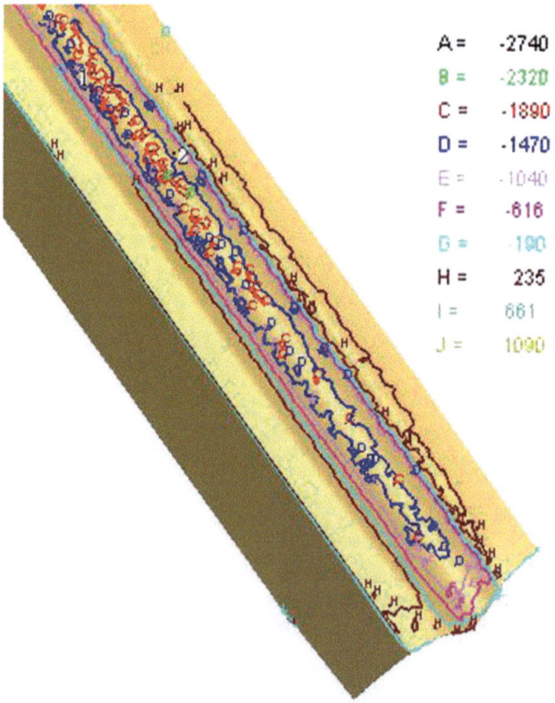

Fig. 4.7 Average stress field contour on the contact surface

Table 4.1 Stress components of the typical points on the contact surface

Position	σ_x/MPa	σ_y/MPa	σ_z/MPa	τ_{xy}/MPa	τ_{yz}/MPa	τ_{zx}/MPa
Point 1	−1976.36	−1107.47	−1483.82	−274.429	83.6146	-6.87734
Point 2	−2509.49	−1894.22	−2108.94	−416.943	18.0315	54.8891

From the table, it is shown that the deviation of σ_z is not very large around the average stress, and the shear stress at the z-direction is very small, which coincides with the analyzed results of the stress state.

The average stress field under different friction coefficients, different materials, and different rolling time periods is shown in Fig. 4.8. Figure 4.8a–c shows the average stress field under different rolling time periods and the same feed rate, the average stress at point a is the maximum, the variation of the average stress in Fig. 4.8b–d is not very large, and the distribution of the stress field is relatively equivalent, which is because the yield stress of AISI-1045 is larger than that of the AISI-1015.

Figure 4.8b, d shows the average stress field under different rolling time periods and feed rates of the same simulation, and the variation of the stress field is not obvious. It indicates that it is feasible to neglect the influence of the feed rate in the solving of the slip-line field and the average pressure.

The stress unit on the contact surface in the numerical simulation is shown in Fig. 4.9. σ_x, σ_y, and τ_{xy} are the stress components on the pitch point obtained by simulation. The average unit pressure can be obtained by simulation according to the geometrical relationship between the stress component and the average unit pressure.

According to the position of the typical points in Fig. 4.7 (point 1 is near $\psi \approx 0$ of zone AB), stress units of $E1_a$, $E1_b$, $E1_c$, and $E1_d$ are taken out from the average stress field shown in Fig. 4.8a–d.

The stress components of the above stress units are collected, respectively, and the unit pressure at each point can be calculated according to the geometrical relationship shown in Fig. 4.8. The theoretical calculation results can be obtained by substitute corresponding parameters into formulas (2.51) and (2.52). The theoretical calculation and numerical simulation results of each point are shown in Table 4.2.

Table 4.2 shows that the error of the theoretical and numerical simulation results is less than 3% in zone AB and less than 5% in zone AC. The mesh density in zone AB is sparse than that in zone AC, which is the main reason of the difference.

The property of the materials (σ_s) has a dominant influence on the average unit pressure. When the friction coefficient is doubled, the difference in the theoretical average unit presses is less than 5% and the difference in the simulated results is less than 3% in the two contact zones; and the influence in zone AC is larger than zone AB.

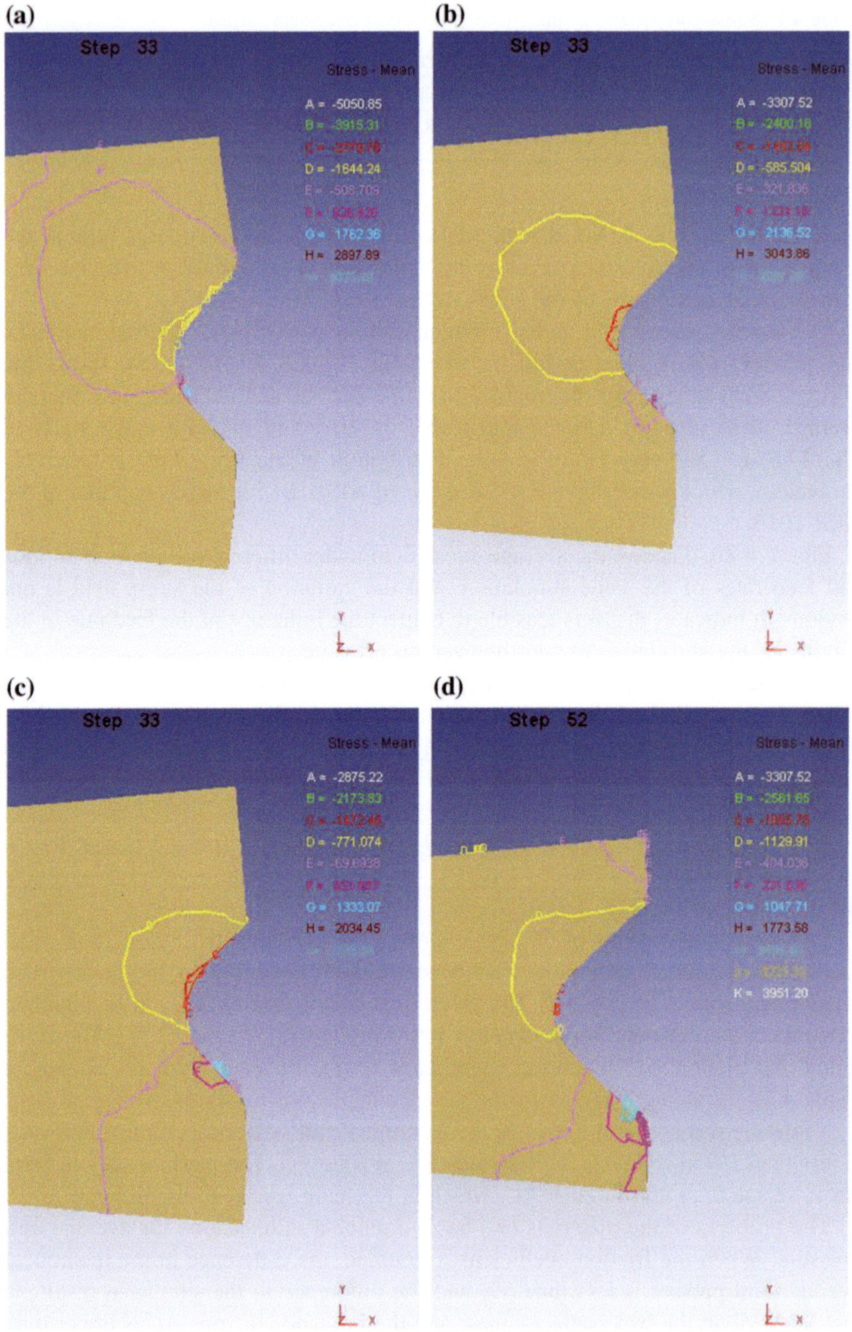

Fig. 4.8 Average stress field on the cross section under different material and friction coefficient, **a** material: 1045, $\mu = 0.1$, **b** material: 1015, $\mu = 0.1$, **c** material: 1015, $\mu = 0.2$, **d** material: 1015, $\mu = 0.1$

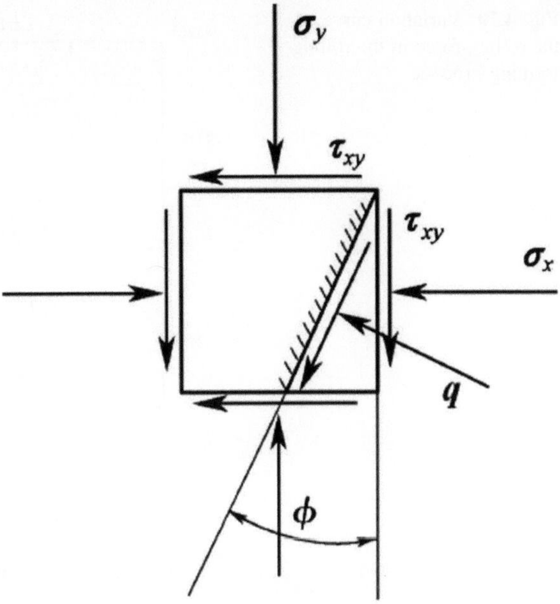

Fig. 4.9 Stress units in the numerical simulation

Table 4.2 Unit average pressure in the theoretical calculation and numerical simulation

Stress unit	Theoretical calculation q_{th}/MPa	Numerical simulation q_{FE}/MPa	Relative error (%)
$E1_a$	1966.79	1939.75	1.37
$E2_a$	1438.02	1477.86	2.69
$E1_b$	1475.09	1515.49	2.66
$E2_b$	1078.52	1125.29	4.15
$E1_c$	1515.80	1541.86	1.69
$E2_c$	1137.60	1098.86	3.40
$E1_d$	1475.09	1507.01	2.11
$E2_d$	1078.52	1110.94	2.91

Table 4.2 shows that the theoretical result coincides well with the numerical simulation one, and the maximum error is less than 5%. The frication coefficient has little influence on the unit average pressure; when the frication coefficient has been doubled, the difference in the calculated value is less than 5% and the simulated result is less than 3%. If the density of the generated meshes is improved, more accurate simulation results can be obtained, and the error of the two results can be further decreased.

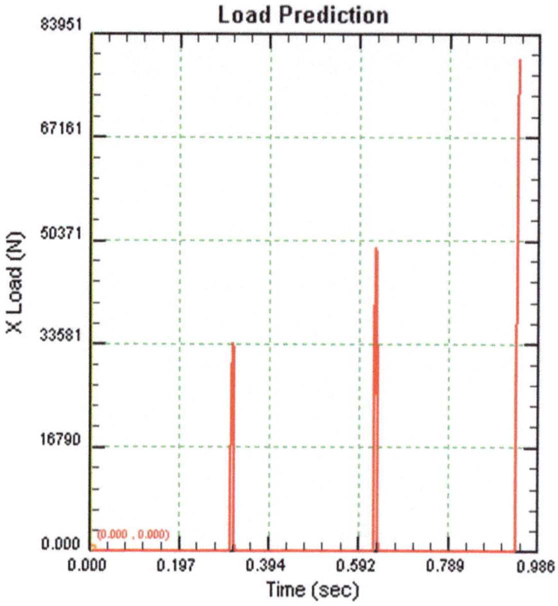

Fig. 4.10 Variation curve of the rolling force in the rolling forming process

4.3.2 Rolling Force in the Rolling Forming Process

The variation curve of the rolling force along *x-axis* (horizontal direction) in the simulation is shown in Fig. 4.10. The roller rolled from the symmetrical single-tooth workpiece three times, which is corresponded to the first-to-third stage of the rolling process. The rolling force rises to the peak value and then decreases in a rolling forming process; the maximum rolling force increases with the increasing feed rate, as analyzed in Sect. 3.2.2.

Under the conditions such as the length of the spline shaft $L = 20$ mm, the material AISI-1045, modulus $m = 1$, pressure angle $\alpha = 45°$, and a friction coefficient $\mu = 0.1$, the stress–strain relationship is shown in Fig. 4.11. It is shown that the initial yield strength of the material is $\sigma_s = 640$ MPa, and the maximum subsequent yield stress in the plastic deformation is $Y_{max} = 910$ MPa. Substitute σ_s and Y_{max} into Eqs. (2.56)–(2.57), and let $k_1 = 1.0$, the maximum rolling force in the rolling forming process can be calculated. Comparing the maximum rolling force with the results shown in Fig. 4.10, the results are listed in Table 4.3.

The theoretical calculated rolling force is smaller than the simulated value applying the initial yield limit, and the error is relatively large, within a scope of 20–30%. While the error is within 1.5% calculated with the maximum yield stress. Therefore, substituting σ_s into Eq. (2.57), the average stress obtained is more suitable for the calculation of the rolling force.

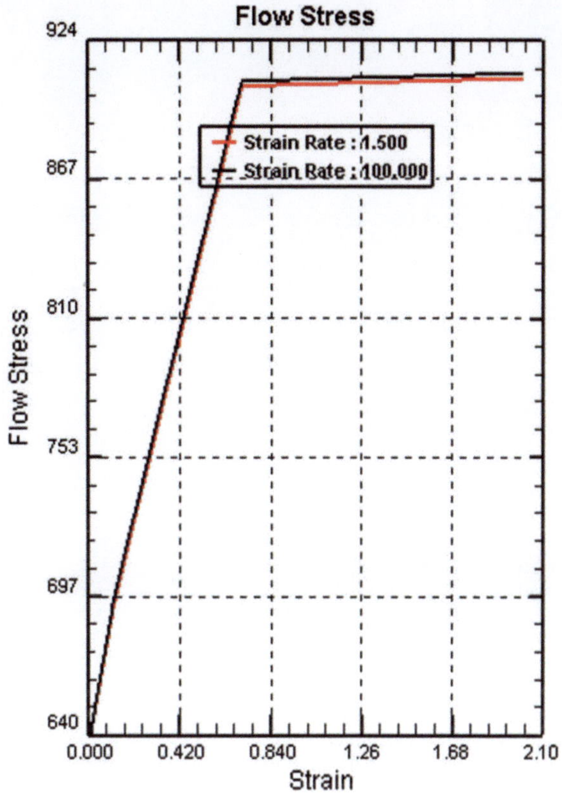

Fig. 4.11 Constitutive relationship of AISI-1045 at room temperature (20 °C) during plastic forming

Table 4.3 Rolling force in the theoretical and numerical simulation results

Y	Theoretical calculation/kN	Numerical simulation/kN	Error/%
σ_s	58.513	79.953	25.815
$Y_{\max}$	78.932		1.277

References

1. Li Guichang. System of Computer Numerical Simulation on Cold Rolling Forming of Splines and Threads [D]. Taiyuan: Taiyuan University of Science and Technology, 2008.
2. Li Rui. The Development of Automatic Modeling and Simulation System on Cold Rolling Process of Spline and Thread [D]. Taiyuan: Taiyuan University of Science and Technology, 2009.
3. Jia Yuejing. Numerical Simulation Software Development of the Involute External Spline Cold Rolling Forming Process [D]. Taiyuan: Taiyuan University of Science and Technology, 2011.
4. Li Guichang, Li Rui, Song Guangxin, et al. Method of Creating Accessory Database by Parameterization Design Gear on UG [J]. Aviation Precision Manufacturing Technology, 2008 (3): 55–57.

References

Chapter 5
Metal Flow Rules and Forming Quality of Cold Rolling Forming Spline Shaft

Abstract Metal flow and deforming rules of the cold rolling spline tooth during the forming process, the accuracy, defects and the forming quality control of the cold rolling forming process have been described and analyzed.

5.1 Metal Flow Rules of the Tooth in Cold Rolling Precision Forming of Spline Shafts

5.1.1 Workpiece Billet and the Forming Parameters

From the principle of spline shaft cold rolling precision forming process, cold rolling forming of spline can be simplified as a plain deformation state and metal flow along the axial direction can be neglected. In order to observe the metal flow conditions in cold rolling forming process, the cold rolling forming spline shaft is divided into several slices; each section of the billet perpendicular to the spline axis is marked with meshes preliminarily and then assembled together. After cold rolling forming on the machine, the sliced billet is separated and metal flow conditions on the sections can be obtained by observing changes of the meshes.

According to Fig. 5.1a, prepare several disk components by turning; series of concentric circles are marked with a spacing distance of 0.20 mm in advance for the reference line of metal flow, and the material flow conditions of the workpiece in the cold rolling precision forming process are recorded. After piercing and drilling of the keyway on each slice, the three round disks are assembled together with a flat key and screw bolt and then the external circles are ground to required external diameters of the spline, as shown in Fig. 5.1b.

Number 45 carbon steel is selected for the cold rolling forming billet materials. The parameters of the cold rolling spline shaft are as follows: modulus $m = 1$, tooth number $Z = 56$, and pressure angle of the pitch circle is 30°. According to Eq. (2.10): $d_z = \sqrt{d_f^2 + \frac{4}{\pi} A_{th} Z} = \sqrt{d_f^2 + \frac{Z}{\pi}[\frac{d_b^2}{3}(\tan^3 \alpha_a - \tan^3 \alpha_f) + d_a s_a - d_f s_f]}$, the calculated external diameter of the billet is 55.45 mm. The cold rolling forming experiment is carried out on the spline cold rolling machine Z28 J-35 of Qingdao Shengjian

J. Song et al., *Cold Rolling Precision Forming of Shaft Parts*,
DOI 10.1007/978-3-662-54048-0_5

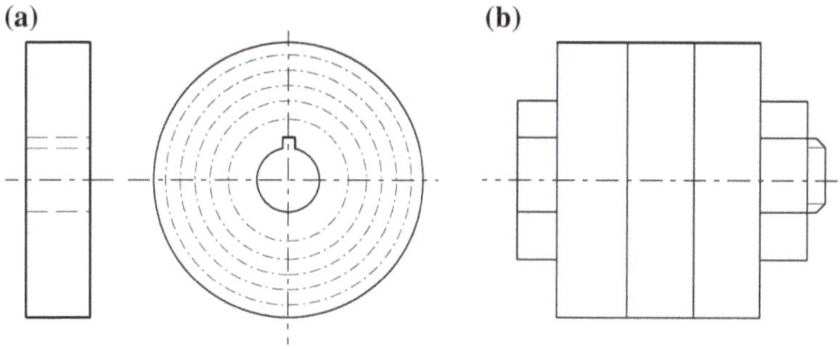

Fig. 5.1 Billets used in the metal flow experiment of cold rolling forming

Machinery Factory; the process parameters include the following: a feed rate of 60 mm/min, rotation speed of the engine driving the spindle 800 r/min, and the tooth number of 280.

5.1.2 Metal Flow Analysis of the Tooth Profile

After cold rolling of the prepared billets on the machine, the concentric circle on the formed disk spline tooth is put under the microscopic for observation. Figure 5.2a is the local amplified picture of the cold rolling forming tooth, and Fig. 5.2b shows the components after cold rolling forming.

Fig. 5.2a shows that after cold rolling forming, metal material on the component surface migrates under the rolling and extrusion of the roller tooth and results in the plastic deformation of the metals. The deformation of the outermost concentric circle is the maximum, and the deformation of the tooth profile decreases with the decreasing of the circle radius; there is almost no deformation occurs on the circle far from the tooth. In cold rolling forming, deformation mainly occurs in a certain

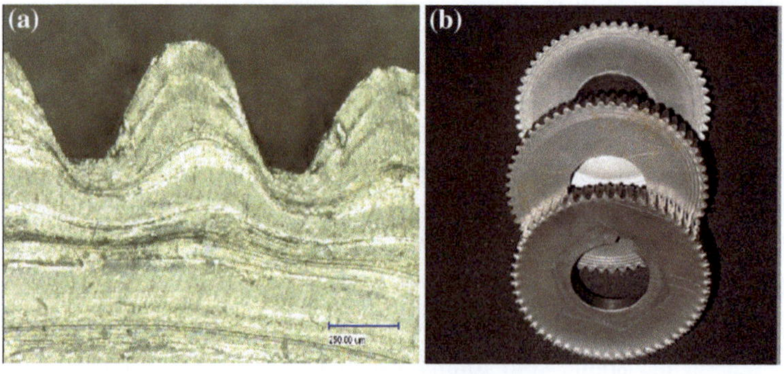

Fig. 5.2 Cold rolling forming spline components and metal flow conditions of the tooth profile

scopes near the workpiece surface, and metal flow on the surface is dramatic than the inner part of the workpiece. Metal of the tooth gullet is concaved under the extrusion of the roller tooth, the distance of the pregrooved circle decreased, and the metal has been extruded out. The metal flow pattern at the addendum of the billet is convex; the distance between the concentric circles is enlarged; the marked lines on both sides of the tooth are not symmetric.

According to the principle of the metal plastic forming, metal flows along the direction of minimum resistance in plastic forming. From the deformation conditions of the concentric circle, it is seen that the concentric circles at the spline dedendum are concaved inside, while the concentric circles of the spline tooth addendum and flank are convex toward outside, and the distance becomes larger near the addendum. Because in the cold rolling forming process, metal of the dedendum part is firstly extruded by the roller addendum and flows to the center of the circle, the density of the metal increased obviously under the inner resistant force of the spline and the pressure of the roller. However, with the enhancement of the metal density, the radial resistance increased, and the metal flow along the radial direction becomes difficult. On both sides of the roller addendum, there is no resistance on the metal surface, and a metal flow pass with the minimum resistance has been formed. Therefore, metal will flow toward both sides of the roller addendum. Under the pressure of the metal flowing tangential to both ends of the addendum, the metal flows toward the free surface on both sides of the roller addendum and bulges have been formed. With the increase in the roller feed rate, the bulged materials become higher and flow toward the dedendum of the roller. Resisted by the pressure and friction force of the roller flank, metal flow of the tooth flank is slowed down, and the materials in the middle of the tooth flow to the dedendum of the roller and form the addendum of the spline. Therefore, the addendum of the spline is formed by free flow of the metals. In the forming process, the dedendum of the tooth is not contacted with the billet material; the addendum of the billet is a free surface not influenced by the pressure of the rollers, and the flow resistance at the radial direction is the minimum. The pressed metal flows to the addendum of the spline and is fully formed to satisfy the dimension accuracy. The deformation of the metal on the spline surface increases under the extrusion of the internal metal flow, and the deformation of the outer layer metal is the maximum and gradually decreases toward the inner layer.

Cold rolling forming experiments on splines of different modulus showed that the tooth height of the spline is related to the deformation affected range in the cold rolling precision forming process, which is within a distance three times of the spline modulus m from the surface of the billet, and there is no deformation outside this area.

On both sides of the formed component tooth, the marked line is not symmetric, which indicates that the metal flow rules are different. In the cold rolling precision forming process, contact on both sides of the workpiece is not symmetric, and the force applied on the tooth surface and the metal flow direction is also different, which verified that the analysis of the contact point sliding direction on the active side and driven side introduced in Sect. 2.5 is correct.

5.2 Precision of the Spline Cold Rolling Forming Process

Taken the external involute spline on the input shaft of a certain gear box for example, billet of the workpiece and rollers are designed according to the requirements of the spline component. As shown in Fig. 5.3a, the involute spline is near the shoulder of the shaft, while machining with cutting method, relief notch must be designed, which will weaken the strength of the shaft. Because the demand of this shaft is relatively large, it is suitable for producing with cold rolling precision forming method.

According to the parameters of the involute spline component shown in Fig. 5.3a, simplified structure design for the shaft is carried out, and the part required for spline cold rolling forming is retained. The structure of the sample billet is shown in Fig. 5.3b. Both sides of the simplified shaft can be used for cold

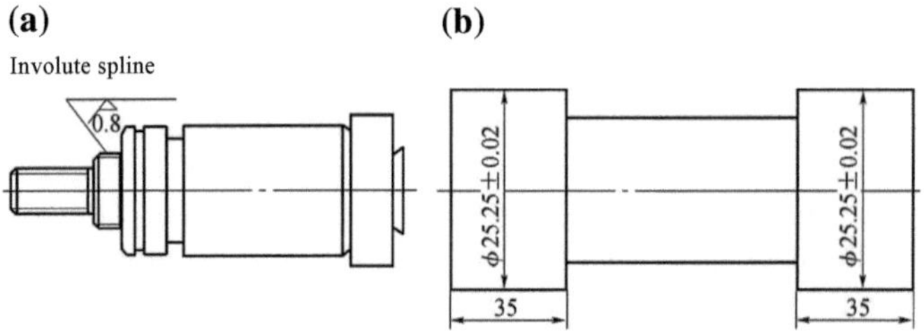

Parameters of the involute spline:

Modulus	0.75	Diameter of the dedendum circle D_f/mm	24.600 minimum
Tooth number Z_1	34	Diameter of the spline initial circle D_i /mm	18.031
Press angle of the pitch circle a_0/(°)	45	Actual maximum tooth width S_{max} /mm	1.178
Diameter of the addendum D_a /mm	$26.100_{-0.12}$	Actual minimum tooth width S_{min} /mm	1.097
Diameter of the dedendum D_1 /mm	25.50	Accumulated pitch error F_P /mm	0.032

Fig. 5.3 Spline parameters of the gear box input shaft and the design of the workpiece billet. **a** structure of the spline shaft; **b** diagram of the spline billet

rolling forming. According to the volume constant principle, the diameter of the billet is calculated for 25.25 mm with the design software for forming involute spline parameters. The external circle of the billet is ground, and the numerical control cold rolling forming machine is Z28 K-20 of the cold rolling forming laboratory in Taiyuan University of Science and Technology; and the tooth number of the roller is 230.

There are many factors influencing the precision of the cold rolling precision forming spline, such as the manufacturing precision and rigidity of the machine, the precision of the roller, adjustment of the machine, manufacturing of the billet, and selection of the process parameters. In the laboratory, the machine is adjusted to the best conditions, and spline shafts manufactured by cold rolling forming are shown in Fig. 5.4.

Single-item measurement method provided in the national standard of China GB/T3478.5-1995-"measurement method for cylindrical straight-tooth involute spline" is used in the precision measurement and analysis of the formed spline. Gauge bar with a diameter of $D_d = 2.00$ mm and a micrometer are used in the measurement of the overpin distance; and radial runout measuring instrument and dial indicator are used in the measurement of the radial runout error of the tooth ring. The measured results are shown in Table 5.1.

The overpin distance of the spline obtained in the cold rolling forming experiments is as follows: $MR\text{emax} = 29.089$ mm and $MR\text{emin} = 29.071$ mm. From GB/T 3478.6-1995 and GB/T3478.8-1995 [1], the precision of the cold rolling forming spline tooth profile can reach IT4. Referring to the value of the radial runout error of the tooth ring, and from GB/T 3478.1-1995, the tolerance grade is IT5 or (b) series. Therefore, the precision of the formed components is higher than the precision shown in Fig. 5.4. In the manufacturing of spline shafts with cutting method, IT7 or IT6 can be obtained by gear hobbing and shaping, and grinding procedure is often needed in the cutting of spline for a precision of IT4 or IT5. Cold rolling forming is an advanced process of higher production rate and forming precision.

Fig. 5.4 Cold rolling forming spline shafts

Table 5.1 Measured dimensions of the sample

Number of the sample	Overpin distance		Radial runout	Number of the sample	Overpin distance		Radial runout
	Maximum	Minimum			Maximum	Minimum	
1	29.088	29.071	0.03	11	29.088	29.076	0.03
2	29.073	29.080	0.03	12	29.073	29.080	0.02
3	29.089	29.086	0.02	13	29.089	29.086	0.03
4	29.088	29.071	0.04	14	29.088	29.077	0.02
5	29.073	29.082	0.04	15	29.073	29.082	0.01
6	29.089	29.086	0.03	16	29.088	29.078	0.03
7	29.088	29.073	0.02	17	29.073	29.080	0.01
8	29.088	29.071	0.04	18	29.088	29.072	0.02
9	29.073	29.084	0.04	19	29.073	29.085	0.02
10	29.089	29.088	0.04	20	29.089	29.084	0.02

5.3 Surface Quality of the Cold Rolling Forming Spline Shaft Parts

5.3.1 Surface Roughness of the Formed Components

In the cold rolling forming of the spline, not only the tooth of the spline are formed, but also through the repeated rolling and finishing on the workpiece surface by the rollers, the roughness of the workpiece can be decreased. The roughness of the component surface before and after rolling is measured with Taylor Hobson Surtronic 25 portable roughness measurement equipment. Taking 10 points on the surface of the experimental billet for roughness measurement, the average value of the roughness is recorded as the roughness before rolling, and the average roughness of the 10 points near the pitch circle of the spline tooth surface after cold rolling is recorded as the measured roughness of the cold rolling spline.

Table 5.2 shows that the surface roughness precision grade of the spline tooth has improved one order compared with the roughness of the ground billet. In the cold rolling forming process, the roller is sliding relatively to the surface of the workpiece, the micrographic geometrics have been modified, the surface roughness has been decreased, and there is a significant influence on the improving of the surface quality. In traditional cutting machining of the spline shafts, because the

Table 5.2 Tooth surface roughness of the cold rolling forming spline

Number of the sample	1	2	3	4	5	6	7	8	9	10
Surface of the billet, Ra/μm	0.746	0.769	0.790	0.768	0.752	0.774	0.764	0.768	0.760	0.759
Formed tooth surface, Ra/μm	0.408	0.412	0.403	0.414	0.401	0.406	0.404	0.407	0.411	0.404

metal fiber of the tooth surface has been cut off, tool marks on the surface increased the surface roughness of the tooth surface, while manufacturing spline with the same precision, grinding procedure should be arranged to meet the requirement of the surface roughness after machine cutting.

After cold rolling forming, the surface roughness of the spline tooth can be obviously decreased, the effective contacting area of the spline coupling increased, and the wear resistance of the component has been enhanced. With the increase in the contacting area, the resistant capability of the contacting deformation and thus the contact rigidity and the connecting strength will be improved. On the other hand, for the spline shafts machined by traditional cutting method, because of the large surface roughness, the components are sensitive to stress concentration, and the fatigue strength will be affected. Industrial application shows that the service life of the spline shaft formed by cold rolling is 2–3 times longer than the spline shafts machined by cutting method. In addition, because the surface roughness of the cold rolling precision forming spline shaft is lower, corrosion gas or liquids can be prevented to infiltrate into the inner layer of the metals through the microconcave of the surface, and the corrosion resistance of the component can be improved.

5.3.2 Tooth Surface Hardness of the Cold Rolling Forming Components

In order to analyze the tooth hardness and strengthening conditions of the cold rolling forming spline shaft, a No. 45 carbon steel spline shaft is manufactured by cold rolling forming with the modulus $m = 1$, tooth number $Z = 56$, and the pressure angle $\alpha = 30°$. Taking a complete tooth from the cold rolling forming spline shaft and slice on the cross section of the shaft, after grinding and polishing, the hardness of the spline cross section is measured with HVS-1000 micro-Vickers hardness tester by impressed into the four diamond pyramid at a load of 15 kg. The measured points are distributed as shown in Fig. 5.5.

Fig. 5.5 Distribution of the microhardness measuring points on the cross section of the tooth

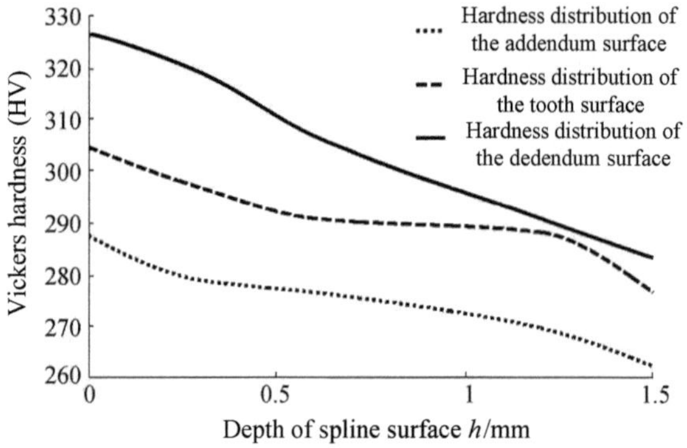

Fig. 5.6 Hardness distribution at different positions of the cold rolling forming spline tooth

In order to analyze the depth of the surface hardening layer, 24 measuring points along the normal direction of the spline tooth surface are selected at the tooth addendum, near the flank pitch circle and the dedendum. The distance between the first point and the external surface is 0.05 mm, and the interval space between each point is 0.05 mm. By fitting of the curve, the trend of hardness on every part of the tooth along the normal direction with the variation of the depth is shown in Fig. 5.6. From the surface of the cross section to the central part of the cold rolling spline shaft, the hardness is decreased. The hardening degree of the addendum, around the flank pitch circle and the dedendum, is as follows:

$$N_{\text{addendum}} = \frac{H - H_0}{H_0} \times 100\% = \frac{280 - 260}{260} \times 100\% \approx 8\%$$

$$N_{\text{flank}} = \frac{H - H_0}{H_0} \times 100\% = \frac{300 - 260}{260} \times 100\% \approx 15\%$$

$$N_{\text{dedendum}} = \frac{H - H_0}{H_0} \times 100\% = \frac{322 - 260}{320} \times 100\% \approx 25\%$$

where

H is the hardness of the external tooth surface and
H_0 is the hardness of the raw material.

The maximum microhardness of the spline addendum surface is HV280, and the thickness of the hardened layer is close to 0.5 mm. The maximum hardness of the flank surface is more than HV300, and the hardening layer of this position is thicker than that of the tooth addendum, which is about 1.5 mm. The hardness of the dedendum is the maximum, which exceeds HV320. The hardness is 25% higher than the hardness before rolling, and the depth of the hardened layer is more than 1.5 mm.

The outmost layer of the spline shaft tooth is the mating surface, the hardness of which has an important influence on the wear resistance of the component. Taking 12 points along the addendum to the dedendum of the tooth, the curve of the

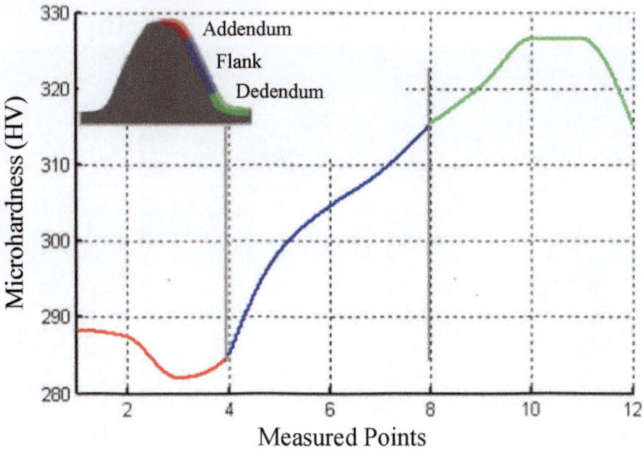

Fig. 5.7 Hardness distribution at different positions of the cold rolling forming spline tooth

measured hardness is shown in Fig. 5.7. The hardness on every part of the tooth surface is about 20% higher than that of the billet. Because the deformation of the tooth dedendum is the maximum, the hardness of the dedendum is the maximum; the hardness of the tooth flank takes the second place, and the hardness of the addendum is rarely improved. Compared with conventional cutting method of spline, in the cold rolling forming process, the wear resistance of the tooth surface is greatly improved due to the strengthening and hardening effect of the rolling process. Therefore, the formed components can satisfy the application requirements of higher strength.

The hardening degree of the cold rolling forming components is related with the composition of the material, the hardness before rolling, tooth depth, and the feed rate. For some kind of aluminum alloys and stainless steel, because the hardening speed is too fast to fill the tooth fully, they can only be used for cold rolling forming of shallow tooth profile gears or involute spline components.

5.3.3 Microstructure of the Cold Rolling Forming Spline Tooth

The enhancement of the cold rolling forming spline tooth hardness is closely related with the variation of the internal microstructure. Quenching and tempering No. 45 carbon steel billet are selected for the cold rolling forming of spline. Intercept a full tooth profile on the formed spline and cut it off along the cross section perpendicular to the tooth axis; after being polished, the sample is etched with 4% HNO_3 alcohol solutions. Microstructure observation is carried out with KEYENCE three-dimensional digital microscopy, and the microstructure on different parts of the spline is shown in Fig. 5.8. From the figure, it is shown that the metals on the surface layer of the spline tooth are distributed with a fibrous form; the flow-line pattern is approximately the same as the external profile of the spline

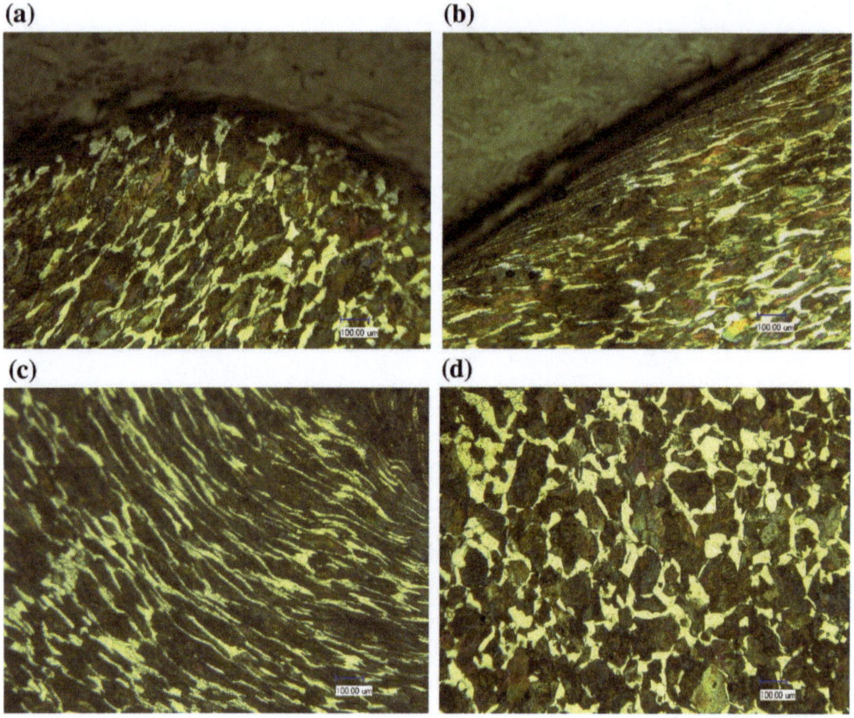

Fig. 5.8 Microstructure and deformation at different positions of the cold rolling forming spline. **a** addendum; **b** tooth flank; **c** dedendum; **d** shaft center

tooth. At the position of the shaft center, the microstructure of the material is the same as the state before rolling.

In the cold rolling forming of spline shafts, because the metal flow is different at the position of tooth addendum, flank, and dedendum, the microstructure is also different at different parts of the formed components. Figure 5.8a is the microstructure of the formed spline addendum, and it is shown that the microstructure is fine and fibrous. Because the addendum of the spline is formed by free flow of the pressed metal, the grains have been crushed, refined, and slightly elongated along the metal flow direction. However, because the applied force is relatively small, there is no obvious fibrous of the microstructure. Figure 5.8b is the microstructure of the spline tooth flank after cold rolling forming. The microstructure of the deformed material is fully refined, fibrous microstructure along the direction of the tooth profile has been formed. Slip bands have been found in the grains, and the grains are extended and distorted along the deformation direction. In the cold rolling forming process, the tooth flank has been rolled for several times by the roller, under the repeated actions of the rolling force and the frictional force, the grains have been repeated and fully refined, which results in the forming of the fibrous microstructure along the tooth profile direction. The microstructure shown in Fig. 5.8c is the most density and presents a strong fibrous pattern that almost coincides with that of the tooth profile. Because the deformation of the tooth dedendum is the maximum, under the extrusion of the roller addendum, the microstructure becomes more compacted. The redundant metal extruded from the dedendum flows along the tooth flank to the minimum resistant direction, and fibrous microstructure shown in the figure has been formed. This kind of fibrous microstructure is favorable for the improvement of the mechanical property, and the hardness of the tooth dedendum is the most highest. Figure 5.8d is the microstructure of the shaft center far from the tooth. It is shown that the microstructure after rolling is the same as the microstructure of the billet because the forming energy is attenuating from the surface of the tooth to the center of the shaft and the deformation is decreased; and there is no plastic deformation at the center of shaft, which is well coincides with the theoretical analysis.

5.3.4 *Tooth Surface Strengthening Mechanism of the Cold Rolling Precision Forming Spline Shaft*

The roughness and hardness of the cold rolling forming spline shaft surface are improved than the property of the billet, which cannot be obtained in the cutting method. Cold rolling precision forming not only can obtain the complex shaping and precision dimension of the spline tooth, but also can improve the property of the components, and the goal of shaping and property modification can be gained at the same time.

After cold rolling forming, the microstructure of the spline tooth has been obviously changed, the grains have been refined, and fibrous microstructure has been obtained. The property of the material is determined by the microstructure, the size of the grain has an important influence on the property of the metal. Relationship between the yield strength and grain size of the metal material can be described with Hall-Petch formula [2]:

$$\sigma_s = \sigma_o + Kd^{-\frac{1}{2}} \tag{5.1}$$

where K is a constant, related with the grain boundary structure, characterizing the influence of the grain boundary on the strength;

σ_o is the yield strength of an actual single crystal and
d is the average diameter of the grains in a multicrystal material.

From Eq. (5.1), it can be seen that the yield strength of the metal material is in an inverse proportional to the grain size; the finer the grain, the higher the yield strength. In the cold rolling precision forming of the spline shaft, grain refinement is caused by the plastic deformation; the average grain size decreased along with the rapid improvement of the dislocation density, which will result in the increase in the spline tooth yield stress σ_s. The equiaxial grain of the spline tooth is crushed and refined, and the shape and internal substructure of the grain have been changed. The segregation, grain boundary, inclusion, and other chemical compound have been changed in the cold rolling plastic forming; new grain and subgrain have been formed; and finally continuously distributed fibrous flow-line pattern microstructure has been formed. The hardness, strength, and corresponding mechanical property of the formed spline have been obviously increased. Under the rolling and extrusion of the roller during the cold rolling forming, part of the works applied on the billet result in the plastic forming of the workpiece, and part of the works converted to the internal residual stress in the formed components. Dislocation of the crystal lattice induced the distortion of the crystal lattice and residual compressive stress on the formed surface of the components, which effectively improved the bending and torsion strength of the spline components. The fibrous microstructure is the most compact and fine at the dedendum of the tooth, which enhanced the comprehensive property of the spline components.

Because of the repeated rolling by the roller tooth, the microconvex and microconcave have been flattened, and a roughness less than Ra0.4 can be obtained. The corrosion resistance, impacting resistance, and fatigue resistance can also be improved, and the life span can be elongated for 2–3 times than the spline shafts machined by cutting method.

5.4 Forming Error and Defects of the Cold Rolling Forming Spline

In the cold rolling precision forming, through the free rolling between the roller and billet, the tooth profile of the spline has been formed to meet the requirements of shape and dimension. By the research of the forming error and defects of the forming components and appropriate design of process parameters, the forming precision and quality of the spline can be controlled and improved, which is also a key factor for the popularization and application of this technology in practical production.

5.4.1 Formation and Control of the Accumulated Pitch Error

5.4.1.1 Variation of the Circular Pitch in the Cold Rolling Forming of Spline Shaft

In the cold rolling precision forming of spline shaft, the two rollers are feeding continuously in the radial direction with a constant speed; from the contacting

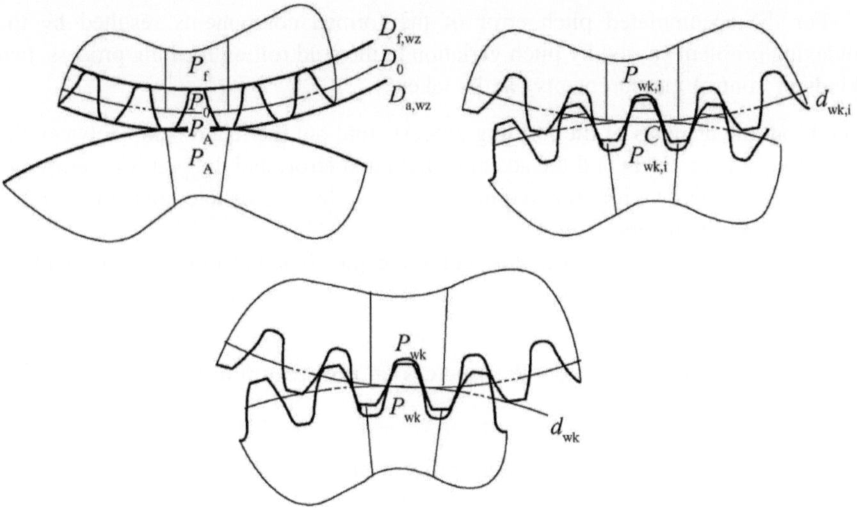

Fig. 5.9 Variation of the pitch circle and the circular tooth pitch of the workpiece

between the roller addendum and the external circle of the workpiece, every time the workpiece turns half of a circle, the tooth of the workpiece will contact with the roller once. With the radial feeding of the roller, the rolling diameter of the pitch circle on the roller and the workpiece has been changed. Figure 5.9 shows the relationship between the roller and the workpiece tooth profile at the three typical positions: the initial contact stage, the feeding and pressing stage, and the finishing stage. In different stages of the forming process, the diameter of the pitch circle changes dynamically; and the circular pitch on the workpiece is different each time contacting with the roller tooth. Therefore, appropriate dynamical indexing should be carried out during the cold rolling forming process. The cold rolling forming precision forming process of the spline shaft is a circular pitch modification process, and the variation of the circular pitch is determined by the change of pitch circle diameter.

In the free indexing cold rolling forming of spline shaft, the rotation speed of the roller is constant, and variation of the pitch circle diameter and the circular pitch results in the indexing error every time the workpiece entering the contacting zone. In order to obtain the accurate circular pitch of the workpiece while the roller tooth impressed into the billet, corresponding circular pitch should be provided to guarantee the requirement of the dynamic indexing at different pitch circle diameters of the rolling forming process. The same problem also exists in the cold rolling forming process of spline with rack dies. Dale J. Garden proposed a kind of variable-pitch rack die design method in an American patent, in which the dynamic indexing problem can be solved and the forming precision can be improved by linearly adjustment of the circular pitch from the first tooth of the roller contacting the workpiece to the first tooth contacting the workpiece in the finishing stage [3].

For the accumulated pitch error of the formed components resulted by the indexing problem caused by pitch variation in the cold rolling forming process, two kinds of control measurements can be taken:

(1) From the analysis of the forming process, find out the relationship between the process parameters and the accumulated pitch error; and the precision error can be controlled within the required category by reasonable matching of the process parameters.
(2) Design new process programs and solve the dynamic indexing problem by movement compensation.

5.4.1.2 Influence of Process Parameters on the Accumulated Pitch Error

In the cold rolling precision forming process of spline shaft, the influence of the accumulated pitch error caused by the variation of the pitch circle diameter and the indexing error is the maximum at the initial indexing stage. The influence of process parameters on the accumulated pitch error can be deduced through the movement analysis of the contacting point.

In the initial indexing stage, the addendum circle of the roller tooth contacts with workpiece at the external circle of the billet. The accumulated pitch error can be analyzed by the movement of the contacting point when the workpiece rotates for half a circle. During the feeding process, the moving trajectory of the contacting points between the roller addendum and the workpiece billet is not a circle, but a reverse Archimedes helix, as shown in Fig. 5.10, the polar coordinate equation of which is:

$$\rho = rz - \frac{v_r}{\omega}\theta \tag{5.2}$$

where

v_r is the radial feed rate of the roller and
ω is the angular rotating speed of the billet.

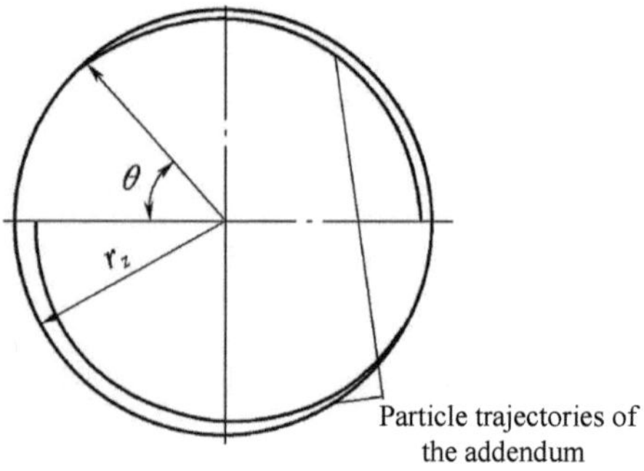

Fig. 5.10 Particle trajectory of the roller addendum on the external circle of the billet

Comparing Eq. (5.2) with the equation of circle, the arc length in the half circle zone is not equal. While the tooth cogging pressed and indexed by the roller of one side rotates to the other side, and the addendum of the roller entering the contacting zone, indexing error will be resulted by the arc length difference, and accumulated pitch error will be formed with a value of:

$$\Delta F_P = \pi r_z - \int_0^{\pi} \sqrt{\rho^2 + \rho'^2(\theta)} d\theta \quad (5.3)$$

The accumulated pitch error F_p of spline has been specified in the standard of involute spline. Therefore, in the cold rolling forming of spline, the accumulated pitch error resulted by the indexing error must be smaller than the standard value F_p.

$$\Delta F_P < F_P \quad (5.4)$$

The rotation angular speed ω of the billet can be determined according to the number of spline teeth Z, the number of roller teeth Z_2, and the rotation speed n of the roller:

$$\omega = 60n\frac{z_2}{z} \quad (5.5)$$

In the cold rolling precision forming of spline shaft, the main process parameters are the radial feed rate v_r and rotation speed n of the roller, and the reduction per rotation is defined as follows:

$$\varepsilon = \frac{v_r}{n} \quad (5.6)$$

The influence of the process parameters on the accumulated pitch error can be obtained by multi-parameter calculation with MATLAB and comprehension of the above equations.

Still taken the cold rolling forming of involute spline shaft for example, the teeth number is 34, modulus is 0.75, and pressure angle is 45°. Four groups of roller teeth number are taken as parametric variables; the rule of ΔF_P varied with the reduction ε is shown in Fig. 5.11. The figure shows that under a determined roller teeth number, accumulated pitch error increases with the variation of the reduction ΔF_P linearly; with the increasing of the roller teeth number, the slope varying curve between the accumulated pitch error ΔF_P and the reduction ε will decrease.

From the relationship between the accumulated pitch error and the process parameters, it is known that the accumulated pitch error can be decreased by decreasing the reduction (the ratio between the feed rate and the main motion speed). Increasing the roller teeth number is favorable for the control of the accumulated pitch error.

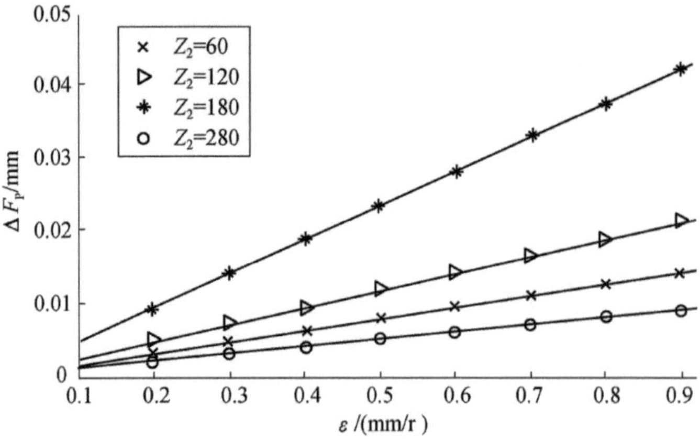

Fig. 5.11 Influence of reduction on the accumulated pitch error

5.4.1.3 Control of the Accumulated Pitch Error by Synchronization of the Roller and Workpiece Rotation Speed

The precision of the cold rolling forming spline tooth profile is influenced by many factors; for example, the contacting position will be affected by the manufacturing precision and rigidity of the machine tool, and the manufacturing precision of the roller and the install precision; therefore, unreasonable accumulated pitch error of the formed components will be resulted. The influence of the manufacturing precision can be solved by improving the manufacturing and assembly precision; the error caused by the variation of the pitch circle diameter due to the principle of the cold rolling forming process can be eliminated by synchronization of the roller and workpiece rotation speed.

From the above analysis, in different stages of the rolling forming, the rotation speed of the roller is constant, and with the radial feeding of the roller and the enhancement of the tooth height, the pitch circle diameter of the roller and workpiece is changing; if the rotation motion of the workpiece is added into the driving movement of the machine, and the free rolling between the roller and the workpiece is changed to a synchronized rotation, the rotation speed can be dynamically compensated through the adjustment of the workpiece rotation speed, and dynamic indexing can be realized according to the variation of the circular pitch.

Firstly, the position of the roller can be determined through the numerical control system of the machine; the forming stage can be calculated and judged; then, the rotation speed of the workpiece at different forming stages (shown in Fig. 5.9) can be controlled according to the following methods:

(1) At the initial contacting indexing stage, contact rolling occurs between the roller addendum and the external circle of the workpiece billet, the addendum circular pitch of the roller indexing the spline with a teeth number of Z on the

external circle of the workpiece billet, and the rotation speed of the workpiece at this stage should be:

$$n_1 = \frac{n_2 d_{a_2}}{dz} \tag{5.7}$$

(2) At the stage of indentation, the pitch circle diameter of the roller and the workpiece varied continuously with the radial feeding of the roller, in order to realize the variable circular pitch rolling forming, the rotation speed $n_{1,i}$ of the workpiece should be dynamically modified according to Eq. (5.8):

$$n_{1,i} = \frac{n_2 d_{2,i}}{d_{1,i}} \tag{5.8}$$

where $d_{1\mathrm{i}}$ and $d_{2\mathrm{i}}$ are the pitch circle diameters of the workpiece and the roller at the contacting instant, which can be calculated according to the center distance between the roller and workpiece and the meshing principle.

(3) At the finishing stage, the relative motion between the roller and workpiece is a kind of meshing movement without top backlash. The rotation speed of the workpiece can be calculated accurately from Eq. (5.9):

$$n_1 = \frac{n_2 d_2}{d_1} \tag{5.9}$$

For non-modified standard spline and rollers, d_1 and d_2 are the pitch circle diameters, respectively.

5.4.2 *Defect Analysis of the Cold Rolling Forming Spline Shaft*

In the cold rolling forming process of spline shaft, because the deformation of the metal is serious, the forming precision is influenced by many factors, and the forming defects and even faulty parts will be produced if the process parameters are not appropriately selected. The main defects of the cold rolling forming spline shaft include the following: disorder of the teeth, mistake of the teeth number, insufficient filling of the tooth, and higher roughness of the tooth surface and chipping of the tooth surface.

5.4.2.1 Disorder of the Tooth and Mistake of the Teeth Number

Disorder of the teeth will happen during the cold rolling forming process; the disorder of the formed workpiece teeth is resulted by the mistake indexing of the

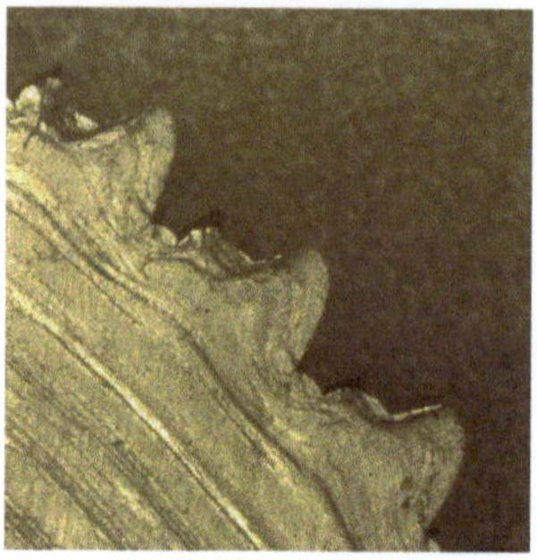

Fig. 5.12 Disorder of teeth in the cold rolling forming of spline shaft

roller, as shown in Fig. 5.12. Disorder of the teeth is a serious defect, which will not only result in the failure of the forming process and damage of the components, but also even result in the blocking faults, roller and machine damage, and serious accident. Therefore, disorder of the teeth should be avoided as possible.

The main reason of the teeth disorder is indexing error, which is related with the billet design and the adjustment of the machine, and can be solved through the following methods:

(1) Unqualified design and manufacture of the billet. Larger diameter of the billet will result in the unequal division of the workpiece circumference in the cold rolling forming.
(2) The circumferential position of the roller has not been well adjusted before rolling, and uncertain position of the two rollers and the workpiece will be resulted by the tooth position error.
(3) The precision of the machine is not qualified, the gap of rotation equipment of the two rollers is too large, or the sliding resulted by the clamping or accident of the machine.

While disorder of the teeth occurs in the forming process, the machine should be stopped for checking; if a redundant tooth has been formed, abnormal sound can be heard and faulty workpiece will be generated. The main reason of the redundant tooth lies in the larger billet diameter. In the same way, if the billet diameter is too small, less teeth number will be obtained. More commonly, unqualified tooth profile precision, such as too sharp or flat addendum, improper overpin distance or waste component will be produced.

5.4.2.2 Tooth Profile Defects

In the double-side non-backlash rolling process, the size and direction of the applied force on both sides of the tooth surface are different. When the forming force is relatively large, the formed tooth will inclined to one side and deduce tooth profile error. The formation of this defect is related with the metal flowing rule in the forming process. It can be seen from Chap. 2 that at the active side of the

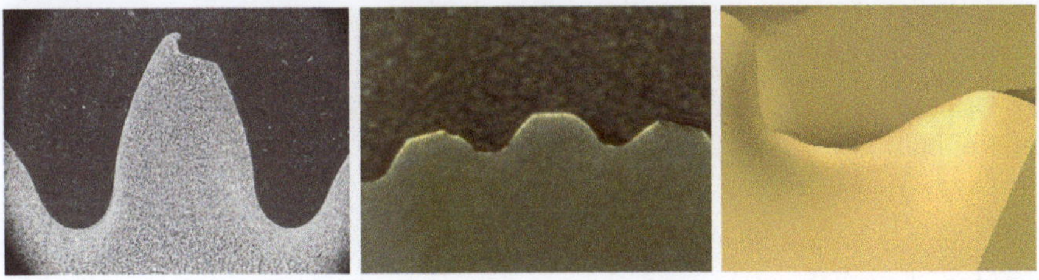

Fig. 5.13 Shape defections of the addendum

workpiece tooth, metal flows from the tooth addendum and dedendum to the nearby of the nodes; at the driven side of the workpiece tooth, metal flows from the node to the addendum and dedendum of the tooth. In the cold rolling precision forming of small modulus spline, the forming force is small and the incline phenomenon of the tooth is not very serious.

Tooth profile defection can be solved with the following methods:

(1) Adjusting the rotation and feed rate of the roller, coordinating the deformation, and reducing the forming force as possible.
(2) At the finishing stage, the roller can be rotated positively and inversely; the active side and driven side of the workpiece tooth can be alternated continuously, and thus, the incline of the tooth can be eliminated.
(3) Modifying the tooth profile of the roller and machining the two sides of the tooth surface to non-symmetrical shape, the influence of metal flow on both side of the tooth shape can be balanced.

The first and second methods can be easily realized by adjustment of the power transmission in the process. The effect of method (1) is not very obvious, and small tooth incline defects can be simply solved with this method. Method (2) is an effective measurement for the eliminating of the tooth incline. In method (3), the modification value of the roller should be determined by experimental and theoretical analysis. The manufacturing of the roller is complex and difficult to realize.

The tooth profile defects also include corner of the tooth addendum, folding, insufficient filling of the tooth, and convex of the dedendum, as shown in Fig. 5.13. The reason of these defections is mainly the disqualification of the billet material, unreasonable design of the billet size, and non-matching of the rolling process parameters.

5.4.2.3 Rough and Chipped Tooth Surface

Sometimes, the cold rolling formed spline shaft surface is very rough, especially at the addendum of the tooth, and there is obvious scratching marks on the surface. Under serious conditions, chips will be extruded on the addendum of the tooth, and the quality of the component will be affected. The reasons of these defects mainly include the following:

(1) Displacement of the roller teeth on both sides;
(2) Rolling for a longer time; and
(3) Large friction coefficient between the roller and billet.

In the two former methods, the defects can be modified by adjusting of the machine, the rough surface and chipped defects on the surface can be solved by variation of the process parameters and shortening of the rolling time. The friction coefficient can be adjusted by changing of lubrication conditions; the rough billet surface and tooth surface defects or damage can also result in the variation of the friction coefficient between the roller and the billet; the reason of the defects can be easily found by observation.

5.5 Quality Control Methods of Cold Rolling Forming Spline Shafts

There are many factors influencing the cold rolling forming spline shaft quality. Precision and rigidity of the machine, and precision of the roller precision all have an important influence on the precision of the formed components. Therefore, it should be considered in the period of equipment selection and die design. In this book, the state of the machine and tools has been ideally simplified. Under condition, the equipment and dies have been determined, and the design of the billet, adjustment of the equipment, and selection of the process parameters should be mainly considered in the control of the forming quality. The following respects should be well considered to guarantee the quality of the forming components and to reduce defects.

5.5.1 Design of the Cold Rolling Billet

There is no necessary for consequent machining of the formed spline shaft parts. The machining requirement of the billet is higher, and the material of the billet, heat treatment station, and geometrical dimension are all very restricted. Key factors of the billet design will be analyzed from the respect of forming quality control in the following context.

In cold rolling forming process, the rolling ability of the material should be considered; theoretically, any plastic materials can be used in the cold rolling forming of the spline shaft. However, cracking is easy to occur in the cold rolling forming of materials with an elongation less than 8%; therefore, this kind of materials is not suitable for cold rolling forming. In addition, the hardness of the material has an important relation with the cold rolling forming process. Higher hardness will result in the rapid decrease in the roller lifetime. If the hardness of the billet material is more than 300 HBS, it is not suitable for cold rolling forming and heat treatment procedure should be arranged to reduce the hardness for cold rolling forming. Comprehensively considering the quality, precision, and strength requirement of the cold rolling forming spline, and the economic lifetime of the roller, the reliability of the machine and the production rate, the elongation rate, and hardness of the component material can be selected according the following demands:

(1) Materials with an elongation δ more than 12% have good rolling property and are suitable for cold rolling precision forming.
(2) Materials with hardness less than 270 HBS are suitable for cold rolling forming because spalling and joint marks on the tooth surface are not liable to occur. In addition, with a higher hardness of the material, the rolling force will be increased, which will reduce the lifetime of the roller and increase the cost of the roller allocated to each component.

Neglecting the convex part at both ends of the components, according to the volume constant principle, the billet diameter before rolling can be deduced. The dimension of the billet can be calculated by the formed spline shaft, but through experiments, it is found that the dimension determined with this method cannot satisfy the requirement of qualified spline regardless of the calculation accuracy. Because the formed spline shaft is related with the parameters and manufacturing precision of the roller, and the rolling process, the dimension of the billet cannot be calculated by the theoretical value of the spline shaft; specified forming machine, roller, and process parameters should also be taken into consideration for the determination of the billet dimensions.

Billet diameter is one of the most important process parameters in the cold rolling forming process of spline shaft; in practical production, it will directly influence the indexing precision and the tooth profile quality of the formed spline. Too large billet diameter will result in the disorder of the teeth and even broken of the roller teeth; too small billet diameter will result in the insufficient filling of the tooth shape and even disorder of the spline teeth.

Many scholars and technicians have studied the calculation method of the billet diameter in the cold rolling forming of spline shafts. Zheng Quangang in Qingdao Shengjian Machinery Factory [4] and Pan Zhiqiang in Jingjiang Wanlima Power Machinery Factory proposed the empirical formula based on the summary of the production experiences [5].

Zhang Dawei deduced the theoretical calculation formula of the billet diameter of round dedendum spline [6], and He Feng in Jianghan Tools Machinery obtained the theoretical formula of the spline billet diameter, ignoring the fillet arc of the dedendum [7]. All these researches are based on the calculation of the nominal billet dimensions. Cold rolling precision forming is suitable for large volume production, and the demands for process stability are strict. Therefore, in the design of the billet, the tolerance must be specially controlled into a reasonable category. The shape errors such as the roundness and cylindricity of the rolling parts on the workpiece billet will be reflected to the formed spline and will influence the precision of the spline.

Authors of this book proposed a determination criterion in the billet dimension design. Firstly, calculating the initial external circle diameter of the cold rolling forming spline billet with the nominal dimension calculation formula (2.10) proposed in this book and combining the indexing condition formula (2.9) and the theoretical reference value calculated with the above methods, the initial value can be used in the roller design and determination of the theoretical value of the roller addendum circle diameter in free cold rolling precision forming, and generally, the roller designed with this theoretical value should be modified. Otherwise, if the theoretical diameter of the workpiece is calculated according to the standard value, then the formed spline shaft will perhaps be unqualified. Therefore, after the design and manufacturing of the cold rolling forming rollers, the billet diameter should be modified combining with the parameters of the rolling machine and the roller based on the theoretical value.

Before determination of the mass production process, according to the theoretical calculation value, cold rolling forming experiments of different billet diameters are carried out, and the final billet diameter in batch production should be determined by the precision measurement of the formed components. In batch production, the external diameter of the billet should be controlled within half of the tolerance band of the formed spline shaft according to the requirement of the precision, and then, satisfied dimension consistency can be obtained. In addition, the geometrical form error of the formed parts on the spline should also be strictly controlled, especially the coaxiality error between the spline external circle and the location datum. The geometrical form error of the billet should be controlled within $\frac{1}{3} \sim \frac{1}{2}$ of the corresponding tolerance of the spline, and higher forming precision and dimension consistency of the spline components can be obtained in the production.

5.5.2 Setting of the Roller and Adjustment of the Tooth Positions

Before cold rolling forming of high-precision spline, the adjustment of the machine is an important step for guarantee of the forming quality. Cold rolling forming machine is a kind of specialized equipment; besides, the adjustments such as common machines, setting of the rollers, and adjustment of the teeth positions are necessary before processing.

Before installation of the rollers, the parallelism of the two spindles, the radial runout error of the spindle external circle to the axis, and the circumferential rotation error of the spindle should be checked; the height of the two fixture center connecting lines should be equal to the height of the spindle. Because the feed rate

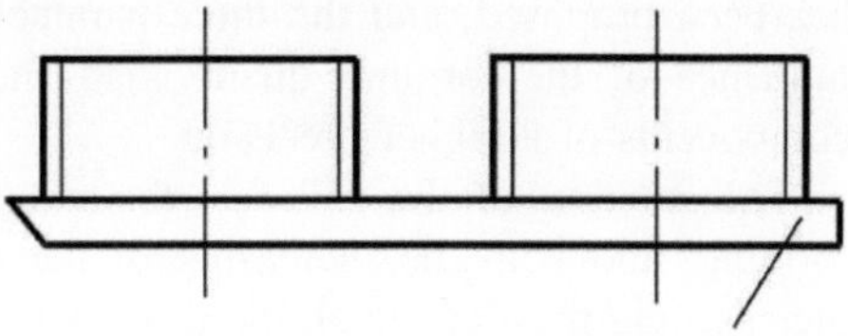

Fig. 5.14 End surface measurement of the roller

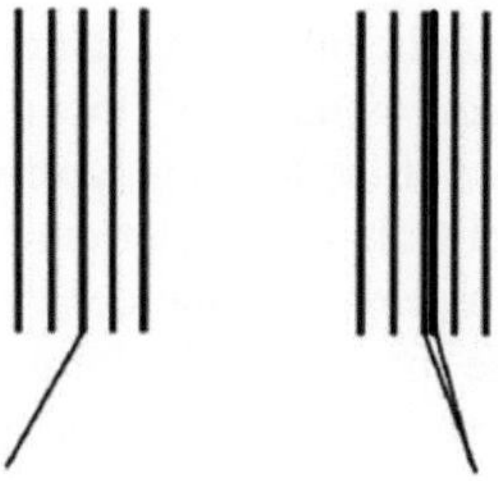

Fig. 5.15 Indentation patterns

of the two rollers is equal in the cold rolling forming process, setting of the cutter is very important. The quality of the formed spline tooth and the lifetime of the roller are depended in a great degree on the installation of the roller and the adjustment of the tooth positions. Before cold rolling forming, the two rollers are installed on the spindle in parallel, and a flat cutter ruler can be used to check if there is a gap between the ruler and the front end surface, as shown in Fig. 5.14.

Indentation method can be used in the teeth adjustment of the roller; the two rollers are driven to rotate relatively; only under conditions, the indentation of the two rollers completely coincides; and the teeth profile of the two rollers can then be coincided with the corresponding positions on the workpiece, as shown in Fig. 5.15 [8].

When oversize occurred in the cold rolling forming spline shaft or abnormal sound is heard, the machine should be stopped immediately for checking, and setting of the roller and adjustment of the teeth position should be carried out every time the machine is adjusted. During the batch production process, setting of the roller and adjustment of the teeth position should be carried out periodically to ensure the precision of the rolling machine.

5.5.3 Reasonable Selection of the Process Parameters

Rotation speed, feed rate of the roller, and rolling time have important influences on the precision, surface roughness of the cold rolling forming spline shaft, and the lifetime of the roller. In Sect. 3.2.2, the selection criterion of the process parameters has been proposed, and the three parameters has a high coupling degree on the influence of the forming quality and should be reasonably matched to obtain components of a higher precision.

The rotation of the roller is the main motion of the cold rolling precision forming, and with the increasing of the rotation speed, the deforming rate will increase and most of the elastic deformation will convert to plastic deformation. If the increasing in the speed is too large, the metal on the workpiece surface will be repeatedly work hardened and peeled off; while the speed is relatively lower, because the recovery of the elastic deformation, the addendum of the spline shaft tooth will not be fully filled. Through the increasing in the roller radial feed rate, the total rotation number of the billet in the forming process will be reduced; the number of the loading cycle will increase, and it is favorable for the transition of the billet from elastic deformation to plastic deformation. But if the feed rate is too large, macro- and microcracking will appear on the billet for lacking of plasticity of the metal, and the forming precision will thus be decreased.

In the free indexing cold rolling forming of spline shafts, the accumulated pitch error is related with the rotation speed and the feed rate of the roller; it is also related with the diameter of the roller and the billet, and the determination of the two parameters is rather complex. Through experiments, it is obtained that with the cold rolling machine Z28 K-20 manufactured by Qingdao Shengjian Machinery Factory, satisfied forming precision can be obtained with a reduction ε of 1–3 mm/r. The smaller the ε, the higher the forming precision is. However, limited by the working parameters of the machine and the moving control precision, too small ε cannot be selected for application.

Under long-time rolling, work hardening layer will be formed on the surface of the workpiece billet, and the property of the components can be improved by work hardening and strengthening. But if the workpiece is rolled for too long a time, the hardness of the hardening layer on the component surface and the central part of the teeth is not the same, which will result in the generation of large quantity of heat and breaking and crushing of the hardened microstructure under long time and repeated loading. Under this condition, the whole surface of the spline tooth will not be a metal surface with complete and regular flow line any more, and peeling and folding will occur on the surface. Therefore, the rolling interval of spline shaft cold rolling forming is commonly among 2–4 s; and the finishing interval should not exceed 2 s.

Common problems and solving methods in cold rolling forming of spline are listed in Table 5.3.

Table 5.3 Common problems and solving methods in the cold rolling forming of spline

Common problems	Reasons	Solving methods
Sharp tooth addendum of the formed components (larger external diameter)	Allowance of the workpiece before rolling is too large	Decrease the allowance of the workpiece before rolling
Flat tooth addendum of the formed components (smaller external diameter)	Allowance of the workpiece before rolling is too small	Increase the allowance of the workpiece before rolling
Burrs on the tooth addendum of the formed components (few burrs on a single side)	Displacement of the teeth positions on both rollers	Adjusting of the teeth position
Elliptical spline after rolling (half of the addendum is sharp and the other half is flat)	The cylindricality error of the workpiece before rolling or the coaxial error with the center hole is too large	Increase the cylindricality of the workpiece before rolling or the coaxiality with the center hole
Folding on the tooth flank of the spline after rolling	Displacement of the teeth positions on both rollers	Adjusting the teeth position
One redundant tooth has been formed after cold rolling	Allowance of the workpiece before rolling is too large	Decrease the allowance of the workpiece before rolling
Axial taper occur on the formed spline	The two rollers are not parallel	Adjusting the parallelism of the two rollers
Serious middle concave and the top convex on the spline addendum after rolling	Allowance of the workpiece before rolling is too large	Decrease the allowance of the workpiece before rolling
Inconsistent overpin distance after batch rolling	(1) Large difference of the workpiece dimension before rolling; (2) large difference of the workpiece hardness	(1) The dimension tolerance of the workpiece before rolling should be controlled to less than 0.04 mm; (2) the difference of the hardness should be controlled within 15–20 HB

References

1. Zhan Zhaoping, Chang Baoyin, Ming Cuixin. Standard Application Manual for Involute Splines [M]. Beijing: China Standards Press, 1997.
2. Zhou Dajun. Technology and Example of Metal Bulk Forming [M]. Beijing: China Machine Press, 2009.
3. Dale J. Garden, Warren Mich. Method of Conditioning the Surface of a Spline Forming Rack [P]. United States Patent. 5509287, 1996-04-23.
4. Zheng Quangang. Cold Rolling Forming Technology of Cylindrical Spur Gear with Small Module Involute Spline [J]. Automobile Technology & Material. 1997, 7: 16–18.
5. Pan Zhiqiang. Cold Rolling Technology of Small Modulus Involute Spline [J]. Machinist Metal Forming, 2003, 12: 43–44.
6. Zhang Dawei, Fu Jianhua, Li Yongtang. Contact Ratio in Process of Spline Cold Roll Forming [J]. China Metal Forming Equipment & Manufacturing Technology, 2008, 4: 80–84.

7. He Feng. Design of Cold Rolling Tools for Small Module Involute Spline Shafts [J]. Tool Engineering, 2001, 35 (2): 23–25.
8. Wu Weihua. The Processing Problem of Involute Spline Rolling [J]. Machinist Metal Cutting, 2004, 12: 26, 37.

Chapter 6
Mechanism and Process Analysis on the Cold Rolling Forming of Threads

Abstract Mechanism and process analysis on two-wheel and three-wheel cold rolling forming of threads have been analyzed. Diameter conditions of three-wheel rolling threads and relative movement between the workpiece and the rolling wheel have also been analyzed.

6.1 Principle of the Thread Cold Rolling Forming Process

In present, commonly used cold rolling methods of threads mainly include flat mold rolling, star wheel rolling, and cylindrical wheel thread rolling. The production rate of flat die rolling is higher; in the production of small diameter threads, the production rate can reach 10000–36000 pieces/h. Automation of the flat die rolling process is liable to realize, but the adjustment of the machine is difficult; precision of the rolling threads is lower than that produced by cylindrical wheel rolling. Therefore, it is widely used in the manufacturing of carbon steel fastener threads, as shown in Fig. 6.1.

Star wheel thread cold rolling is the most efficient thread processing method in present, and it is easy to realize automation; and the production rate is limited by the material loading speed, not the rolling speed. Therefore, while material loading is difficult or the production volume is not very large, star wheel cold rolling is not the most advantageous method; it is suitable for rolling simple shaped threads of large batches.

Cylindrical roller can be used for rolling various kinds of threads, worm wheel, gears, stud, screw, locating screw, and hollow components. It is most widely used in the manufacturing of high-strength, high-precision, and complex threads.

According to the number of the rollers, cold rolling forming of thread with cylindrical rollers can be classified as two-wheel rolling and three-wheel rolling; According to the rolling principle, it can be classified as radial rolling, tangent rolling, and axial feeding rolling, as shown in Fig. 6.2. Two-wheel rolling and three-wheel radial rolling of threads with cylindrical roller are mainly studied in this book [1].

J. Song et al., *Cold Rolling Precision Forming of Shaft Parts*,
DOI 10.1007/978-3-662-54048-0_6

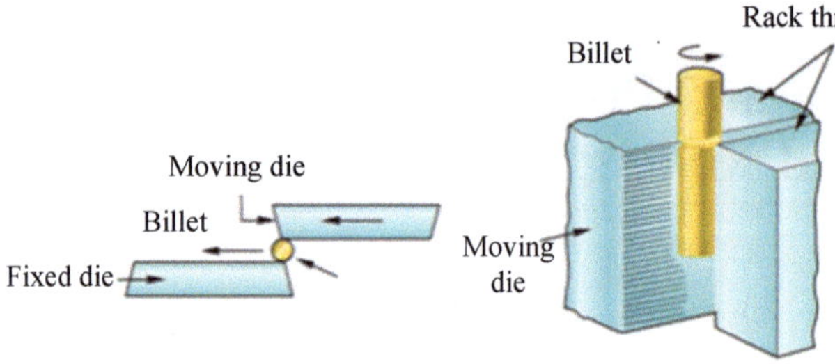

Fig. 6.1 Flat die rolling

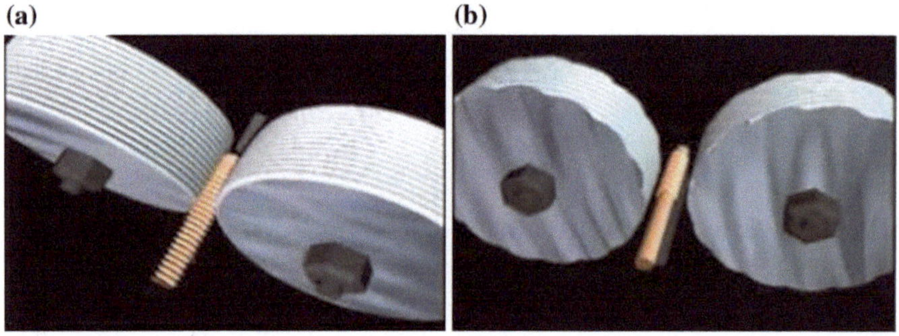

Fig. 6.2 Thread rolling methods with cylindrical roller **a** two-wheel radial feeding rolling; **b** two-wheel axial feeding rolling

6.1.1 Two-Wheel Rolling of Threads

Two-wheel rolling principle of threads is shown in Fig. 6.3. The two wheels are a couple of dies installed on the rolling machine. The helical direction of the two rolling wheels is the same and opposite to the direction of the thread on the workpiece. The rolling wheels are installed on the two parallel spindles; the end surface of the two wheels should be in the same plane, and the threads should be staggered for half a pitch. In the rolling process, the workpiece is mounted between the two wheels and can be supported by the support pad underside it or the centers at both ends. Two spindles rotate with the same speed and direction, and one or both of the spindles are feeding radially relative to the workpiece, besides rotating on its own axis. The workpiece then rotates with the wheel, and the thread can be formed under the pressure gradually. After rolling to the predetermined dimensions, feeding should be stopped and a period of finishing rolling is necessary. The feeding wheel will withdraw; and the workpiece can be unloaded, and the cold rolling process of the threads is finished.

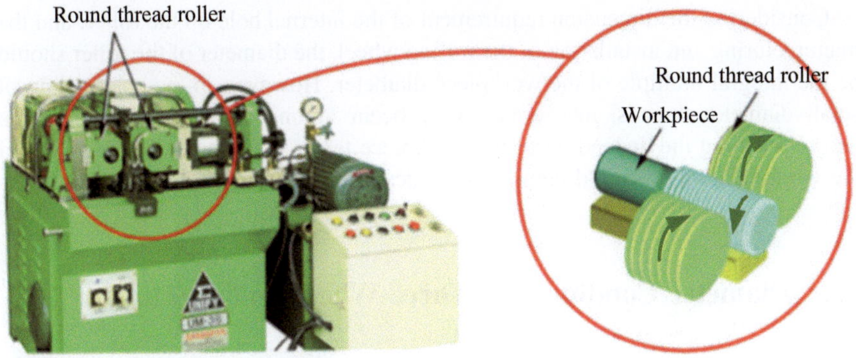

Fig. 6.3 Two-wheel rolling principle of threads

6.1.2 Three-Wheel Rolling of Threads

Three-wheel thread rolling is carried out on the three-axial rolling machine. The three rolling wheels are mounted on the spindles with an angle of 120°, and the three wheels are rotating at the same direction (opposite to the rotating direction of the workpiece) and speed, and radially feeding toward the workpiece at the same speed simultaneously. Threads are generated under the pressure of the rolling wheels. After processing to the predetermined dimension, then the finishing procedure will be carried out without any feed rate, and the feeding rollers are detached with the workpiece, as shown in Fig. 6.4.

Compared with two-wheel rolling, three-wheel rolling has the advantages of higher efficiency, suitable for rolling hollow thread components, and there is no need for location and sustaining. However, the adjustment of the teeth profile is difficult, and the size of the rolling thread is limited in a certain degree in three-wheel rolling.

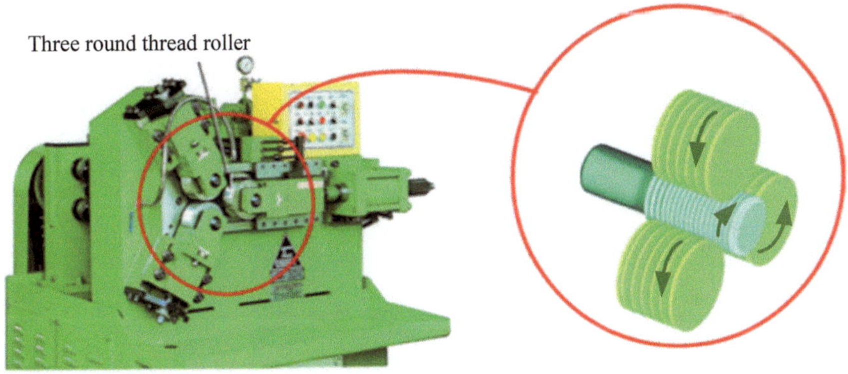

Fig. 6.4 Principle of three-wheel thread rolling

Considering the dimension requirement of the internal hole on the roller, and the manufacturing and installation of the rolling wheel, the diameter of the roller should be the integral multiple of the workpiece diameter. However, in the cold rolling of small diameter threads, interference may occur among the rolling wheels even before reaching the dedendum of the workpiece teeth. Therefore, the dimension of the workpiece billet should be accurately designed.

6.2 Diameter Conditions of Three-Wheel Rolling Threads

Figure 6.5 is the positional relationship between the rolling wheel and the workpiece in three-wheel thread rolling. It is shown that while cold rolling with three-wheel equipment, the relation exists during the finishing stage of the rolling.

$$O_1O_2 > 2R$$

And: $O_1O_2 = \sqrt{3}(r+R)$
Then: $\sqrt{3}(r+R) > 2R \Rightarrow r > \frac{2-\sqrt{3}}{\sqrt{3}}R$
That is,

$$r > 0.155R \tag{6.1}$$

where

R is the radius of the rolling wheel addendum, mm and
r is the small radius of the rolling thread, mm.

Therefore, the small radius of the rolling thread must be 0.155 times larger than the external radius of the wheel. In practical operation, gaps should also be left. The

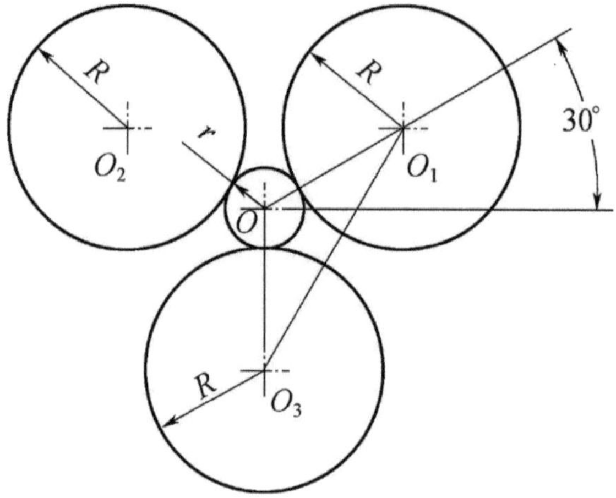

Fig. 6.5 Positional relationship between the wheel and workpiece in three-wheel thread rolling

minimum internal radius of the rolling thread in three-wheel rolling should be as follows:

$$r = 0.155R + \delta$$

where δ is the gap, mm; according to actual conditions, δ should be within the range of 1–1.5 mm.

It can be seen from formula (6.1) that the diameter of the workpiece should be increased with the increase in the wheel diameter. In conditions, the process parameters are satisfied and rolling wheel of small diameters should be selected (decreasing the head number of the rolling wheel).

6.3 Cold Rolling Forming Process of Threads

During cold rolling forming of threads, under a constant feed rate S of the rolling wheel (radial displacement of the rolling wheel during each rotation of the workpiece), the reduction Δs of the wheel relative to the workpiece (the difference of the workpiece radius before and after a single wheel rolling through the workpiece) is dynamically changed. The cold rolling process can be divided into four stages according to the difference of the rolling wheel reduction.

The first stage is called as the wedging stage, as shown in Fig. 6.6. From the starting of the rolling to the stage the workpiece rotating for $1/m$ circle (m is the number of the wheels), the rolling wheel rolls directly on the cylindrical surface before the wheel arriving at the position the previous wheel has rolled through. According to the geometrical relationship, the contact point between the wheel addendum and the workpiece is an Archimedes spiral line, the depth of which increases from zero to form a V groove on the workpiece gradually. The thread rolling process is analyzed, taking three-wheel rolling as an example.

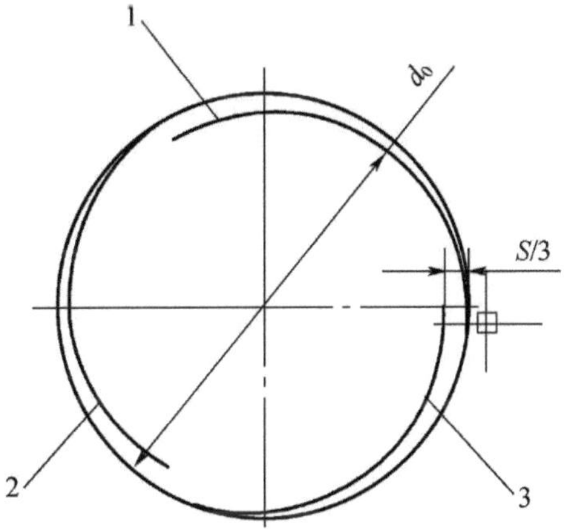

Fig. 6.6 First stage of the thread cold rolling process

In Fig. 6.6, lines 1, 2, and 3 are the curves formed by the contacting points between the addendum of the three wheels and the workpiece, the trajectory of which can be obtained as (expressed with polar coordinate, and the axial center of the workpiece is taken as the polar point):

For line 1:

$$\rho_1 = \mathrm{r}_0 - \frac{v_r}{\omega}\theta,\ \theta \in \left(0, \frac{2}{3}\pi\right) \tag{6.2}$$

where

r_0 is the initial radius of the workpiece, mm;
v_r is the radial feed rate of the rolling wheel, mm/s; and
ω is the angular speed of the workpiece, rad/s.

It is obvious that the reduction Δs of each rolling wheel increases from zero to $S/3$ gradually, where S is the preset feed rate of the rolling wheel.

The second stage is called as the steady rolling stage, as shown in Fig. 6.7. The workpiece rotates 1/*m* circle to the predetermined stroke of the workpiece, which is the most important stage of the rolling process. During this stage, each wheel is rolling on the trajectory of the former wheel; the reduction in the process is a constant *S*/*m*. During this stage, the rolling force increases gradually with the increase in the work hardening of the workpiece and the contacting area.

The number of cycles the workpiece rotates in the second stage is as follows:

$$m = \frac{S_{total}}{v_r} \cdot \frac{\omega}{2\pi} - \frac{1}{3}\ (\text{cycle}) \tag{6.3}$$

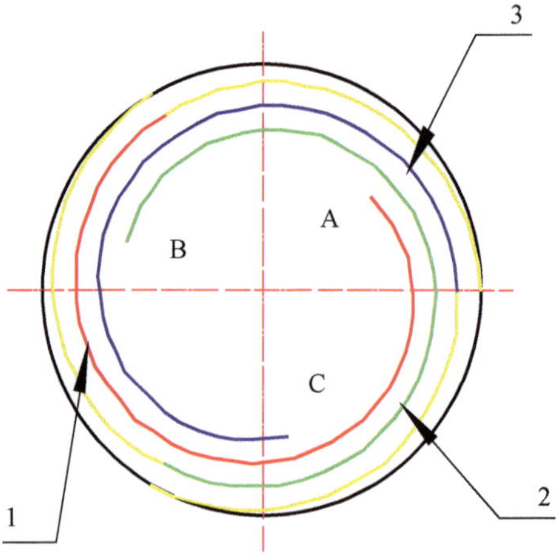

Fig. 6.7 Trajectory in the second stage of thread cold rolling

where

S_{total} is the total feed rate of the rolling wheel after the finishing of the cold rolling forming of the Archimedes spiral threads.

Similarly, the trajectory line can be obtained as follows:
Line 1:

$$\rho_1 = \mathrm{r}_0 - \frac{v_r}{\omega}\theta,\ \theta \in \left(\frac{2}{3}\pi, \frac{\omega}{v_r} \cdot S_{total}\right) \quad (6.4)$$

Line 2:

$$\rho_2 = \mathrm{r}_0 - \frac{v_r}{\omega}\left(\theta - \frac{2}{3}\pi\right),\ \theta \in \left(\frac{4}{3}\pi, \frac{\omega}{v_r} \cdot S_{total} + \frac{2}{3}\pi\right) \quad (6.5)$$

In order to obtain the reduction in the rolling wheel at the moment, subtract line 2 from line 1, then:

$$z = \rho_1 - \rho_2 = \left(r_0 - \frac{v_r}{\omega}\theta\right) - \left(r_0 - \frac{v_r}{\omega}\left(\theta - \frac{2}{3}\pi\right)\right) = \frac{2\pi}{3} \cdot \frac{v_r}{\omega} \quad (6.6)$$

It is obvious that the feed rate of the wheel is a fixed value, which is independent with the angle θ.

In addition, according to the definition of the feed rate:

$$S = \frac{2\pi}{\omega} \cdot v_r \quad (6.7)$$

Substituting Eq. 6.7 into Eq. 6.6, the reduction in this period can be obtained as follows: $\Delta s = \frac{S}{3}$.

The third stage is called as the prefinishing stage, as shown in Fig. 6.8. At the predetermined stroke, the rolling wheels will not feeding any more; wheel 1 will move from point A to point B, wheel 2 will move from point B to point C, and wheel 3 from point C to point A.

Theoretically, the trajectory of the stage is as follows:
Line 1:

$$\rho_1 = r_0 - S\theta \in \left(\frac{\omega}{v_r} \cdot S_{total} \cdot \frac{\omega}{v_r} \cdot S_{total} + \frac{2\pi}{3}\right) \quad (6.8)$$

Line 2:

$$\rho_2 = r_0 - S\theta \in \left(\frac{\omega}{v_r} \cdot S_{total} + \frac{2\pi}{3} \cdot \frac{\omega}{v_r} \cdot S_{total} + \frac{4\pi}{3}\right) \quad (6.9)$$

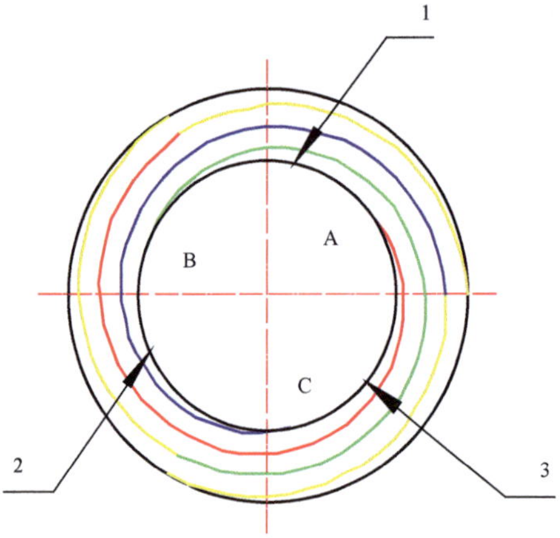

Fig. 6.8 Third stage of the thread cold rolling

The reduction in this stage can be obtained by substituting line 2 into line 1, that is:

$$z = \rho_1 - \rho_2 = (r_0 - S_{total}) - \left(r_0 - \frac{v_r}{\omega}\left(\theta - \frac{2\pi}{3}\right)\right) = \frac{v_r}{\omega}\theta - \frac{S}{3} - S_{total}$$
$$\theta \in \left(\frac{\omega}{v_r} \cdot S_{total} \cdot \frac{\omega}{v_r} \cdot S_{total} + \frac{2\pi}{3}\right) \tag{6.10}$$

Therefore, the reduction in the wheel in this stage will decrease from $\frac{S}{3}$ to zero.

The fourth stage is called as the finishing stage. The feed rate and the reduction of the rolling wheel are zero, which is an ideal station. In practical production, because the roundness error of the workpiece, small irregular reduction in the wheel will be produced, which will play a finishing role to the workpiece threads, and the roughness of the surface can be reduced. Therefore, this stage is of a great influence on the final quality of the workpiece.

6.4 Relative Movement Between the Workpiece and the Rolling Wheel

6.4.1 Axial Motion Relationship

Usually, the helix angle of the rolling wheel pitch diameter is equal to that of the rolled thread. Therefore, while the pitch cylinder of the rolling wheel is pure rolling relative to the pitch cylinder of the rolled thread, there is no relative motion between the roller and the workpiece. While axial feeding rolling method is adopted, the helix angle of the rolling wheel pitch diameter is not equal to that of the rolled thread; under conditions, the pitch cylinder of the rolling wheel is pure rolling

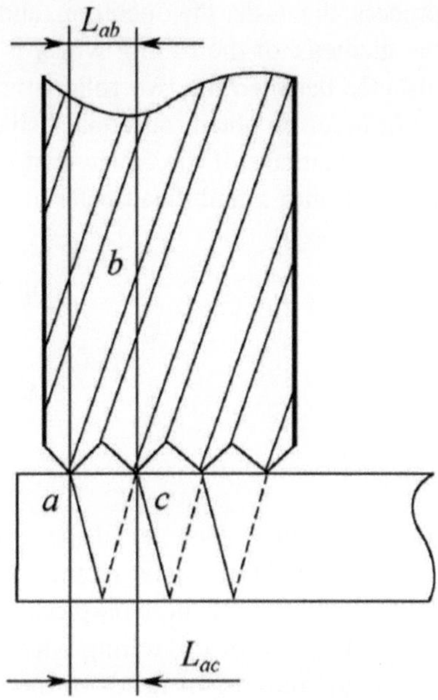

Fig. 6.9 Schematic diagram of thread rolling with rolling wheel

relative to the pitch cylinder of the rolled thread, and axial relative motion will generate between the rolling wheel and the workpiece.

As shown in Fig. 6.9, in the thread rolling process, in order to eliminate the axial relative motion between the roller and the workpiece, while the rolling thread rotates for one circle, axial moving distance of the contact point of the billet L_{ac} is equal to the moving distance L_{ab} of the contact point on the rolling wheel. Assuming that the helix angle of the processing thread is $\tau_{\text{workpiece}}$ and the helix angle of the roller is τ_{roller}, there is:

$$L_{ab} = \pi d \tan \tau_{\text{workpiece}} \tag{6.11}$$

$$L_{ac} = \pi d \tan \tau_{\text{roller}} \tag{6.12}$$

where d is the rolling diameter of the workpiece.

According to $L_{ab} = L_{ac}$, there is:

$$\tau_{\text{workpiece}} = \tau_{\text{roller}} \tag{6.13}$$

Therefore, in order to avoid the axial movement between the workpiece and the roller, the helix angle of the roller should be equal to that of the rolled thread.

Actually, influenced by the factors such as inner hole of the roller, strength, machine tool, and manufacturing, the pitch diameter of the rolling wheel is several integral multiple larger than that of the workpiece. Therefore, the rolling wheel is often manufactured in a pattern of multihead thread (there is also some equal head thread). The larger the diameter of the rolling wheel, the smoother the rolling process, the easier the operation, and the longer the roller lifetime will be. However, the diameter of the rolling wheel is limited by factors such as the adjustable limit distance between the two roller spindles.

In order to obtain an equal helix angle of the roller and the workpiece, and a bigger diameter of the roller than the workpiece, the rolling wheel can be manufactured with a multihead pattern. Therefore

$$\tan \tau_{\text{workpiece}} = \frac{P}{\pi d_2} \tag{6.14}$$

$$\tan \tau_{\text{roller}} = \frac{T}{\pi D_2} \tag{6.15}$$

where

d_2 is the pitch diameter of the rolled thread;
D_2 is the pitch diameter of the rolling wheel;
P is the pitch of the workpiece;
T is the lead of the rolling wheel; and
n is the number of heads of the rolling wheel.

According to formula (6.13), there is:

$$\frac{\mathrm{P}}{d_2} = \frac{T}{D_2} \tag{6.16}$$

Substituting $T = nP$ into the above formula, there is:

$$D_2 = nd_2 \tag{6.17}$$

That is, when the pitch diameter of the rolling wheel is n times of the roller thread pitch diameter (n is the value of the rolling wheel heads number divided by that of the workpiece thread); there is no relative axial movement between the roller and the workpiece.

6.4.2 Rotational Motion Relationship Between the Workpiece and the Rolling Wheel

There is an important significance of the rotational motion relationship between the roller and the workpiece for the recognization of the relative motion between them, the forming of the component, the adjustment of the rolling wheel, and the reason of the roller abrasion and the workpiece deformation.

Figure 6.10 is the schematic diagram of the thread rolling. The rolling wheel is rotating anticlockwisely, and the workpiece is driven to rotate clockwisely.

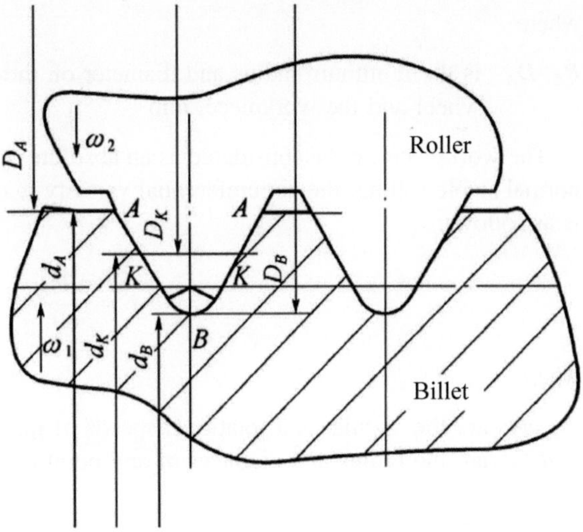

Fig. 6.10 Schematic diagram of the thread rolling

The circumferential velocity v of any point on the rolling wheel is as follows:

$$v = \omega_1 R = \frac{\pi D n_1}{60} \tag{6.18}$$

where

ω_1 is the angular velocity of the roller, rad/s;
n_1 is the rotation speed of the roller, r/min;
v is the linear velocity of any point on the roller, mm/s; and
R, D is the radius and diameter of any point on the roller.

The maximum circumferential velocity on the surface of the roller is at point B (the addendum of the tooth), and the velocity V_{B} is as follows:

$$V_B = \omega_1 R_B = \frac{\pi D_B n_1}{60} \tag{6.19}$$

where

R_B, D_B is the radius and diameter of the roller addendum surface.

The minimum circumferential velocity on the surface of the roller is at point A, and the velocity V_A is as follows:

$$V_A = \varpi_1 R_A = \frac{\pi D_A n_1}{60} \tag{6.20}$$

where

R_A, D_A is the minimum radius and diameter on the contact surface of the rolling wheel and the workpiece, mm.

The workpiece can be considered as an absolute rigid body. Under conditions of normal stable rolling, the circumferential velocity ω of any point on the workpiece is as follows:

$$\omega = \omega_2 r = \frac{\pi d n_2}{60}$$

where

ω_2, n_2 are the angular and rotational speeds of the workpiece and
r, d are the radius and diameter of any point on the workpiece, mm.

$$V_{A\text{workpiece}} = \omega_2 r_A = \frac{\pi d_A n_2}{60} \tag{6.21}$$

where

r_A, d_A are the maximum radius and diameter of the workpiece surface, mm.

The minimum circumferential velocity on the surface of the workpiece is at point B (refers to the contacting surface), and the velocity $V_{B\text{workpiece}}$ is as follows:

$$V_{B\text{workpiece}} = \omega_2 r_B = \frac{\pi d_B n_2}{60} \tag{6.22}$$

where

r_B, d_B are the radius and diameter on the contacting surface of the workpiece and the wedge top surface of the rolling wheel, mm.

6.4.3 *Relative Sliding Between the Rolling Wheel and the Workpiece*

(1) Under conditions of stable rolling, the workpiece can be considered as an absolute rigid body, and the relative motion between the workpiece and the roller can be expressed as follows: At a point K, the circumferential velocity of the workpiece and the roller is the same, and there is no relative sliding between them. Therefore

$$V_K = \omega_1 R_K = \omega_2 r_K = V_{K\text{workpiece}} \tag{6.23}$$

$$\frac{\omega_1}{\omega_2} = \frac{r_K}{R_K} = \frac{n_1}{n_2}$$

$$n_2 = \frac{R_K}{r_K} n_1$$

where

V_k, W_k are the linear velocity of the roller and the workpiece at point K, mm/s and
R_k, r_K are the rolling radius of the roller and the workpiece, mm.

(2) At section KB, the circumferential velocity V of the roller is always larger than that of the workpiece W, and there is a relative sliding between the roller and the workpiece.

The circumferential velocity difference is at point B, and the relative sliding speed ΔV_B is as follows:

$$\Delta V_B = V_B - V_{BI} = \omega_1 R_B - \omega_2 r_B = \frac{\pi}{30}(R_B n_1 - r_B n_2) = \frac{\pi n}{30}\left(R_B - r_B \frac{R_K}{r_K}\right) \tag{6.24}$$

where

W_B, V_B are the linear velocities of the workpiece and rolling wheel at point B, mm/s.

Because the pressure and relative sliding at point B are relatively large, the location of B on the roller is the most liable to abrasive.

(3) At section KA, the circumferential velocity V of the roller is smaller than that of the workpiece $V_{workpiece}$, and there is a relative sliding between them.

The circumferential velocity difference is at point A, and the relative sliding speed ΔV_A is as follows:

$$\Delta V_A = W_A - V_{A\text{workpiece}} = \omega_2 r_A - \omega_1 R_A = \frac{\pi}{30}(r_A n_2 - R_A n_1) = \frac{\pi n}{30}\left(r_A \frac{R_K}{r_K} - R_A\right) \tag{6.25}$$

where

V_A, $V_{A\text{workpiece}}$ are the linear velocity of the roller and the workpiece at location A, mm/s.

Although the relative sliding at location A is relatively large, the pressure is not very large; therefore, the abrasion is as serious as location B. The relative sliding of section *KA* and *KB* is existing during the whole rolling process, which is the sliding produced in the pure rolling between the rolling wheel and the rolled thread.

In addition, during the rolling process, even if the helix angle of the roller and the workpiece is equal, there is still a small axial movement between them, which will not influence the quality of the rolling thread. However, while the sliding is very serious, the production rate will be affected, and the length of thread can be rolled with specified roller will also be shortened, because the workpiece will be moved outside of the rolling wheel before the complete formation of the thread.

Reference

1. http://www.kinefac.Com [OL].

Chapter 7
Process Parameters in the Cold Rolling Forming of Threads

Abstract Parameters of thread cold rolling forming process have been narrated, including calculation of the workpiece billet diameter, selection of the rolling force, rolling speed, and the feed rate, and determination of the minimum wall thickness criteria in cold rolling of hollow threads.

Cold rolling forming of threads is a kind of plastic processing method, in which the strain of the deforming metal is overlapped by plastic strain and elastic strain. When the rolling wheel rolling and unloading at a certain place, the material is recovered, the elastic strain disappears, plastic strain has been retained, and then, the thread can be formed. The better the plasticity of the material, the easier the forming of the thread is. Therefore, plastic deformation is the theoretical basis for the forming of the rolling thread and the basis for the analysis of the process parameters in the cold rolling forming of threads.

7.1 Thread Billets Before Cold Rolling Forming

7.1.1 Commonly Used Calculation Formulas of the Workpiece Billet Diameter

In practical production, theoretical and empirical formulas formed in the last many years have often been used in the calculation of the billet diameter, and the calculation methods are different in various reference materials.

The billet diameter of the rolling workpiece can be classified into three types: round addendum and flat dedendum, flat addendum and dedendum, and round addendum and dedendum. And the billet diameter can be calculated with the following formulas, respectively [1]:

J. Song et al., *Cold Rolling Precision Forming of Shaft Parts*,
DOI 10.1007/978-3-662-54048-0_7

(1) Round addendum and flat dedendum of the rolling wheel tooth is as follows:

$$d_0 = \left(d_{2\text{actual}} + \frac{1.37r^2 - 1.155h_0^2}{t} + K\frac{P^{2.5}}{d^{0.8}}\right)_{-\delta} \tag{7.1}$$

where

- d_0 is the billet diameter of the workpiece before rolling, mm;
- $d_{2\text{actual}}$ is the required actual pitch diameter of the thread after rolling, mm;
- r is the arc radius of the rolled thread dedendum or the arc radius of the roller addendum, mm;
- P is the pitch diameter of the thread, mm;
- K is a coefficient. For steel, $K = 0.1$; For nonferrous metals, $K = 0.13$;
- d is the nominal diameter of the rolled thread, mm;
- h_0 is related with the actual external and pitch diameter of the rolling thread, while the dedendum of the roller thread is not participate in the forming of the rolled thread addendum, $h_0 = 0.433P - \frac{1}{2}(d_{\text{actual}} - d_{2\text{actual}})$; And h_0 is related with the height of the thread tooth dedendum, while the dedendum of the roller thread is participate in the forming of the rolled thread addendum, $h_0 = 0.433P - h_2$;
- D_{actual} is the required actual external diameter of the rolled thread according to the deviation of the external diameter, mm;
- δ is the diameter tolerance of the rolled thread billet, mm.

(2) Flat addendum and dedendum of the rolling wheel tooth is as follows

$$d_0 = \left(d_{2\text{actual}} + \frac{1.155\left(h_3^2 - h_0^2\right)}{t} + K\frac{t^{2.5}}{d^{0.8}}\right)_{-\delta} \tag{7.2}$$

where $h_3 = 0.433\text{t} - h_1$ (the height of the roller tooth).

(3) Round addendum and dedendum of the rolling wheel tooth is as follows:

$$d_0 = \left(d_{2\text{actual}} + \frac{1.37\left(r^2 - r_1^2\right)}{t} + K\frac{t^{2.5}}{d^{0.8}}\right)_{-\delta} \tag{7.3}$$

where r_1 is the arc radius of the rolled thread addendum or the arc radius of the roller tooth dedendum, mm.

There are also some empirical formulas, such as:

$$d_{\text{billet}} = \sqrt{d^2 - 1.3dp - 0.6p^2} \tag{7.4}$$

where

d is the external diameter of the thread;

P is the pitch diameter of the thread [2].

$$d_{billet} = d_{2\text{actual}} + 0.0212p + 0.1952\frac{p^2}{d_2} \tag{7.5}$$

where $d_{2\text{actual}} = d_{2\text{theoretical}} - \delta(\text{tolerance})$, d_2 is the pitch diameter of the thread [3].

$$d_{\text{billet}} = d - a - 0.624p \tag{7.6}$$

where a is the tolerance of the pitch diameter [3].

$$d_{\text{billet}} = d_{2\min} + 0.3TD_2 \tag{7.7}$$

where

$d_{2\min}$ is the minimum pitch diameter;
TD_2 is the tolerance of the pitch diameter [2].

The above formulas are all billet diameter calculation equations summed up by the production corporations according to their actual conditions, which is suitable for the production conditions of them.

7.1.2 *Solution of the Billet Diameter of the Cold Rolling Forming Thread*

7.1.2.1 Flat Addendum and Dedendum of the Thread Tooth

According to the constant volume principle during plastic forming, as shown in Fig. 7.1, thc mctal material moved to part A from part B. Assume that the large diameter of the thread is d_W, the small diameter is d_1, the pitch diameter is d_2, the pitch is P, the billet diameter is d_0, and corresponding radius is r_w, r_1, r_2, and r_0.

Solving volume A first and taking a differential element d_r, we obtain:

$$dV_A = \left(\frac{P}{2} - 2\cot\alpha(r - r_2)\right) \cdot 2\pi r \cdot dr \tag{7.8}$$

The volume of part A is as follows:

$$V_A = \int_{r_0}^{r_W} \left(\frac{P}{2} - 2\cot\alpha(r - r_2)\right) \cdot 2\pi r \cdot dr \tag{7.9}$$

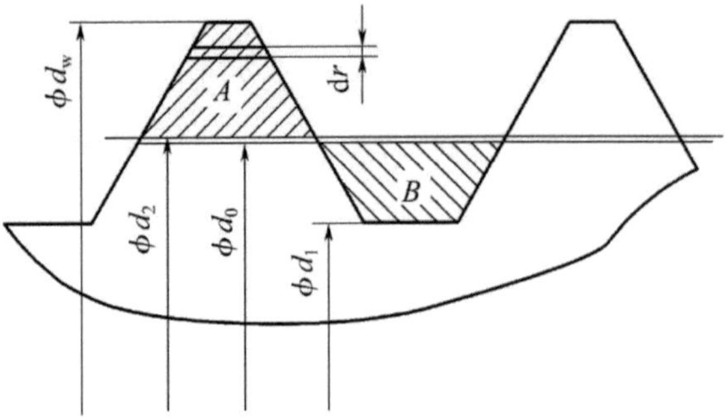

Fig. 7.1 Schematic diagram of thread rolling with a flat addendum and dedendum rolling wheel

It can be obtained as follows:

$$V_A = \left(\frac{P}{2}\pi + 2\pi \cdot \cot\alpha \cdot r_2\right) r^2 - \frac{4}{3}\pi \cot\alpha \cdot r^3 \Big|_{r_0}^{r_W}$$
$$V_A = \left(\frac{P}{2}\pi + 2\pi \cdot \cot\alpha \cdot r_2\right)\left(r_W^2 - r_0^2\right) - \frac{4}{3}\pi \cot\alpha \cdot \left(r_W^3 - r_0^3\right) \tag{7.10}$$

In the same way:

$$dV_B = \left(\frac{P}{2} - 2\cot\alpha(r_2 - r)\right) \cdot 2\pi r \cdot dr \tag{7.11}$$

The volume of part B is as follows:

$$V_B = \int_{r_1}^{r_0} \left(\frac{P}{2} - 2\cot\alpha(r_2 - r)\right) \cdot 2\pi r \cdot dr \tag{7.12}$$

$$V_B = \left(\frac{P}{2}\pi - 2\pi\cot\alpha \cdot r_2\right) r^2 + \frac{4}{3}\pi\cot\alpha r^3 \Big|_{r_1}^{r_0}$$
$$V_B = \left(\frac{P}{2}\pi - 2\pi \cdot \cot\alpha \cdot r_2\right)\left(r_0^2 - r_1^2\right) + \frac{4}{3}\pi\cot\alpha \cdot \left(r_0^3 - r_1^3\right) \tag{7.13}$$

$V_A = V_B$, and then:

$$-Pr_0^2 = -\left(\frac{P}{2} + 2\cot\alpha \cdot r_2\right) r_W^2 - \left(\frac{P}{2} - 2\cot\alpha \cdot r_2\right) r_1^2 + \frac{4}{3}\cot\alpha\left(r_W^3 - r_1^3\right)$$

$$r_0 = \sqrt{\left(\frac{1}{2} + \frac{2}{P}\cot\alpha \cdot r_2\right) r_W^2 + \left(\frac{1}{2} - \frac{2}{P}\cot\alpha \cdot r_2\right) r_1^2 - \frac{4}{3P}\cot\alpha\left(r_W^3 - r_1^3\right)} \tag{7.14}$$

7.1.2.2 Round Addendum and Flat Dedendum of the Rolling Wheel Tooth

As shown in Fig. 7.2, round addendum and flat dedendum rolling wheel is one of the most commonly used rollers, and the derivation method of volume A is the same with that mentioned above.

$$V_A = \left(\frac{P}{2}\pi + 2\pi \cdot \cot\alpha \cdot r_2\right)\left(r_W^2 - r_0^2\right) - \frac{4}{3}\pi \cot\alpha \cdot \left(r_W^3 - r_0^3\right) \tag{7.15}$$

$$V_B = \int_{\frac{d_1}{2}+R\left(1-\frac{\sqrt{3}}{2}\right)}^{r_0} \left(\frac{P}{2} - 2\cot\alpha(r_2 - r)\right) \cdot 2\pi r \cdot dr \tag{7.16}$$

$$\begin{aligned} V_B = &\left(\frac{P}{2}\pi - 2\pi \cdot \cot\alpha \cdot r_2\right)\left(r_0^2 - [r_1 + R(1-\cos\alpha)]^2\right) \\ &+ \frac{4}{3}\pi \cot\alpha \cdot \left(r_0^3 - [r_1 + R(1-\cos\alpha)]^3\right) \end{aligned} \tag{7.17}$$

For the volume of the arc area, assume that R is the radius of the addendum circle, still taken a differential element d_r for consideration, and then:

$$dV = 2\pi r \cdot dr \cdot 2\sqrt{R^2 - \left(R - r + \frac{d_1}{2}\right)^2} \tag{7.18}$$

$$V_C = \int_{r_1}^{r_1+R(1-\cos\alpha)} 2\pi r \cdot 2\sqrt{R^2 - (R - r + r_1)^2}dr \tag{7.19}$$

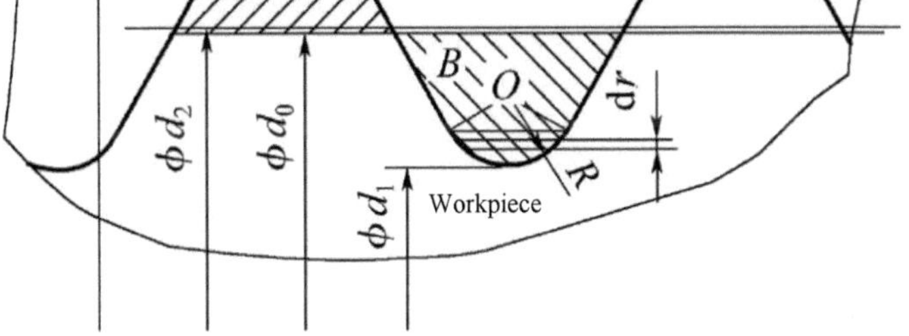

Fig. 7.2 Schematic diagram of thread rolling with a round addendum and flat dedendum roller

Let $R - r + r_1 = R\cos t$, then $r = R + r_1 - R\cos t$, $dr = R\sin t dt$, and it can be obtained as follows:

$$V_C = 4\pi R(R + r_1)\left(\frac{1}{2}\alpha - \frac{\sin 2\alpha}{4}\right) - \frac{4}{3}\pi R^3 \sin^3\alpha \tag{7.20}$$

According to the volume constant principle, $V_A = V_B + V_C$

$$-\mathrm{P}r_0^2 = -\left(\frac{P}{2} + 2\cot\alpha \cdot r_2\right) r_W^2 - \left(\frac{P}{2} - 2\cot\alpha \cdot r_2\right)(r_1 - R(1 - \cos\alpha))^2$$
$$+ \frac{4}{3}\cot\alpha\left(r_W^3 - (r_1 - R(1 - \cos\alpha))^3\right) + 4R(R + r_1)\left(\frac{\alpha}{2} - \frac{\sin 2\alpha}{4}\right) - \frac{4}{3}R^3\sin^3\alpha$$

$$r_0 = \sqrt{\begin{array}{c}\left(\frac{1}{2} + \frac{2}{P}\cot\alpha \cdot r_2\right) r_W^2 - \left(\frac{1}{2} - \frac{2}{P}\cot\alpha \cdot r_2\right)(r_1 - R(1 - \cos\alpha))^2 \\ \hline -\frac{4}{3P}\cot\alpha\left(r_W^3 - (r_1 - R(1 - \cos\alpha))^3\right) - 4\frac{R}{P}(R + r_1)\left(\frac{\alpha}{2} - \frac{\sin 2\alpha}{4}\right) - \frac{4}{3}R^3\sin^3\alpha\end{array}} \tag{7.21}$$

Although the above equation is relatively complex, the calculation precision is higher when all of the factors influencing the external diameter of the billet is considered, and the results can be obtained by programming calculation after each parameters have been substituted into the equation in the utilization.

7.1.3 Chamfer of the Billet

The tooth type of the rolling wheel is rather thin, under large pressures, both ends of each tooth are in a forced state, and the axial force has been counteracted; therefore,

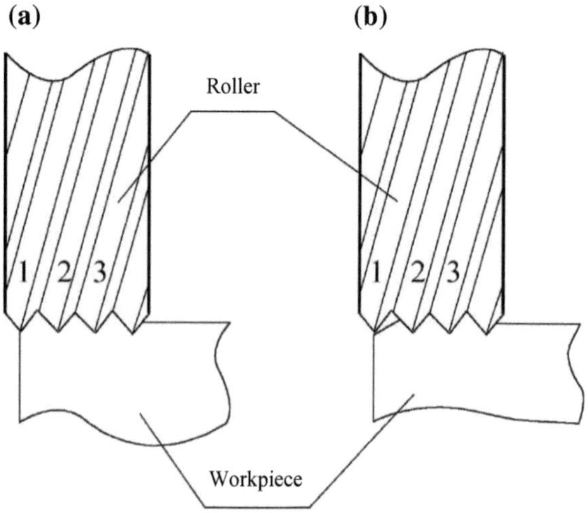

Fig. 7.3 Force diagram of the billet end on the rolling wheel **a** there is a chamfer on the end; **b** there is no chamfer on the end

the bending moment of the tooth is very small or even can be neglected, shown as tooth 2 and tooth 3 in Fig. 7.3. In Fig. 7.3b, when the end of the billet is chamfered, tooth 2 has not completely entered the internal of the billet; therefore, the forces applied on both sides are unequal, which will lead to the generation of axial force and bending moment. However, compared with tooth 1 in Fig. 7.3a, the axial force is relatively small.

From the aspect of rolling wheel protection, the smaller the chamfer, the best is the effect, but the effective number of the screws will be reduced in such conditions. With a chamfer of 15°–30°, the chamfer of the rolled thread is close to 45°, which can satisfy most of the requirements of thread components.

7.1.4 Material of the Rolling Billet

Considering the stability of the rolling process and the quality, precision, and strengthening requirements for thread components, not only reasonable and economical lifetime of the rolling wheel but also the reliability and production rate of the rolling machine should be guaranteed. Therefore, besides of certain plasticity, there are still some detailed demands for the rolling billet.

Most ductile metal materials with an elongation of more than 12% are suitable for rolling threads because of their good rolling performance. For materials with an elongation among 8–12%, because the elongation of the material is located near the limit of the rolling performance, they are suitable for rolling metric thread. For materials with an elongation of less than 8%, because defects such as cracks will generate on the addendum of the thread, they are not suitable for rolling forming.

The hardness of the billet has a close relationship with the rolling process of thread. The lifetime of the roller will reduce with the increase in the billet hardness; therefore, considering from economical aspect, the hardness of the billet material should be less than HRC37. A hardness of the billet more than HRC37 has exceed the reasonable category of the rolling hardness, because the shortening of the roller lifetime will increase the cost of the roller apportioned to each thread component; it will also increase the radial rolling force. In addition, the surface of the billet should be defectless; otherwise, cracks will generate during the rolling process.

The rollability, peeling resistance, and gaps of the material are three important factors to evaluate and select the rolling materials. Rollability includes ductility and deforming resistance during the cold rolling forming. Peeling resistance is a characteristic related with the shearing strength of the rolled metal. During the rolling process, if the metal flow at the flank of the rolling wheel is faster than that of the middle part, with the filling of the thread tooth, a seaming will be formed by the folding metal. The forming of the gaps is firstly depended on the rolled metal and then on the shape of the rolling thread.

Table 7.1 shows the rollability index of 17 commonly used steel and several nonferrous alloys. At a certain rolling speed, the thread rolling force is in an inverse proportion to the rollability index of the metal. With the decrease in the material rollability index, the production rate of unit energy will be reduced. For example, to manufacture threads of 1/2-13UNC-2A with an axial feeding rolling machine of an output power of 7.5 kW, the production rate will be 11.5 m/min under the rollability index of 1.00, and it will reduce to 6.9 m/min with a rollability index of 0.6 under the same conditions.

The rollability of the materials also provides a kind of method to compare the radial force and the expected lifetime of the rolling dies. Under the same rolling conditions, the required radial force of thread rolling is in an inverse proportion with the rollability index of the material. If the failure reason of the roller is cracking of the addendum, then the lifetime of the roller will be approximately in a direct proportion to 1/3–1/4 of the rollability index. Therefore, under the same conditions, if the rollability index is 1.00, the lifetime of the roller is 30000 m, while if the rollability index is 0.5, then the expected lifetime of the roller will be 1800–3700 m.

The lead and sulfur elements in brass and steel will decrease the peeling resistance of the rolling thread. And the carbon content will also decrease the peeling resistance of the carbon steel. Generally, the peeling performance will increase with the increase in the cold working time, especially those work hardening alloys such as stainless steel of 300 series, copper and some aluminum alloys. The peeling resistance can be increased by annealing of the materials before rolling.

Table 7.1 Rollability of different materials

Metals	Hardness (HB)	Rollability index	Peeling	Gaps
Carbon steel and low alloy steels				
1010	137	1.11	B	C
1018	148	1.08	B	C
1020	156	0.96	B	C
1095	260	0.47	B	B
1095	320	0.42	B	B
1112	198	1.00	C	C
1117	173	1.03	C	C
1144	225	0.78	B	C
4140	205	0.93	B	C
4140	234	0.57	B	C
4140	300	0.42	B	B
4340	235	0.45	B	B
8620	215	0.60	B	C
Stainless steel				
303	174	0.46	C	B
316	150	0.45	B	B
416	221	0.58	C	B
430	225	0.56	C	B
Nonferrous alloys				
Aluminum 2017, 2024	135	1.40	B	C
Phosphor bronze	130	1.28	C	B

Table 7.1 shows the peeling performance of commonly used thread rolling materials. Copper and copper alloys with a large ductility can be classified into type A under the annealing state; metals contain large quantity of high-speed cutting additions, for example, thread steel can be classified into type D.

When the dimension of the billet is insufficient, unclosed gaps are easy to form on the addendum of the thread, which will decrease the lifetime of the components under corrosion environment. Metals with a lower hardness and better flowability are more liable to form deeper gaps. Figure 7.4 shows the schematic diagram of the gaps on the addendum of the thread, and Fig. 7.5 shows the four kinds of degree of gaps. The gap forming degrees of several commonly used materials are listed in Table 7.1.

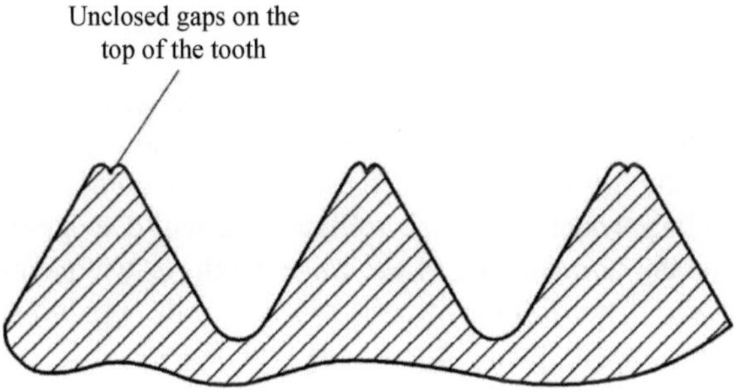

Fig. 7.4 Gaps on the tooth addendum of the rolled thread

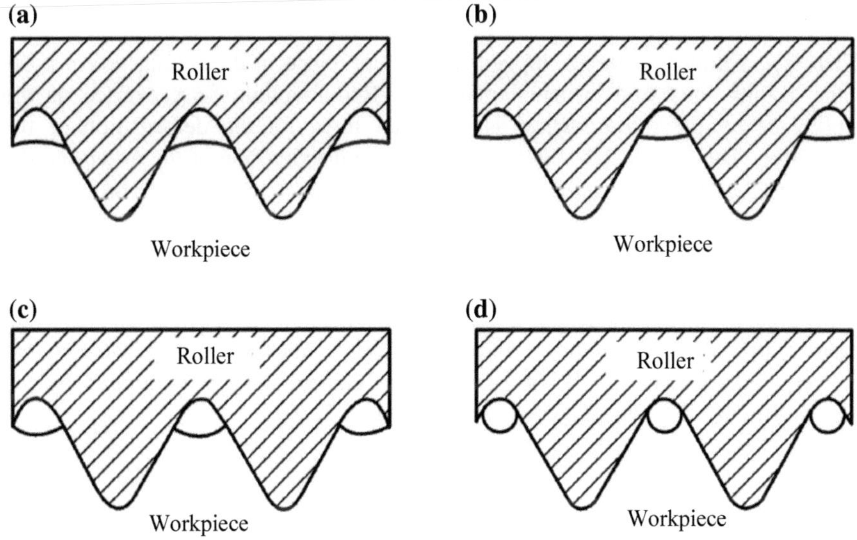

Fig. 7.5 Schematic diagram of the rolled thread addendum types

7.2 Parameters in the Cold Rolling Forming Process of Threads

Rolling parameters in thread cold rolling forming include rolling force, rolling speed, and feed rate and time. Reasonable selection of rolling parameters is an important premise to ensure the reliable and durable working of the rolling machine, stable rolling, rolling precision, and increasing roller lifetime. Selection of the rolling parameters is also related with the dimension of the rolling thread and the mechanical properties of the rolling materials.

7.2.1 Rolling Force

Rolling force is an important factor to apply plastic deformation to the billet material and for the forming of the threads. Appropriate selection of the rolling force will shorten the rolling period, increase the production rate, and prolong the lifetime of the rolling wheel. Too large rolling force will damage the rolling tools, for example, cracking and damage of the roller tooth, or overheating of the components. Otherwise, if the rolling force is too small, the production rate and the stability of the processed thread dimension will be affected, and threads with complete teeth can even not be rolled. Therefore, in actual production, reasonable selection of the rolling force is of great importance.

7.2.1.1 Mechanical Analysis of the Rolling Process

There are mainly two forces applied on the workpiece in the rolling process: the radial rolling force and the tangential force. Radial rolling force exerted on the workpiece by the rolling wheel is the main force for the deformation of the workpiece. The force is parallel to the central connecting line of the roller and the workpiece, while the tangential rolling force will rotate the workpiece and complete the deformation. Radial force is the pressure applied by the hydraulic cylinder to the workpiece, which is an important basis for the calculation of the hydraulic system in the calculation of the equipment. Tangential force is formed by the moment transferred by the motor to the roller through the spindle, which is an important basis for the calculation of the motor power. It can be approximately considered that the required moment can be obtained by the tangential force multiplied by the radius of the rolling wheel. As shown in Fig. 7.6.

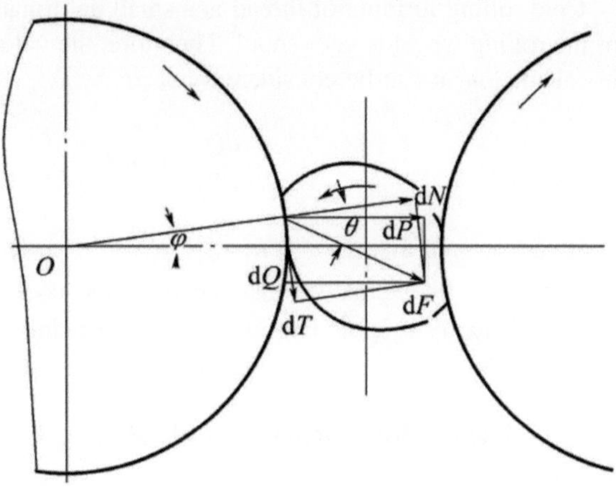

Fig. 7.6 Mechanical analysis of the thread cold rolling forming process

In the cold rolling forming of threads, there are two forces applied to the workpiece, that is, the normal force dN and the induced frictional force dT, the direction of which is tangent with the circle of the rolling wheel and the billet. According to coulomb's law, the value of dT is as follows:

$$dT = \kappa dN$$

Because the shape of the thread tooth, the above formula cannot be substituted by $dT = \mu dN$.

where

κ is the calculation coefficient;
μ is the frictional coefficient;

κ is always smaller than μ and is related with the tooth angle of the thread tooth. The resultant force is $dF = \sqrt{dT^2 + dN^2}$

where $dF = dN\sqrt{1+\kappa^2}$.

The resultant force dF can be decomposed into the horizontal force dP and the perpendicular force dQ. dP is parallel to the central line of the two rollers, which is the rolling force applied by the roller to the billet, while the perpendicular force dQ will drive the billet to rotate around its axis and complete the rolling process, which is the force produced by the torque applied by the spindle to the roller. Therefore,

$$dP = dF\,\cos(\theta - \varphi) = dN\sqrt{1+\kappa^2}\cos(\theta - \varphi)$$

$$dQ = dF\,\sin(\theta - \varphi) = dN\sqrt{1+\kappa^2}\sin(\theta - \varphi);\ \tan\theta = \kappa.$$

$$\frac{dP}{dN} = \frac{\cos(\theta - \varphi)}{\cos\theta} \tag{7.22}$$

Cold rolling forming of thread is a small deformation process, and the feed rate of the rolling wheel is very small. Therefore, the value of φ is also very small, and in calculation, it can be considered that $dP \approx dN$

$$\frac{dQ}{dP} = \tan(\theta - \varphi) \tag{7.23}$$

It can be seen that $\tan\theta = \kappa$; $\frac{dQ}{dP} < \kappa$.

Therefore, $\frac{dQ}{dP}$ is smaller than the frictional coefficient, and the torque can be obtained directly by μdP multiplied with the radius of the roller.

7.2.1.2 Factors Influencing the Cold Rolling Force of the Thread

The value of the rolling force is decided by the product of unit rolling force and the contacting area, and the main influencing factors include:

1) Influencing factors of the deforming resistance (i.e., the unit rolling force):

   (1) The performance of the materials:
   The chemical composition, microstructure, and mechanical properties all have a great influence on the unit rolling force and are the basic factors determining the unit rolling force. The contents of carbon and chromium in the steel have a great influence on the extrusion force; the unit rolling force will increase with the increasing of the tensile strength, yield strength, and hardness of the materials. Because the cold rolling forming of the threads is a continuous small deformation process, the bigger the cold working sensitivity of the materials, the larger the deforming force will be required in the rolling process.

   (2) Deforming degree: The bigger the deforming degree, the larger the deforming resistance of the materials will be. In the cold rolling forming of threads, the deforming degree can be measured with the pitch of the thread. The bigger the pitch is, the larger the unit deformation force will be. In each rolling pass, the deforming degree can be measured by the reduction in the roller, and the unit deforming force is increased with the reduction.

   (3) Deforming speed: It has no obvious influence on the unit rolling force. In the cold rolling process, besides the plastic deformation of the billet, part of the deforming energy is converted into temperature rise. The larger the deforming speed is, the more obvious the temperature rise will be. Therefore, during the rolling process, the rolling force will slightly decrease with the increasing rolling speed, but it is not obvious.

   (4) Influence of the rolling thread teeth contour: If the teeth contour is favorable for metal flow, the unit rolling force will be smaller. For example, the rolling force of metric thread is smaller than that of the trapezoidal thread.

(5) Geometrical shape of the rolling dies: The shape and size of the rolling wheel has a great influence on the unit rolling force in the cold rolling process of threads. Reasonable design of the roller will promote the metal flow in the tooth profile, improve the frictional conditions, and reduce the flowing resistance, which will not only improve the lifetime of the roller but also reduce the unit rolling force. For example, in the rolling of metric threads, the roller should be round addendum flat dedendum as possible, and flat addendum and dedendum of the rolling wheel should be avoided to adopt in order to improve the metal flow property, reduce the stress concentration, and prolong the service time of the rolling wheel.

(6) Lubrication conditions
For better lubrication conditions, the frictional resistance can be greatly reduced and thus decrease the unit rolling force.

2) Influencing factors affect the contacting area of the roller and the workpiece.

The contacting area of the roller and the workpiece in the cold rolling of threads is mainly determined by the length of the workpiece and the enveloping angle of the roller and the workpiece. In Fig. 7.7, $\angle AOB$ is the enveloping angle. The larger the enveloping angle is, the larger the corresponding length c and the contacting area will be.

In the solving of the enveloping length c corresponding to the enveloping angle, as shown in Fig. 7.7, assume that the workpiece is a rigid-plastic body and there is no elastic deformation; therefore, point B should be on the connecting line of the center points of the roller and the workpiece. According to the geometrical relationship of triangular ΔOO_1A, there is:

$$OA^2 = OO_1^2 + O_1A^2 - 2OO_1 \cdot O_1A \cdot \cos\phi \tag{7.24}$$

And then: $\cos\phi = 1 - 2\sin^2\frac{\phi}{2}$

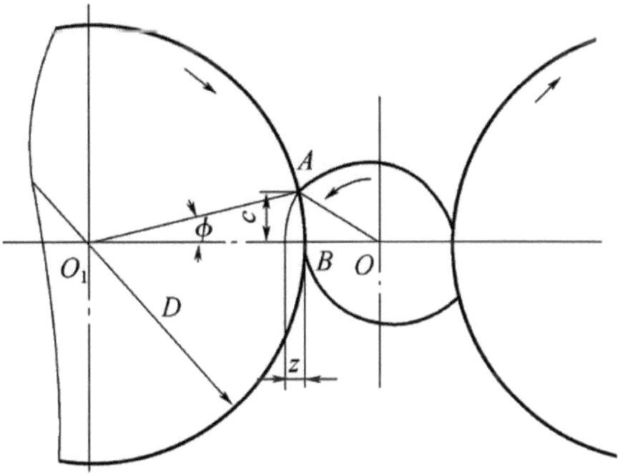

Fig. 7.7 Schematic diagram of the thread cold rolling process

Substitute into the above formula, then:

$$OA^2 = OO_1^2 + O_1A^2 - 2OO_1 \cdot O_1A + 4\sin^2\frac{\phi}{2}(OO_1 \cdot O_1A)$$

Therefore

$$2\sin\frac{\phi}{2} = \sqrt{\frac{OA^2 - (OO_1 - O_1A)^2}{OO_1 \cdot O_1A}} \tag{7.25}$$

It can be obtained from the figure that

$$OA = \frac{d_1}{2} + \Delta s,\ OO_1 = \frac{d_1 + D_1}{2},\ O_1A = \frac{D}{2}$$

where

D is the external diameter of the rolling wheel;
d_1 is the bottom diameter of the rolling thread;
Δs is the reduction of the workpiece, namely the value of the feed rate of the roller per round divided by the number of the roller.

Therefore

$$2\sin\frac{\varphi}{2} = \sqrt{\frac{\left(\frac{d_1}{2} + \Delta s\right)^2 - \left(\left(\frac{D+d_1}{2}\right) - \left(\frac{D}{2}\right)\right)^2}{\frac{D+d_1}{2} \cdot \frac{D}{2}}}$$

Simplified, then

$$\begin{aligned}
2\sin\frac{\varphi}{2} &= \sqrt{\frac{4 \cdot (d_1 + \Delta s)\Delta s}{(D + d_1)D}} \\
\cos\frac{\varphi}{2} &= \sqrt{1 - \frac{\Delta s(d_1 + \Delta s)}{D(D + d_1)}} \\
\sin\varphi = 2\sin\frac{\varphi}{2}\cos\frac{\varphi}{2} &= \sqrt{\frac{4(d_1 + \Delta s)\Delta s}{D(D + d_1)} \cdot \left(1 - \frac{\Delta s(d_1 + \Delta s)}{D(D + d_1)}\right)} \\
\sin\varphi &= \sqrt{1 - 4\left(\frac{\Delta s(d_1 + \Delta s)}{D(D + d_1)} - \frac{1}{2}\right)^2}
\end{aligned} \tag{7.26}$$

Therefore, the enveloping line is as follows:

$$c = \frac{D}{2} \cdot \sin\phi = \frac{D}{2}\sqrt{1 - 4\left(\frac{\Delta s(d_1 + \Delta s)}{D(D + d_1)} - \frac{1}{2}\right)^2} \tag{7.27}$$

From Eq. 7.27, the three parameters influencing the length of the enveloping line are as follows: the diameter of the roller, the diameter of the rolling thread, and the reduction of the roller.

With a roller of M30 and an external diameter of 120 mm, the influence of the reduction on the length of the enveloping line can be drawn with software MATLAB, as shown in Fig. 7.8. It is shown that the length of the enveloping line is the increasing function of the roller reduction. And the reduction in the roller has an obvious influence on the length of the enveloping line.

Taken a M30 thread, for example, with a feed rate of 0.15 mm/r, the curve the enveloping line length varying with the diameter of the roller is shown in Fig. 7.9. It is shown that when the diameter of the roller is less than 100 mm, the length of the enveloping line increases obviously with the diameter of the roller; otherwise, the enveloping line will tend to be gentle, that is, the influence of the roller diameter on the length of the enveloping line is small. Because the diameter of the roller is generally larger than 100 mm, the influence of the roller diameter variation on the contact area and the rolling force is not important.

Figure 7.10 shows the variation curve of the enveloping line length to the diameter of the billet under the roller diameter of 120 mm and a reduction in 0.15 mm. The length of the enveloping line increases obviously with the increasing thread diameter.

From the above analysis, it can be seen that the length of the enveloping line is the monotonic function of the roller diameter D, the diameter of the rolling thread d,

Fig. 7.8 Variation curve of the enveloping line length with the reduction of the roller

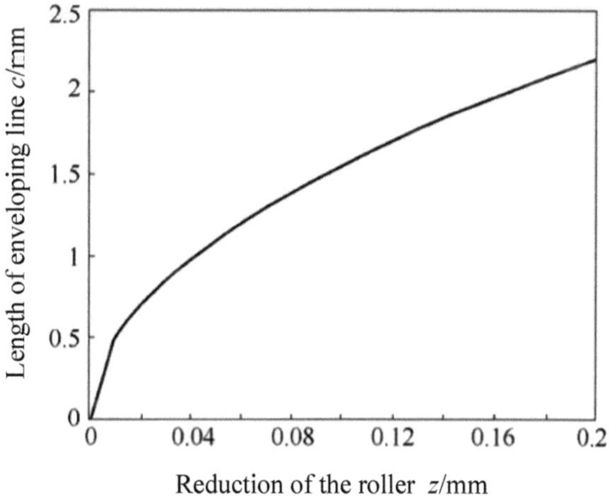

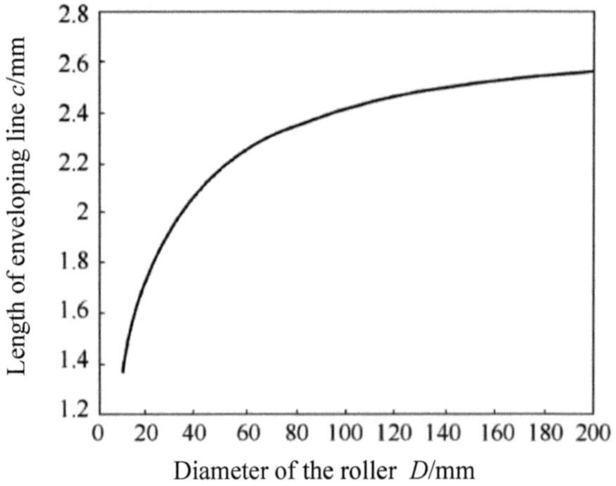

Fig. 7.9 Variation curve of the enveloping line length with the diameter of the roller

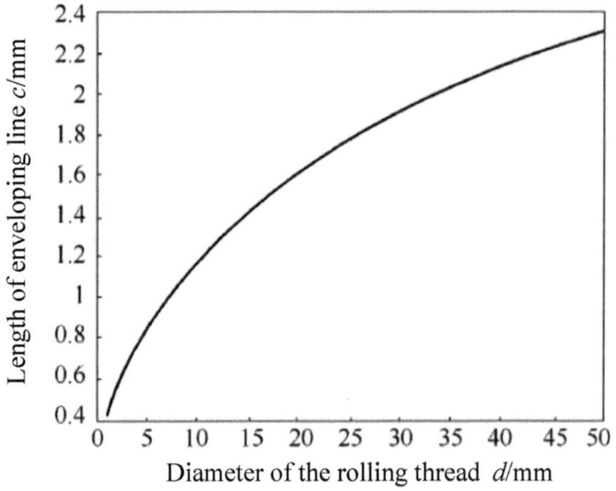

Fig. 7.10 Variation curve of the enveloping line length with the diameter of the rolling thread

and the reduction of the roller Δs. Therefore, the factors influencing the contacting area include the following:

(1) The length l of the rolling thread, which is in a direct proportion with the rolling area.
(2) Reduction Δs: The larger the roller reduction and the enveloping angle of the roller and the billet are, the larger the contacting area will be.
(3) Diameter of the roller: The larger the roller diameter is, the larger the roller and workpiece enveloping angle and the contacting area will be. Therefore, when the roller diameter is larger than 100 mm, the influence of the roller diameter variation on the contacting area is not very important.

(4) The larger the diameter of the workpiece is, the larger of the contacting area will be, and the influence of the workpiece diameter on the contacting area cannot be neglected.

Affected by the factors mentioned above, in the rolling process, the first stage can be called as the wedging stage, in which the reduction in the roller increases to $1/m$ (m is the number of the rollers) from zero gradually. With the increasing reduction, the deforming degree and contacting area also increase and the rolling force will increase rapidly from zero. The second stage is called as the stable rolling stage, in which the reduction in the roller is invariant. With the increasing deforming degree in the rolling process, work hardening will occur and the rolling force will increase gradually. In the third stage of the rolling process, the roller will not feed any more, and the reduction will reduce from $1/m$ to zero. In this stage, the rolling force will decrease with the reducing reduction, but because of the influence of work hardening and the deformation degree, the rolling force will decrease slowly. The fourth stage is called as the finishing stage, in which the variation of the rolling force is related with the roundness of the workpiece especially for hollow threads. When the wall thickness of thc thread is small, irregular shape of the workpiece is easy to occur in the shaping stage; the rolling force is not stable and will vary with the diameter of the workpiece until finally unloading. Therefore, the function of this stage is shaping and finishing of the workpiece and round rolling of the hollow workpiece to modify the elliptic and irregular shapes of the workpiece and obtain qualified components.

According to the analysis, the rolling force can be expressed as the following function:

$$P_0 = f[\sigma_S(HRC, HB, \delta), P, D, d, l, K, \mu]$$

where

σ_S, HRC, HB, δ	is the characteristic parameters of the material, which are the yield stress, rockwell hardness, brinell hardness and elongation, respectively;
l	is the length of the rolling thrcad, mm;
P	is the pitch of the rolling thread, mm;
D	is the diameter of the roller, mm;
d	is the diameter of the rolling thread, mm;
K	is the coefficient of the tooth profile;
μ	is the frictional coefficient.

7.2.1.3 Calculation of the Rolling Force

Cold rolling of threads is a local deforming process, and the exact pressure force is very difficult to obtain. There are various kinds of methods to solve the rolling force obtained by researchers all over the world, but all of them are approximate solutions, and there are great differences of different formulas. Several typical solutions are introduced in the following.

(1) M. и.писареθский in the former Soviet derived a rolling force calculation formula of two-roller rolling of metric threads:

$$P_0 = 34.3\sqrt{\frac{\sigma_T^3}{E} \cdot \frac{D}{D + d_1}} \cdot d_1 \left(a + \frac{d_0 - d_1}{\cos\,\alpha/2} \right) K_P \tag{7.28}$$

where

σ_T is the yield limit of the metal, and $\sigma_T = 3 - 4\sigma_S$, MPa;
E is the elastic modulus of the metal;
D is the external diameter of the roller, mm;
d_1 is the inner diameter of the thread, mm;
a is the width of the top and bottom thread teeth, mm;
d_0 is the diameter of the billet, mm;
α is the profile angle of the thread teeth, °;
K_P is the heads number of the rolling thread.

This formula is a theoretical equation, and the tooth profile, material characteristics, and the diameter of the roller and the rolled thread have been considered.

(2) According to the experimental materials of the former Soviet, the radial rolling force can be calculated with the following formula [4]:

$$P_0 = 0.536A(HB + 22) \tag{7.29}$$

where

HB is the brinell hardness of the material;
A is the contacting area of the roller and billet, $A = \frac{2}{3} \cdot \frac{L_0}{P} \cdot b \cdot h \frac{1}{\cos\,\alpha/2}$;
L_0 is the thread length of the billet, mm;
P is the pitch of the thread, mm;
b is the chord length, mm, $b = \frac{2}{\cos\,\omega}\sqrt{r^2 - (r - h)^2}$;
r is the radius of the billet, mm;
h is the height of the tooth, mm.

(3) For the manufacturing of high-quality threads used in locomotive, according to the introduction in relative materials, following empirical formula has been used to solve the rolling force [1]:

$$P_0 = 58.86 L_0 P^{1.2} (HRC)^{0.7} d^{0.3} \tag{7.30}$$

where

- P_0 is the rolling force, N;
- L is the length of the rolling thread with a complete screw, mm;
- P is the pitch of the rolled thread, mm;
- HRC is the rockwell hardness of the rolled workpiece;
- d is the metric diameter of the rolled thread, mm.

(4) Rolling force calculation formula introduced in other materials has also been used sometimes [5]:

$$P_0 = 250P\sqrt{\frac{HB}{4\delta}} + Z \cdot P\left(\frac{1}{2K} + 0.95\right) LF \tag{7.31}$$

where

- P_0 is the rolling force, N;
- P is the pitch of the workpiece, mm;
- HB is the brinell hardness of the workpiece;
- δ is the elongation of the workpiece;
- K is the heads number of the roller threads;
- L is the thread length of the workpiece, mm;
- F is the modified coefficient of the rolling force, for triangular threads of 60°, $F = 1$; for semicircular threads, $F = 1.5$; and for rectangular threads, $F = 2$;
- Z is the material related coefficient, for carbon steel, $Z = 0.1$–0.5; for low alloy steel, $Z = 0.5$–1.0; for high strength alloy steel, $Z = 1$–2; and for special steel such as stainless steel and copper alloys, $Z = 2$–5.

(5) The calculation formula provided by Qingdao Shengjian Machinery Factory is as follows:

$$P_0 = P_{10} \cdot K\sqrt{\frac{d}{P} \cdot \frac{L}{10}} \tag{7.32}$$

where

- P_{10} is the rolling force needed by every 10 mm of thread length. For thread of M6, $P_{10} = 0.42$; for thread of M16, $P_{10} = 0.70$; and for thread of M20, $P_{10} = 0.84$;
- K is the deforming resistant coefficient. For 35 steel, $K = 1$; for 45 steel, $K = 1.65$, and for 40Cr steel, $K = 1.8$;
- L is the thread length of the rolling workpiece.

The above formula is suitable for cold rolling forming of threads with an external diameter of 140 mm.

Rolling force can also be found from the chart drawn with experience data directly according to rolling parameters. These different formulas and charts are all obtained from calculation and experiences, but they are calculated for the same rolling condition, because parameters or coefficients are different, the obtained rolling force may be different greatly. In practical production, the initial rolling force can be calculated with selected formula or chart, and then modified according to actual conditions.

7.2.2 Selection of the Rolling Speed

Rolling speed includes the circumferential velocity ω_1 of the rolling wheel and the radial feed rate v_r. Factors influencing the deformation are linear velocity v_t of the roller and workpiece contacting area and the radial feed rate of the roller relative to the workpiece. Selection of the rolling speed is one of the important factors determining the deformation degree of the material, which will influence the rolling quality of the thread and must be selected appropriately.

The theoretical linear velocity is as follows:

$$v_t = |\omega_1| \cdot \frac{D}{2} = |\omega_0| \cdot \frac{d}{2} \tag{7.33}$$

where

ω_1 is the pure rolling radius of the roller;
D is the external diameter of the roller;
ω_0 is the angular velocity of the workpiece;
d is the pure rolling radius of the workpiece.

In calculation, D and d can be chosen as the external diameter of the roller and the inner diameter of the thread.

From the simulation results, it can be seen that there is a sliding phenomena between the workpiece and the roller; therefore, the formula is not always correct. Thus, the rolling speed can be calculated with the following equation:

$$v_t = |\omega_1| \cdot \frac{D}{2} \cdot \eta \tag{7.34}$$

where η is the sliding coefficient; according to the actual conditions, it is in the category of 0.83–0.93.

The circumferential velocity of the roller or the rotating velocity per minute is related with the physical–mechanical property of the material and the geometrical

Table 7.2 Recommended rolling speed in the rolling of threads with rolling wheel (m/min)

Pitch (P/mm)	Material		
	Aluminum alloy and steel	Alloy steel and stainless steel	Heat-resistant material and titanium alloy
0.35, 0.5, 0.6, 0.7	30	25	20
0.8, 1, 1.25, 1.5	20	20	15
1.75, 2	20	15	12

parameters and precision of the rolling thread. In practical production, optimized speed should be selected to roll out qualified thread components with small rolling force and obtain higher production rate. Table 7.2 is the recommended rolling speed in the rolling process of threads with rolling wheel [4].

According to the materials introduced by Soviet scholars and thread rolling experiments, for materials of $\sigma_b < 588.6\,\text{MPa}$ and an elongation more than 10 m/min, favorable circumferential rolling velocity should be within 20–25 m/min; for alloy steel, the optimized velocity should be about 12 m/min.

7.2.3 *Selection of the Feed Rate in Thread Rolling*

Selection of the feed rate S in the rolling process is determined by the production rate, quality of the rolling thread, load of the transmission components on the machine, and the lifetime of the rolling wheel. With the increasing feed rate per revolution, the production rate of the machine will be enhanced, but the rolled components will be heated, and the roundness error of the pitch diameter, the external diameter, and the inner diameter will generate. Simultaneously, while rolling the thread components with higher hardness, the lifetime of the roller will be reduced. Therefore, the feed rate should be selected from the economic point of view, and factors should also be considered to satisfy the demands of the transmission torque.

In thread rolling with rolling wheel, the value of feed rate is equal to the rolling depth on the billet in a revolution of the workpiece. Therefore, the feed rate is relative to the rotation speed of the billet, that is, relative to the rotation speed and the radial feed rate of the rolling wheel. On the premise of ensuring the service time of the roller and the precision of the thread, higher rotation speed of the roller should be selected to improve the production rate.

The approximate value of the feed rate can be decided by the following equation:

$$S = \mu^2 \frac{D \cdot d_0}{D + d_0} \cdot \left(1 + \sqrt{\frac{p}{E}}\right) \tag{7.35}$$

where

μ is the external friction coefficient;
p is the unit rolling force;
E is the elastic modulus of the material.

When rolling of metric threads, $p = 2.5\text{–}3\sigma_T$;

σ_T is the yield stress of the material, MPa.

Experimental study shows that the radial feed rate obtained from Eq. (7.35) is approximately an optimized value. Too large feed rate will result in the cracking in the internal part of the billet, but smaller feed rate will lead to the decrease in the production rate.

For soft materials with a yield stress of 243.3 MPa and elongation more than 16%, and hollow components, the feed rate should be decreased to 0.4–0.5 of the original value.

After the billet rotating for one revolution, the radial feed rate of the billet is related with the thread pitch and the mechanical properties of the material. For the materials recommended by Soviet, the radial feed rate should be selected according to Table 7.3 [4].

The radial feed rate of the roller can be solved by the circumferential velocity and feeding amount of the roller. Feed rate v_r of the roller increases with the increase in the circumferential velocity and radial feeding amount, which will result in the tangential movement of the metal surface and the peeling on the external surface of the thread. The deformation of the metal will be uniformed, and defects of peeling and adhesion of the metal on the roller will be reduced by increasing the circumferential velocity and decreasing the feed rate of the roller at the same time. Recommended feed rate of commonly used materials has been listed in Table 7.4.

Table 7.3 Radial feed rate/(mm/r)

Materials and the tensile strength	Pitch P/mm				
	1.0	1.5	2.0	2.5	3.0
Steel $\sigma_b < 392.4$	0.043–0.15	0.06–0.0175	0.73–0.2	0.08–0.25	0.083–0.26
$\sigma_b = 392.4 \sim 490.5$	0.03–0.1	0.43–0.15	0.06–0.175	0.073–0.225	0.08–0.25
$\sigma_b > 539.6$	0.023–0.09	0.033–0.12	0.053–0.16	0.06–0.20	0.063–0.225
Brass	0.03–0.175	0.03–0.20	0.06–0.23	0.07–0.27	0.08–0.30

Table 7.4 Feed rate of different materials in thread rolling/(mm/s)

Materials	Feed rate
Aluminum	6.0–8.0
Brass	3.3–3.0
Low-carbon steel	2.0–3.5
High-carbon steel	0.9–1.7

7.3 Cold Rolling Forming of Hollow Thin-Walled Thread Components

Some thread components with holes such as tube joints and thread bushes can be manufactured with rolling methods. The rolling of thick-walled hollow thread components is similar to the rolling of solid thread components. But for thin-walled thread components, because the radial pressure during the rolling process is very large, there is a great difference from the rolling of solid threads.

7.3.1 Cold Rolling Forming of Hollow Threads

The difference between cold rolling forming of hollow threads and solid threads lies in the whole elastic and plastic deformation applied on the hollow components.

7.3.1.1 Elastic Deformation in the Cold Rolling Forming of Hollow Threads

As mentioned above, in the rolling process, elastic and plastic deformation on the components will always occur. Elastic deformation of the cold rolling threads can be classified into two kinds: One is the local elastic deformation on the tooth profile that will recovery after unloading, and the other is the elastic deformation of the whole circle that does not exist in the cold rolling of solid threads. All of the elastic deformations will recover after the exiting of the rolling wheel, and the actual deformation of the tooth profile is smaller than that of the solid thread. Therefore, the forming of hollow thread is very difficult, even under the same rolling conditions. The thinner the thread wall, the larger the elastic deformation will be, and the forming process will be more inadequate.

7.3.1.2 Plastic Deformation in the Cold Rolling of Hollow Threads

For the processing of some thin-walled hollow threads, plastic deformation will occur on local part or the whole circle of the workpiece; under the conditions, the rolling force has increased to a certain degree, which will result in the failure of the component. There are several types of the failure: In the first circumstance, the thread will be fractured directly, longitudinal cracks will occur on the thread, as shown in Fig. 7.11a; in the second circumstance, the thread has been pressed to an ellipse (in two-roller rolling) or other irregular shapes (in multi-roller rolling), and the components cannot be completely formed, as shown in Fig. 7.11b; in the third circumstance, the roundness of the pitch diameter surpassed the tolerance range of the rolled thread. The three failure types are all resulted by the cracks on the hollow

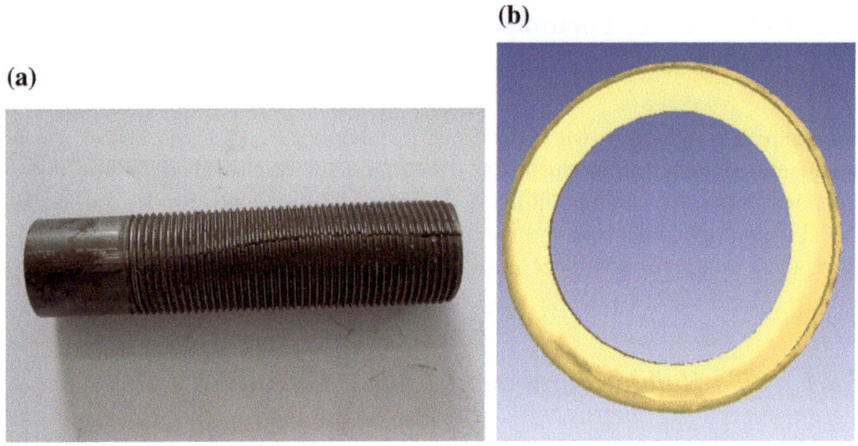

Fig. 7.11 Failure types of the rolled hollow thread **a** longitudinal cracking; **b** irregular deformation obtained in simulation

components deduced by plastic deformation; therefore, rollability should be firstly ensured before rolling of the hollow threads.

7.3.1.3 Repairing of the Deformation in the Finishing Stage of the Hollow Thread Rolling

In the finishing stage of the rolling process, only the tooth profile is finished of the solid thread, while for hollow thread, finishing will also has an effect on the roundness error resulted by the small plastic deformation in the rolling of the circle; therefore, the stage is very important to the rolling of hollow thread. The rolling wheel will not feed any more, the workpiece is rotating with the roller, and the distance between the two rollers is constant; thus, the deformed workpiece can be modified gradually to be qualified round components to meet the requirements of the rolling thread. However, in the three-roller cold rolling forming of hollow thread components, the stress formed in the workpiece under the same radial pressure is greatly smaller and the repairing effect will be better than that in the two-roller rolling process.

7.3.2 Wall Thickness of the Cold Rolling Hollow Thread Components

According to the plastic mechanics, stress will generate when external force applied on an object. Under certain deforming conditions (such as deforming temperature and speed) of each stress components according to a certain relationship, the particles will enter into the plastic state, which can be described with a formula: $f(\sigma_1, \sigma_2, \sigma_3) = C$.

According to Tresca yield criteria, when the tangent stress on the object (particle) reaches a certain value, yield will occur on the object. That is:

$$\sigma_1 - \sigma_3 = \sigma_S \tag{7.36}$$

where

σ_1 is the maximum principal stress;
σ_3 is the minimum principal stress;
σ_S is the yield strength of the material.

The above formula is the rolling strength condition of the hollow thin-walled thread components. Formula 7.29 can be used to discuss the rollability condition of thin-walled threads, and the stress state of the dangerous points can be solved accordingly.

In order to determine the relationship between the rolling force and the inner stress of the component, the hollow thread in the rolling process can be considered as a round circle applied with the rolling force. The minimum wall thickness conditions of two-wheel and three-wheel rolling will then be analyzed as follows.

7.3.2.1 Wall Thickness Condition in Two-Wheel Rolling of Thin-Walled Thread Components

The same as rolling of solid threads, in the rolling process, a radial force and a tangent force are applied on the hollow workpiece through the rolling wheel of the rolling equipment. The workpiece is deformed plastically to form the threads under the radial force, and the workpiece rotates under the tangent force to complete the rolling process. For hollow threads with a large wall thickness, the rolling process is similar to that of the solid components. But for thin-walled components, because the hollow thread component is under the station of a very large radial force and the workpiece is under a free rotation condition, the tangent force is very small. For convenience of calculation, the tangent force can be neglected and only the influence of the tangent force on the thin-walled workpiece is calculated.

In the force analysis of the billet circle in the two-wheel rolling, the friction force is neglected, and the force applied on the thread is simplified as two symmetrical radial forces to obtain the inner force of a random point. It can be simplified as a plane curve rod under the symmetrical force P and can be expressed with the axial line, as shown in Fig. 7.12a.

Therefore, the inner force of the circle can be solved with the method solving statically indeterminate structures, and the circle should be cut along the horizontal

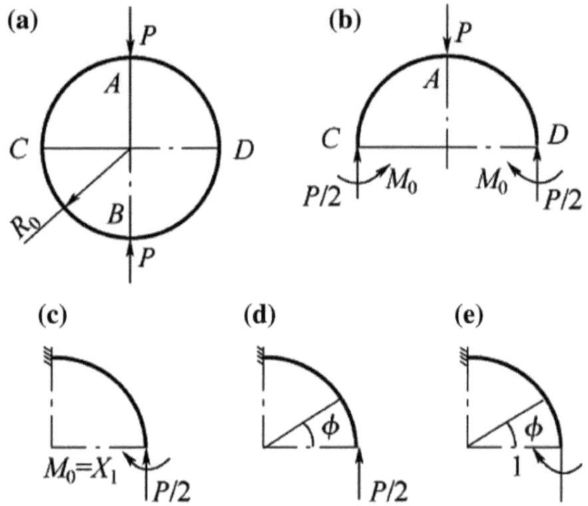

Fig. 7.12 Schematic diagram of the decomposition force in the two-wheel thread rolling

diameter, as shown in Fig. 7.12b. From the symmetrical characteristic of the load, the shearing force Q on the cross section C and D is equal to zero, and there only exists the axial force N and the bending moment M_0. According to the balance condition, it can be easily obtained that $N = P/2$, and only M_0 is a redundant restriction force, which can be recorded as X_1. Because of the symmetry of the circle, only 1/4 of the circle is studied. Because the rotation angle on the symmetrical cross section A and D is equal to zero, section A can be considered as a fixed end, and the zero rotation angle of the section D can be seen as the deformation compatibility condition, as shown in Fig. 7.12c, and can be expressed as follows:

$$\delta_{11} X_1 + \Delta_{1p} = 0 \tag{7.37}$$

where

Δ_{1p} is the rotation angle in cross section D, when only $N = \frac{P}{2}$ (Fig. 7.12d) is applied in the basic statically determinate system;

δ_{11} is the rotation angle in cross section D, when $X_1 = 1$ (Fig. 7.12e) and is the only force;

M and $\overline{M}$ is the bending moment in the cross section of angle ϕ, when $N = \frac{P}{2}$(Fig. 7.12d) and $M_0 = 1$ (Fig. 7.12e) in the basic statically determinate system.

$$M = \frac{PR_0}{2}(1 - \cos\phi),\ \overline{M} = 1. \tag{7.38}$$

Therefore

$$\Delta_{1p} = \int_0^{\frac{\pi}{2}} \frac{M\overline{M}R_0 d\phi}{EI} = \frac{PR_0^2}{2EI}\int_0^{\frac{\pi}{2}}(1-\cos\varphi)d\phi = \frac{PR_0^2}{2EI}\left(\frac{\pi}{2}-1\right) \tag{7.39}$$

$$\delta_{11} = \int_0^{\frac{\pi}{2}} \frac{\overline{M}\overline{M}R_0 d\phi}{EI} = \frac{R_0}{EI}\int_0^{\frac{\pi}{2}} 1 d\phi = \frac{\pi R_0}{2EI} \tag{7.40}$$

Substituting Eq. (7.31), it can be obtained as follows:

$$X_1 = -PR_0\left(\frac{1}{2}-\frac{1}{\pi}\right) \tag{7.41}$$

Under the action of $\frac{P}{2}$ and X_1, the bending moment in any cross section can be expressed as follows:

$$M(\phi) = -\frac{PR_0}{2}(1-\cos\phi) + PR_0\left(\frac{1}{2}-\frac{\pi}{2}\right) = -PR_0\left(\frac{1}{\pi}-\frac{\cos\phi}{2}\right) \tag{7.42}$$

When $\phi = 90°$, the maximum value of M is $\frac{1}{\pi}PR_0 = 0.318PR_0$.

The value is the actual bending moment of the 1/4 circle. According to the symmetrical principle, the bending moment of the whole circle can be obtained, as shown in Fig. 7.13.

Therefore, the inner force of any cross section on the 1/4 circle can be expressed as follows:

$$\left.\begin{array}{l} M(\phi) = -PR_0\left(\frac{1}{\pi}-\frac{\cos\phi}{2}\right) \\ N(\phi) = \frac{P}{2}\cos\phi \\ Q(\phi) = \frac{P}{2}\sin\phi \end{array}\right\} \tag{7.43}$$

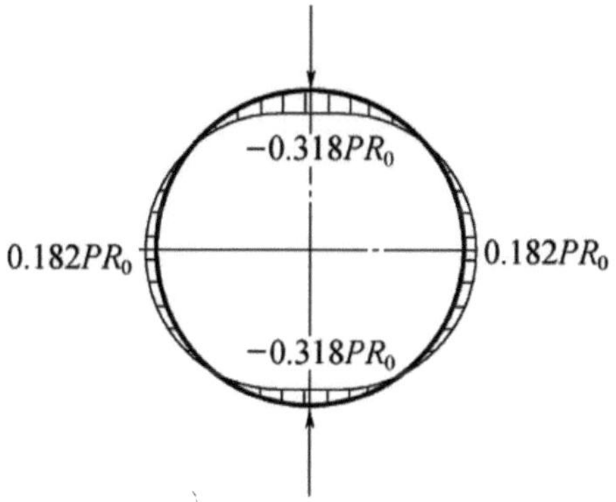

Fig. 7.13 Bending moment of the billet in three-wheel rolling

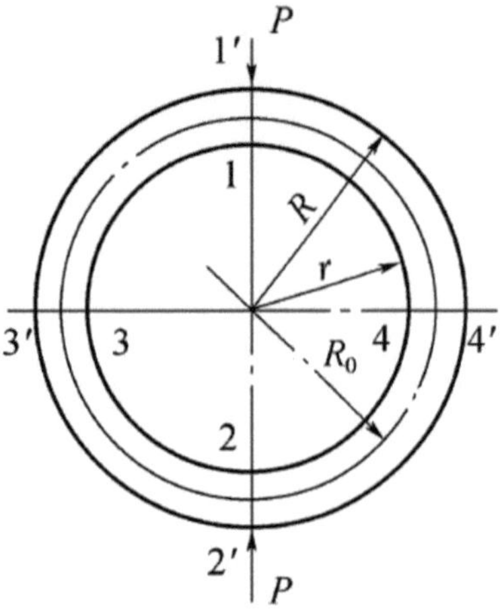

Fig. 7.14 Schematic diagram of the rolling forces applied on the billet in three-wheel rolling

The deformation of the workpiece can be analyzed according to the inner force state of the workpiece in the process of two-wheel rolling. As shown in Fig. 7.14, the plastic deformation of the circle began from the contacting point 1′, 2′ between the rolling wheel and the workpiece. When the stress increased to the yield limit of point 1 and point 2, local plastic deformation will occur. Therefore, for the workpiece requiring higher precision, the rolling condition of the hollow thin-walled thread can be decided by the stress of points 1 and 2 which cannot exceed the yield stress. At the limitation of the yield stress of points 1 and 2, local plastic deformation only occurred on the cross section of 11′ and 22′, while global plastic deformation has not occur on the whole circle. Also, because the cold rolling process of the thread is a procedure of continuous small deformation, the rolling process can thus be continued. When the other two dangerous cross section 33′ and 44′ enters the plastic state, the whole circle will deform plastically and become an irregular shape, and the rolling process cannot even be completed. Therefore, for materials with a good plasticity, the rolling capability of the hollow thin-walled thread can be verified by the stress of point 3 and point 4 which cannot surpass the yield stress of the material. But for some brittle materials, the yield stress and tensile stress are close to materials such as 40Cr and Cr18Ni9Ti, and the tensile stress of points 1 and 2 will reach the limitation before the whole thread reaches the yield limitation. Therefore, cracking will occur from point 1 and point 2. The tensile stress of points 1 and 2 can be used to judge the rolling capability. Therefore, the stress state of points 1, 2, 3, and 4 should be solved firstly, and then, appropriate conditions can be selected to judge whether the hollow thread is capable of rolling.

The stress state of points 1 and 2 can be solved firstly. On the dangerous cross section 11′ and 22′, according to the stress state of any point in the circle of formula (7.43), the axial force N on these two cross sections is zero and the shearing force is $P/2$. According to the knowledge of material mechanics, the maximum normal stress is on the farthest position from the neutral axis, while the shear stress is parabolically distributed on the rectangular cross section, and the shear stress is zero at the top and bottom edges of the section. Therefore, on points 1 and 2, the shear stress is zero, and only normal stress is generated by the bending moment applied on points 1 and 2. At points 3 and 4, on the dangerous cross section of 33′ and 44′, according to the formula (7.43), the shearing force is $Q = 0$, and the axial force is $N = P/2$; therefore, the normal stress generated by the bending moment and the compressive stress generated by the axial stress are applied on points 3 and 4.

Taking an element at point 1 and 2, cylindrical coordinates have been used. The stress components are shown in Fig. 7.15. According to the above analysis, the shear stress of this position is zero, $\sigma_r = 0$, σ_θ is the normal stress generated by the bending moment.

Taking an element at points 3 and 4, the stress component on the element is shown in Fig. 7.16. Here, the shear stress is zero; then, $\sigma_r = 0$, σ_θ is the resultant force of the compressive stress generated by the normal stress and the axial stress.

When the cold rolling thread is longer enough, the inner surface of the circle is in the state of plane strain. Also, because the shear stress at these points is zero, the maximum normal stress is the normal stress generated by the bending moment, and the minimum stress is zero. Therefore, the yield criterion can be simplified as follows:

$$\sigma_1 = \sigma_S \tag{7.44}$$

where $\sigma_\theta = \sigma_1$.

Fig. 7.15 Stress components in element 1

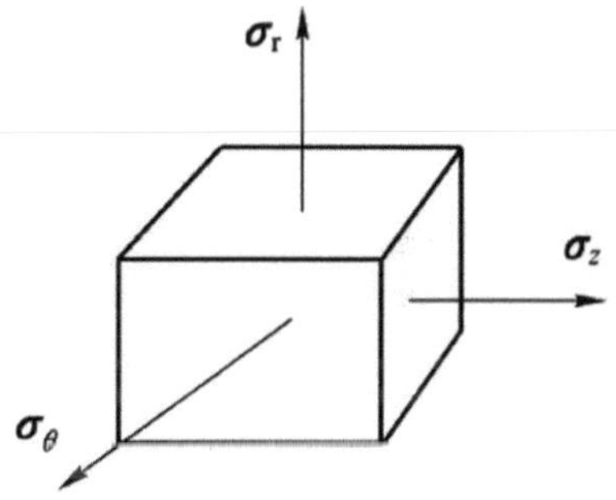

Fig. 7.16 Stress components in element 2

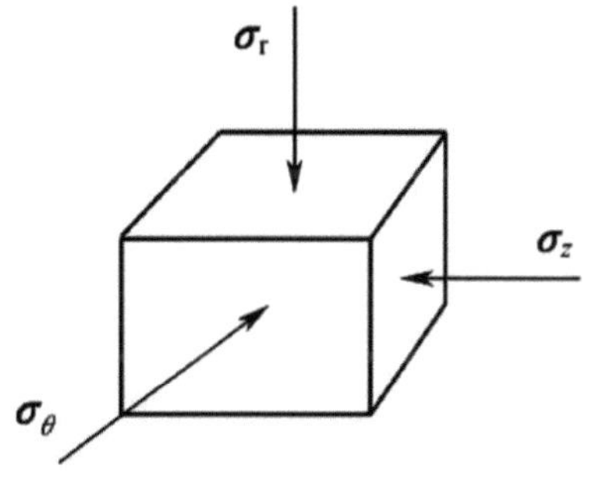

Therefore, the stress components σ_θ of these points should be solved only.

The calculation formula of any point on the cross section of a plane curved bar can be expressed as follows in the state of pure bending:

$$\sigma = \frac{My}{T\rho}$$

where

y is the distance of the point to the neutral axis;
ρ is the distance from the center of the curvature;
T is the static moment of the whole cross section to the neutral axis;
M is the bending moment in the cross section.

Assuming that the outer radius of the billet is R, the inner radius is r, the wall thickness is δ, and the length is l, $R_0 = \frac{1}{2}(R + r)$;

The area of the cross section is $A = l\delta$;

The radius of the neutral layer is $r_0 = \frac{\delta}{\ln\left(\frac{R}{r}\right)}$; and

$$T = A \cdot (R_0 - r_0).$$

At points 1, 2, 3, and 4, $\rho = r$, $y = r_0 - r$.

The tensile stress of points 1 and 2 can be calculated as follows:

$$\sigma_{1,2} = \frac{0.318PR_0(r_0 - r)}{l\delta(R_0 - r_0)r} \tag{7.45}$$

While the compressive stress generated by the bending moment and axial force at points 3 and 4 can be calculated with the following formula:

$$\sigma_{3,4} = \frac{0.182PR_0(r_0 - r)}{l\delta(R_0 - r_0)r} + \frac{P}{2l\delta} \tag{7.46}$$

After the solution of the stress state of points 1, 2, 3, and 4, corresponding formula can be used to judge whether the hollow thread can be rolled. Then, there is:

$$\sigma_{1,2} \leq \sigma_s \tag{7.47}$$

$$\sigma_{3,4} \leq \sigma_s \tag{7.48}$$

$$\sigma_{1,2} \leq \sigma_b \tag{7.49}$$

where σ_b is the tensile strength of the material.

For threads of higher precision, strength, and safety coefficient, formula (7.47) can be used to discriminate the capable of rolling; for thread components of plastic material, formula (7.48) can be used; and for brittle material, formula (7.49) should be used to discriminate whether the hollow thread component can be directly rolled.

The above formula is complex for the calculation of the wall thickness of the hollow component, and it can be approximately calculated with the formula of a straight beam. For example, the stress of points 1 and 2 can be calculated with the following formula:

$$\sigma_{1,2} = \frac{M}{W} = \frac{0.318PR_0}{\frac{1}{6}l\delta^2} < \sigma_s \tag{7.50}$$

And:

$$\delta \geq \sqrt{\frac{1.908PR_0}{l\sigma_s}} \tag{7.51}$$

where

- l is the length of the rolling thread, mm;
- P is the rolling force, N;
- δ is the wall thickness, mm.

7.3.2.2 Wall Thickness Condition in Three-Wheel Rolling of Thin-Walled Components

The same analysis method can be used as above. It can be simplified as a circle applied with three force P with a angle of 120° each other. Cutting the circle along the line of 120°, as shown in Fig. 7.17a, according to the symmetrical property of the load, the shearing force on sections A and B, only the axial force N, and the bending moment M_0 exist; according to the balance condition, it is easy to obtain that $N = \frac{\sqrt{3}}{3}P$. Therefore, only M_0 is the redundant restraining force, which can be recorded as X_1, as shown in Fig. 7.17b. According to the symmetrical property, only 1/6 of the circle is taken for study.

Because the rotation angles of symmetrical sections A and B are both zero, section A can be considered as the fixed end and section B can be selected as the coordinate deforming condition, as shown in Fig. 7.17c, which can be expressed as follows:

$$\delta_{11}X_1 + \Delta_{1p} = 0 \tag{7.52}$$

where Δ_{1p} is the rotation angle of section B while only $N = \frac{\sqrt{3}}{3}P$ applied on the basic static determinate system, as shown in Fig. 7.17d; δ_{11} is the rotation angle of section B as $X_1 = 1$, and acted alone, as shown in Fig. 7.17e.

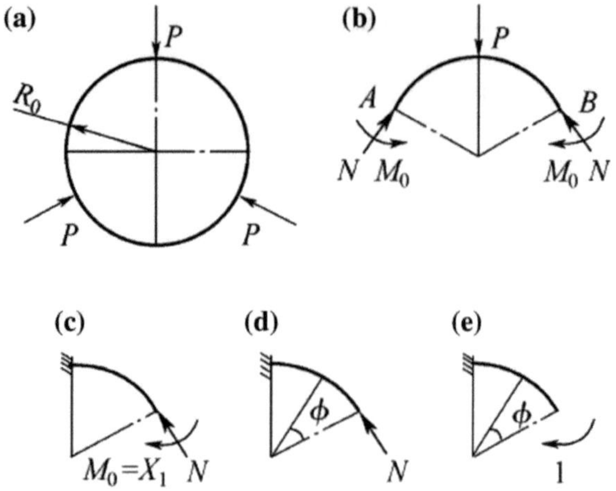

Fig. 7.17 Force decomposition diagram of the three-wheel rolling threads

M and $\overline{M}$ is, respectively, the bending moment of the section at angle ϕ, while only $N = \frac{\sqrt{3}}{3}P$ (Fig. 7.17d) and the unit moment $M_0 = 1$ (Fig. 7.17e) applied on the section, and there has:

$$M = -\frac{\sqrt{3}}{3}PR_0(1 - \cos\,\phi),\ \overline{M} = 1; \tag{7.53}$$

Therefore

$$\Delta_{1p} = \int_0^{\frac{\pi}{3}} \frac{M\overline{M}R_0 d\phi}{EI} = -\frac{\sqrt{3}PR_0^2}{3EI}\int_0^{\frac{\pi}{3}}(1 - \cos\,\phi)d\phi = -\frac{\sqrt{3}PR_0^2}{3EI}\left(\frac{\pi}{3} - \frac{\sqrt{3}}{2}\right);$$
$$\delta_{11} = \int_0^{\frac{\pi}{3}} \frac{\overline{M}\overline{M}R_0 d\phi}{EI} = \frac{R_0}{EI}\int_0^{\frac{\pi}{3}} 1^2 d\phi = \frac{\pi R_0}{3EI}.$$

Substituting into formula (7.53), it can be obtained as follows:

$$X_1 = -\frac{\sqrt{3}}{\pi}PR_0\left(\frac{\pi}{3} - \frac{\sqrt{3}}{2}\right).$$

Under the combined action of $\frac{\sqrt{3}}{3}P$ and X_1, the bending moment of any section is as follows:

$$\begin{aligned} M(\phi) &= -\frac{\sqrt{3}PR_0}{3}(1 - \cos\,\phi) + \frac{\sqrt{3}}{\pi}PR_0\left(\frac{\pi}{3} - \frac{\sqrt{3}}{2}\right) \\ &= -\left(\frac{3}{2\pi} - \frac{\sqrt{3}}{3}\cos\,\phi\right)PR_0 \end{aligned} \tag{7.54}$$

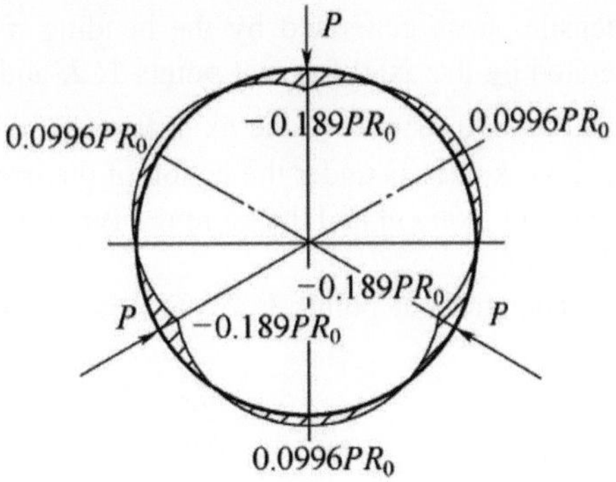

Fig. 7.18 Bending moment diagram of the workpiece in three-wheel thread rolling

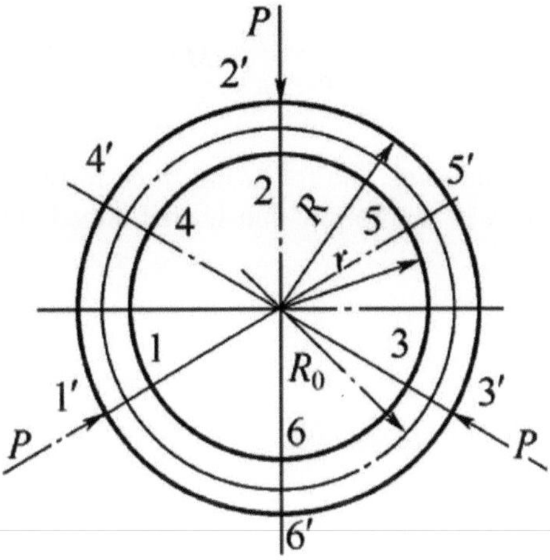

Fig. 7.19 Force diagram of the workpiece in three-wheel thread rolling

This is the actual bending moment of the 1/6 circle, and the bending moment of the whole circle can be obtained together with the symmetrical relationship of the circle, as shown in Fig. 7.18.

When $\phi = 60°$, the maximum value of M is $M - 0.189\,PR_0$. And the inner force of any section of the 1/6 circle can be expressed as follows:

$$\left.\begin{aligned} M(\varphi) &= -\left(\frac{3}{2\pi} - \frac{\sqrt{3}}{3}\cos\varphi\right)PR_0 \\ N(\varphi) &= \frac{\sqrt{3}}{3}P\cos\varphi \\ Q(\varphi) &= \frac{\sqrt{3}}{3}P\sin\varphi \end{aligned}\right\} \tag{7.55}$$

According to the analysis of the two-wheel rolling, the stress state of points 1, 2, and 3 and points 4, 5, and 6 should also be solved. On section 11′, 22′, 33′ (Fig. 7.19), and $M = 0.189\,PR_0$, the axial force is $\frac{\sqrt{3}}{6}P$, and the shearing force is $\frac{P}{2}$. It can also be obtained that the shear stress at points 1, 2, and 3 farthest to the neutral axis is zero. Therefore, the workpiece is under the action of the normal tensile stress generated by the bending moment and the compressive stress generated by the axial force at points 1, 2, and 3. While on the section of 44′, 55′, 66′, and $M = 0.0996\,PR_0$, the axial force $N = \frac{\sqrt{3}}{3}P$ and the shearing force $Q = 0$. And the workpiece is under the action of the normal compressive stress generated by the bending moment and the compressive stress generated by the axial force at points 4, 5, and 6.

The stress at points 1, 2 and 3 is as follows:

$$\sigma_{1,2,3} = \frac{My}{S\rho} - \frac{N}{F} = \frac{0.189PR_0(r_0 - r)}{l\delta(R_0 - r_0)r} - \frac{\sqrt{3}}{6}\frac{P}{l\delta} \tag{7.56}$$

And at points 4, 5, and 6, the stress is as follows:

$$\sigma_{4,5,6} = \frac{My}{S\rho} + \frac{N}{F} = \frac{0.0996PR_0(r_0 - r)}{l\delta(R_0 - r_0)r} + \frac{\sqrt{3}}{3}\frac{P}{l\delta} \tag{7.57}$$

The same, the following three formulas can be used to judge the rolling capability of the hollow thin-walled thread components:

$$\sigma_{1,2,3} \le \sigma_s \tag{7.58}$$

$$\sigma_{4,5,6} \le \sigma_s \tag{7.59}$$

$$\sigma_{1,2,3} \le \sigma_b \tag{7.60}$$

For the requirement of threads of higher precision, strength, and safety coefficient, formula (7.58) is suitable; for threads of plastic materials, formula (7.59) and for brittle materials formula (7.60) should be used to discriminate whether the hollow thread component can be directly rolled.

Also, the wall thickness of the hollow thread can be calculated with the approximate straight beam formula. For example, the stress at points 1, 2, and 3 can be solved according to the following formula:

$$\begin{aligned}\sigma_{1,2,3} &= \frac{M}{W} - \frac{N}{A} = \frac{0.189PR_0}{\frac{1}{6}l\delta^2} - \frac{\sqrt{3}P}{6l\delta}\\ &= \frac{P}{l\delta^2}(1.134R_0 - 0.289\delta)\end{aligned} \tag{7.61}$$

When the hollow thread is a thin-walled component, $R_0 \gg \delta$, and the influence of the axial force can be neglected, there has [6]:

$$\delta \ge \sqrt{\frac{1.134PR_0}{l\sigma_s}} \tag{7.62}$$

References

1. Cui Changhua. Thread Rolling Process [M]. Beijing: China Machine Press, 1978.
2. Zhao Junjie, Ma Zhenhai, Hu Zhenhuan. Numerical Simulation of the Thread Wedge Rolling Process [J]. Journal of Iron and Steel Research, 2002, 14(1): 22–25.
3. Ma Qin. On the Calculation of the Diameter of the Outer Screw Thread again [J]. Technology Fastener of China, 2000, 2: 11–13.
4. Wang Xiulun. Cold Rolling Processing Technology of Threads [M]. Beijing: China Railway Publishing House, 1990.
5. Chen Zhaozhi. Simple Analysis on the Process and Application of Rolling Technology [J]. Morden Components, 2004 (5): 72–73.
6. Qi Huiping, Li Yongtang, Fu Jianhua, Liu Zhiqi. Minimum Wall Thickness of Hollow Threaded Parts in Three-die Cold Thread Rolling [J]. Physics Letters B. 2008, 22: 6112–6117.

Chapter 8
Numerical Simulation of the Thread Cold Rolling Forming Process

Abstract The rolling force, velocity field, and stress and strain field in the cold rolling forming of threads are analyzed with numerical simulation methods. The failure process of thread components has also been analyzed.

Cold rolling is a kind of local deformation plastic forming process, and parameters such as the theoretical solution of the rolling force are difficult to obtain. Therefore, numerical simulation method is adopted to substitute the trial-and-error method to reduce the waste of materials and shorten the calculation time. In this book, software DEFORM is used to carry out the numerical simulation of the thread cold rolling process.

8.1 Basic Procedure and Assumptions

8.1.1 Numerical Simulation Procedure of the Thread Cold Rolling Process

(1) Set up of the finite element model. Three-dimensional solid prototyping software Unigraphics is used to set up the 3D solid model and generate the STL file of the model.
(2) The model is imported into the preprocessing module of DEFORM; three-dimensional and tetrahedron elements are used for mesh dividing and the contacting condition between the billet and the dies; and constraining, loading, and motion conditions can also be generated. Calculation results such as rotation speed of the rolling wheel, radial feed rate, frictional conditions of the billet, and the roller can be saved in the database file.
(3) The processing module of DEFORM is used for the calculation, because the cold rolling forming process of threads is a small deformation process, the deformation is complex, and the calculation process is a bit slow.

J. Song et al., *Cold Rolling Precision Forming of Shaft Parts*,
DOI 10.1007/978-3-662-54048-0_8

(4) According to the stress and strain conditions obtained from the simulation, the post-process module of DEFORM is used to analyze the simulation results. In the model, the material is 35 steel, and the thread cold rolling simulation process is carried out at room temperature (20 °C). The elastic modulus and Poisson's ratio are input from the database of the materials. The elastic modulus and Poisson's ratio is E = 206754 MPa and $v = 0.3$, respectively.

8.1.2 Simplification of the Model

In order to improve the calculation speed, necessary assumption has been made to reduce the calculation workload:

(1) In actual rolling process, the frictional coefficient between the roller and the workpiece varies with the changing various conditions, and the frictional coefficient is assumed to be a constant.
(2) The rolling wheel is assumed to be a full rigid body, and the workpiece is a plastic body without elastic deformation.
(3) The temperature of the workpiece is kept at room temperature during the rolling process.

8.2 Simulation of the Cold Rolling Forming Process of a M10 × 1.5 Solid Thread

Firstly, the finite element models of the billet and the rolling wheel are set up with three-dimensional prototyping software Unigraphics, as shown in Fig. 8.1, which includes two rolling wheels, the workpiece, and the bracing structure. The two rolling wheels are rotating actively counterclockwise, and the workpiece is rotating clockwise under the driven by the rolling wheels. Simultaneously, the two rolling wheels are feeding in the radial direction to form the threads of the workpiece. Simulation parameters are shown in Table 8.1.

Firstly, the workpiece is meshed appropriately with the preprocessing module. Too fine meshes will increase the time of calculation and even failure of the calculation, and too sparse meshes will result in lower precision of the calculation and even difficult for convergence. Because cold rolling forming of threads is a kind of local deformation process, large stress will generate on the contacting surface of the rolling wheel and the workpiece, and the stress and strain are both a little small far from the contacting point, or even be zero. Therefore, the meshes of the billet can be locally refined, and the refinement should be carried out on the surface of the billet, as shown in Fig. 8.2 [1].

From the above analysis, it can be seen that the dies will exert radial and tangential force to the workpiece which can be expressed by the radial and tangential load of the rolling wheel. Therefore, the analysis of the force applied on the workpiece can be analyzed by analyzing the load and moment of the rolling wheel.

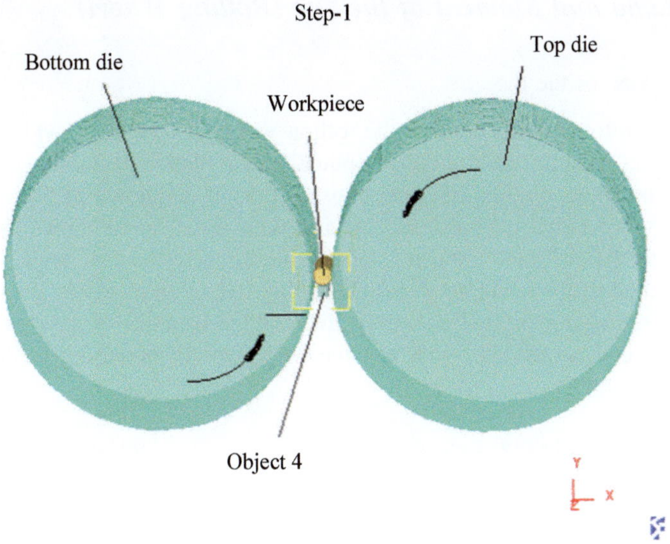

Fig. 8.1 Numerical simulation model of the two-wheel rolling process

Table 8.1 Simulation parameters in the cold rolling forming of solid threads

Items	Parameters	Items	Parameters
Material of the workpiece	AISI1020 (be equivalent to 20 steel)	Model of the rolling wheel	Rigid body
Model of the workpiece	Plastic body	Rotation speed of the rolling wheel	5.02655 rad/s
Initial diameter of the workpiece	9.04 mm	Feed rate of the rolling wheel	3 mm/s
Length of the workpiece	20 mm	Temperature	20 °C
Bracing structure	Rigid body	Frictional coefficient	0.15

Fig. 8.2 Meshing diagram of the workpiece

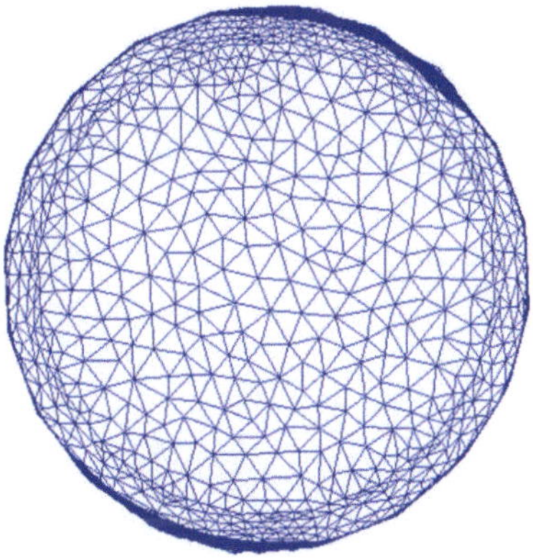

8.2.1 *Load and Moment of the Die (Rolling Wheel)*

(1) Radial load of the die

Because the information stored on the rolling wheel is very large, while the database surpassed the limitation of the computer, the simulation database will be saved as 4 files automatically, and discontinuous will occur, as shown in Fig. 8.3.

The curve of radial rolling force varying with time can be drawn as Fig. 8.4 according to the collected data. It is shown that during the first stage, 0–0.36 s, which is called as the wedging stage, the rolling reduction increases gradually, till to the 1/3 feed rate of the rolling wheel. Because the feed rate of the roller and the contacting area between the roller and the workpiece is increasing gradually, the

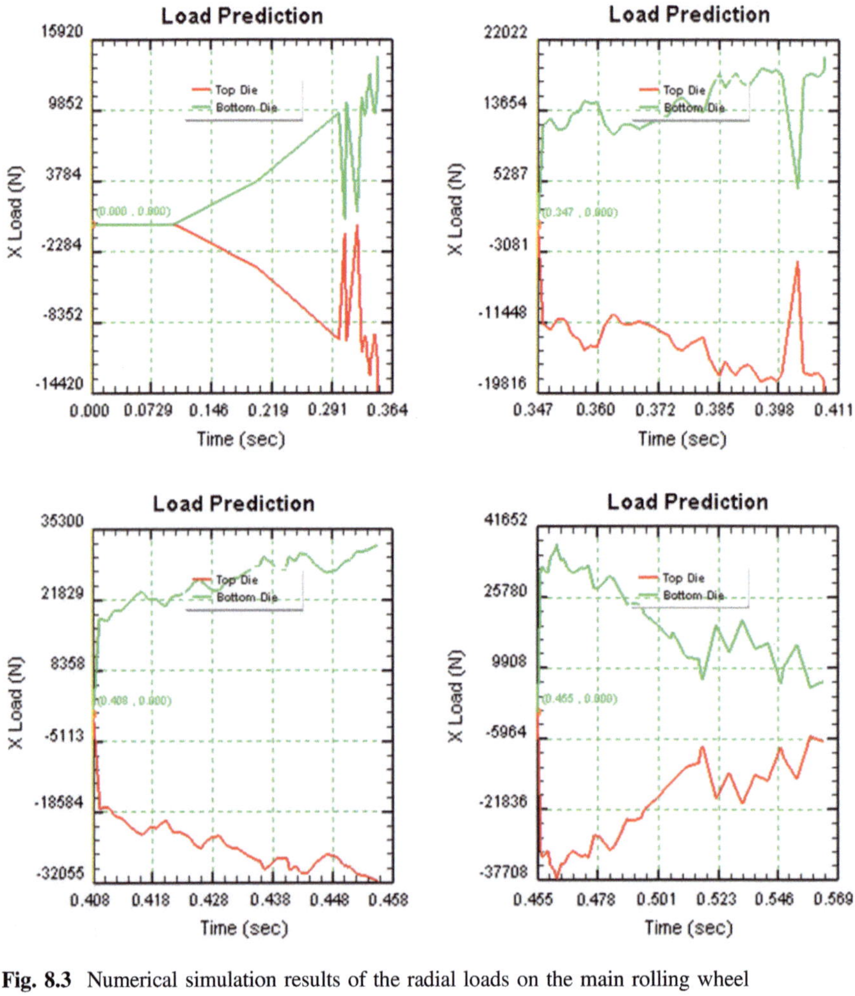

Fig. 8.3 Numerical simulation results of the radial loads on the main rolling wheel

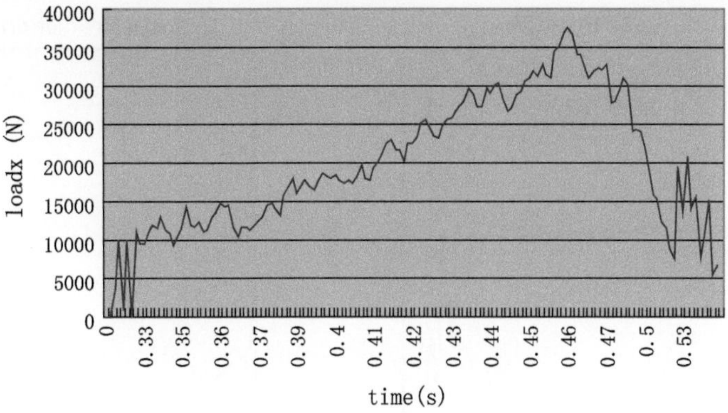

Fig. 8.4 Curve of the radial load on the roller varying with time

radial rolling force will improve rapidly with the feeding of the rolling wheel, and there is no load generated at the initial stage before the contact between the roller and the workpiece.

The second stage of the rolling is from 0.36–0.463 s, which is called as the stable rolling stage. The reduction in the rolling wheels remains the same, but the contacting area is increasing gradually, and each roller is rolling once more on the position the previous rolling wheel has rolled through, which will result in the work hardening on the surface of the material. Therefore,the radial force will also increase gradually in this stage. The stage after 0.496 s is called as the finishing stage, in which the feed rate and the reduction in the rolling wheel are both zero. The rolling force reduces rapidly and then fluctuates within a certain range.

(2) Tangential load of the rolling wheel

Figure 8.5 shows the tangential load diagram of the rolling wheel. The curve of tangential load varying with time drawn according to the acquisition data is shown in Fig. 8.6. Compared with the radial load, the up-and-down tendency of the tangential load is completely the same with that of the radial load, only with different values. Here, the tangential load is actually the force Q mentioned in the above mechanical analysis.

The curve of the ratio between the tangential force and the radial force varying with the rolling time is shown in Fig. 8.7 according to the acquisition data. It is shown that the value is quite stable and is smaller than the frictional coefficient 0.15, which verified the above theoretical analysis: During the rolling process of threads, the ratio between the tangential and the radial load is increasing with time, but the increasing degree is very small. When the rolling wheel does not feeding anymore, the ratio will decrease abruptly and fluctuate around a certain value.

(3) Axial load of the rolling wheel

Figure 8.8 shows the curve of rolling wheel axial load varying with time drawn according to the acquisition data. It is shown that there is no obvious rule of the

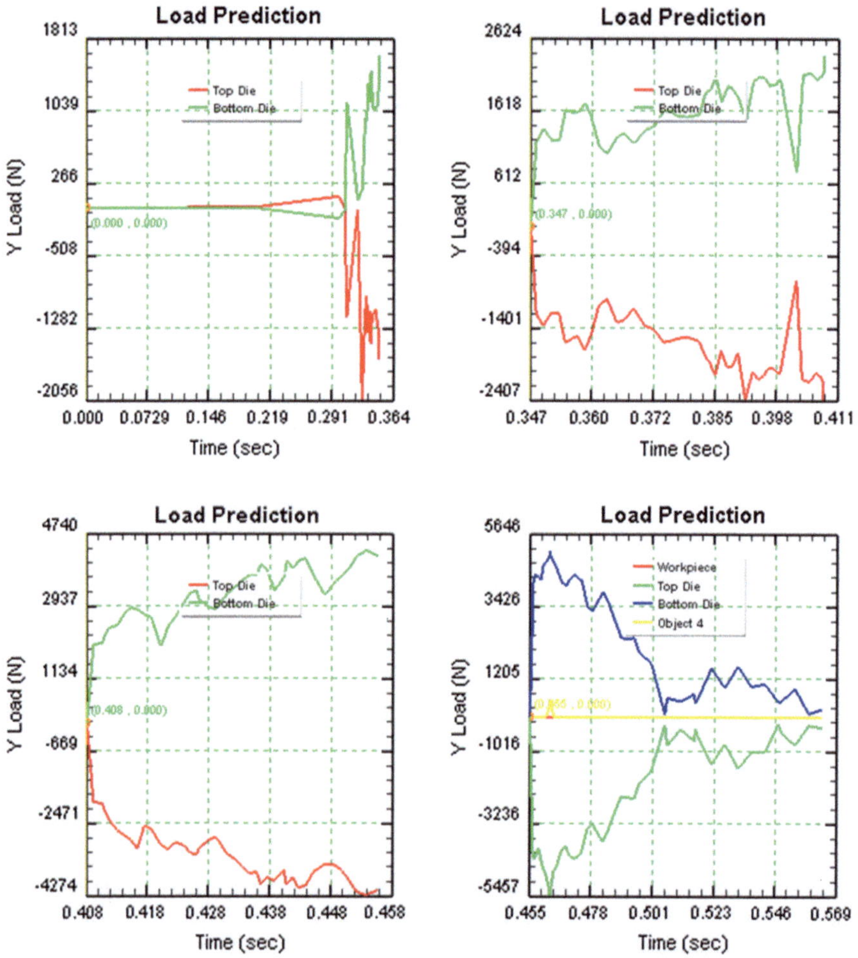

Fig. 8.5 Simulation results of the tangential load of the main rolling wheel

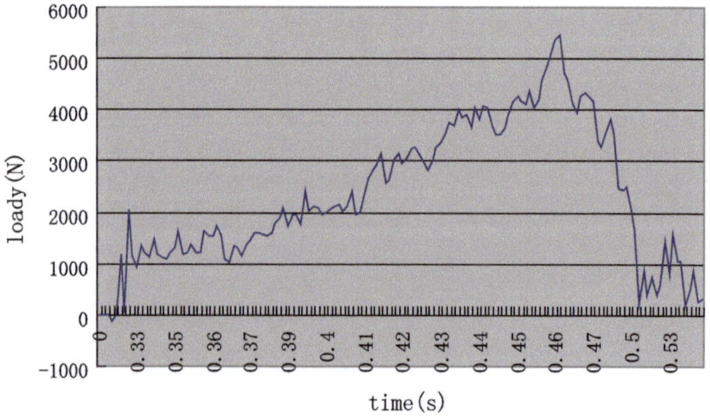

Fig. 8.6 Curve of tangential load varying with time of the roller

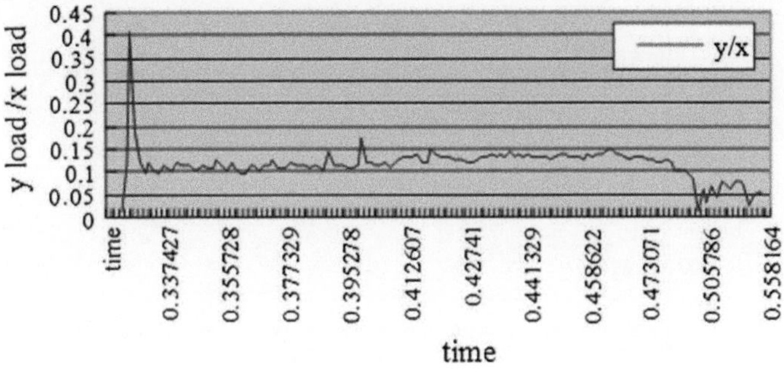

Fig. 8.7 Curve of the ratio between the tangential and radial roller load varying with time in the cold rolling of threads

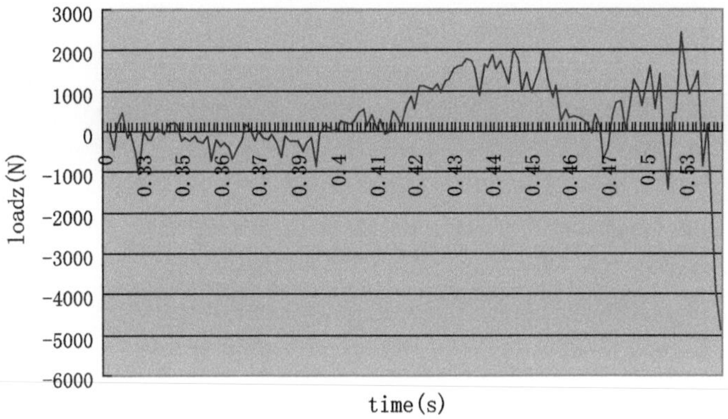

Fig. 8.8 Curve of the axial load of the roller varying with time

axial force of the roller; the direction and value are varying continuously. But with the cold rolling process, the axial load has a tendency of increasing, and it has not decreased during the finishing stage.

(4) Torque

Figure 8.9 shows the curve of the rolling wheel torque varying with time. It is shown that the changing tendency of the torque is the same as that of the rolling force, and the torques of the two rollers are both positive and along the same direction.

Figure 8.10 shows the curve of the ratio between the torque and the tangential load varying with time. It shows that the ratio is almost a horizontal line with a value of about 75, which happens to be the radius of the rolling wheel. It indicates that the torque of the workpiece is the product of the tangential load and the radius of the roller as described in the above analysis.

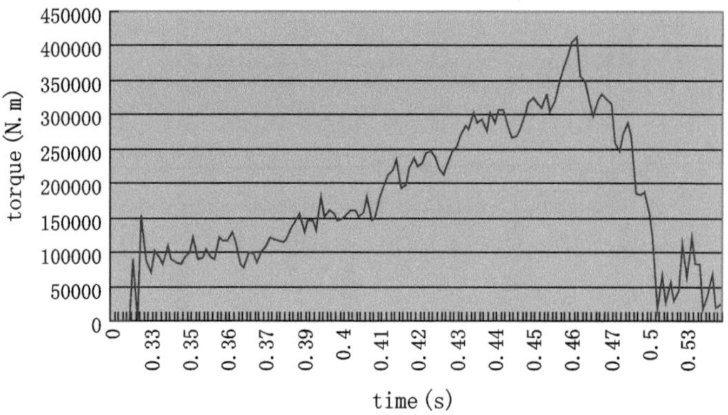

Fig. 8.9 Curve of the rolling wheel torque varying with time

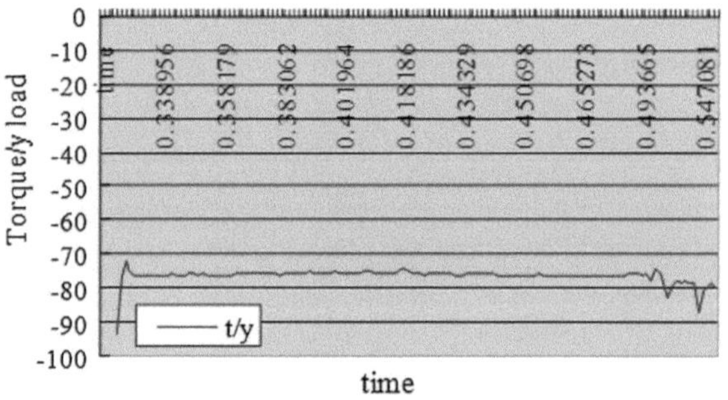

Fig. 8.10 Curve of the ratio between the torque and the tangential load varying with time

8.2.2 Stress State Analysis of the Workpiece

Figures 8.11 and 8.12 show the stress in the cold rolling forming solid thread at different stages of the process. Through comparing, it is shown that the plastic forming of step 100 is much larger than that of the step 10, but it is still a local deformation. The effective compressive stress at the contacting point of the roller and the workpiece is the maximum; the farer the contacting point, the smaller the effective stress will be, until it decreases to zero. Therefore, in the cold rolling forming process of solid thread, the workpiece is always undergoing a local deformation near the contacting points.

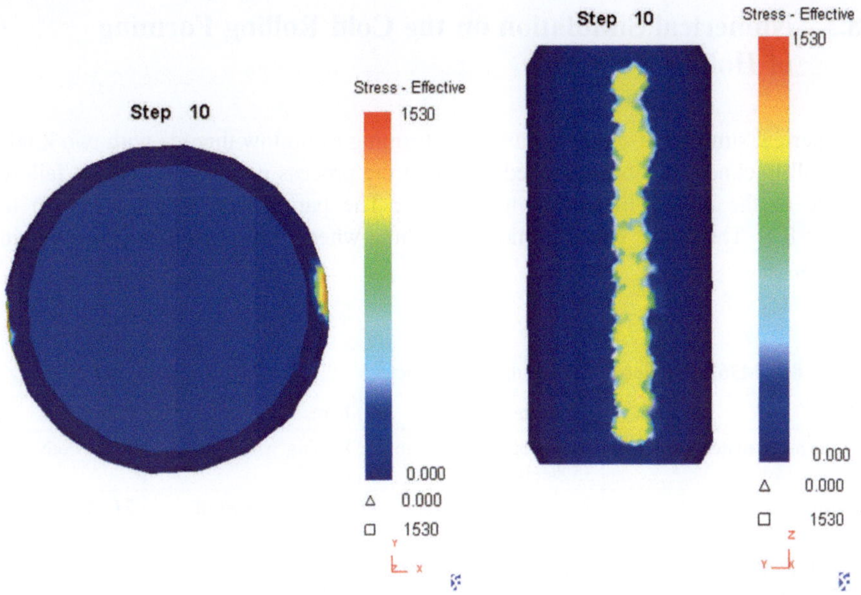

Fig. 8.11 Effective stress state at step 1

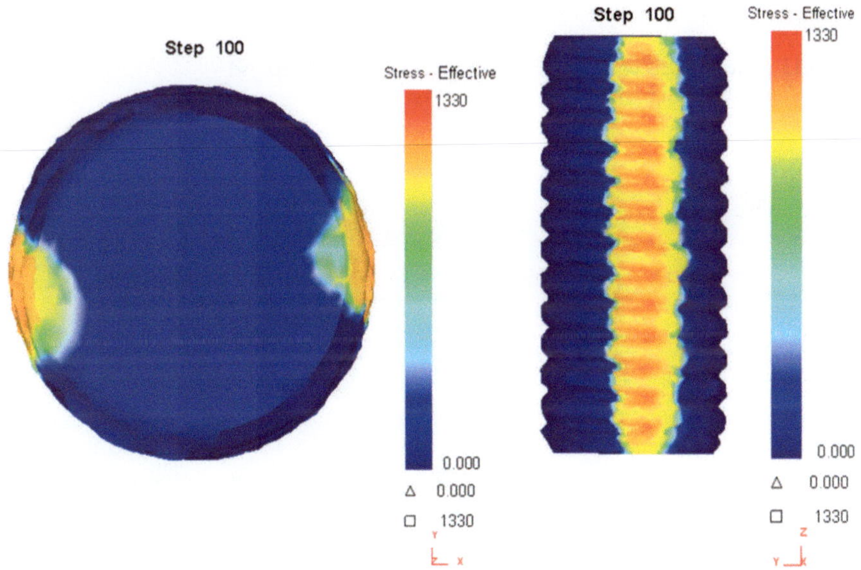

Fig. 8.12 Effective stress state at step 2

8.3 Numerical Simulation on the Cold Rolling Forming of Hollow Threads

Numerical simulation on the cold rolling forming of hollow threads with two kinds of wall thickness has been studied to obtain the process parameters and the failure mode of the cold rolling forming process. The parameters used are shown in Table 8.2. The finite element model of three-wheel thread rolling is shown in Fig. 8.13.

Table 8.2 M36X4-50 numerical simulation parameters

Items	Parameters	Items	Parameters
Workpiece materials	AISI1035 (effective to 35 steel)	Rolling wheel model	Rigid body
Workpiece model	Plastic body	Rotation speed of the roller	5.02655 rad/s
Initial dimension of the workpiece	φ33.483/φ24–50 mm	Feed rate of the roller	3 mm/s
Length of the workpiece	50 mm	Frictional coefficient	0.1

Fig. 8.13 Three-dimensional mode of three-wheel thread rolling

8.3.1 *Analysis of the Rolling Force*

8.3.1.1 Radial Load of the Rolling Wheel

Figures 8.14 and 8.15 show, at 0–0.51 s, the first wedging stage in the rolling process, in which the load of the rolling wheel is increasing rapidly. Period 0.51–2.6 s is the second stage of the rolling, in which the load curve is relatively flat, but still presents a rising trend. 2.6–2.67 s is the third stage of the rolling, in which the load of the rolling wheel decreases gradually. From 2.67 s to the end of the rolling is called as the finishing stage, in which the rolling force decreases rapidly to a very small value. The variation trends of the radial rolling force in the cold rolling of hollow and solid components are basically the same. While in the cold rolling of

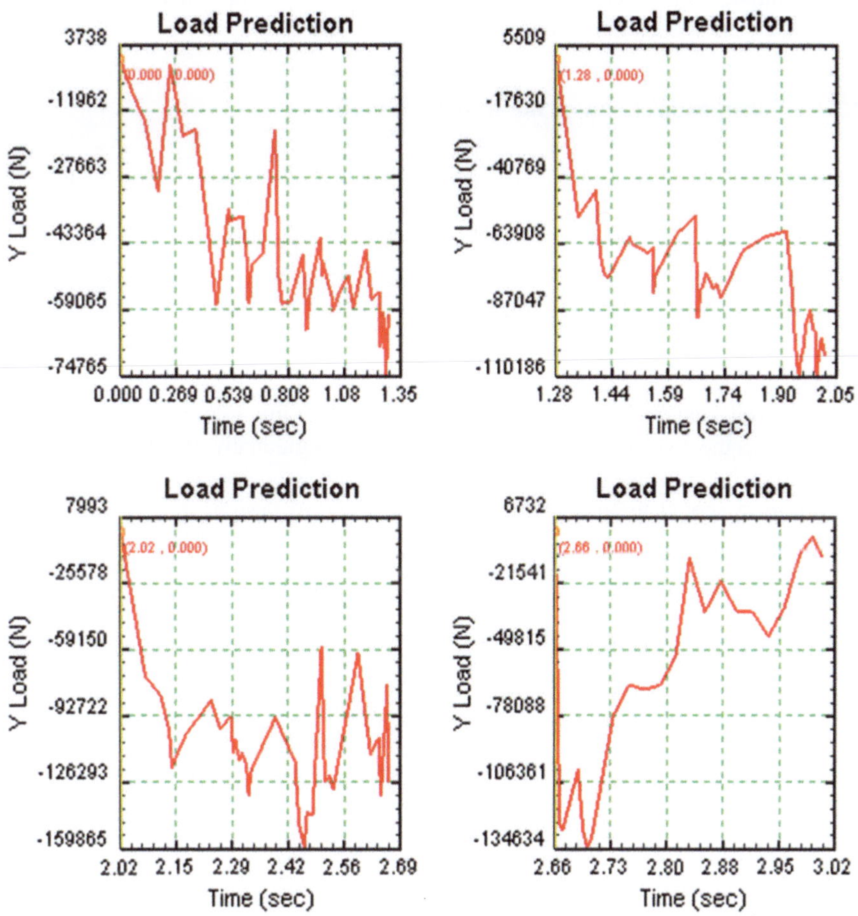

Fig. 8.14 Radial load of the rolling wheel

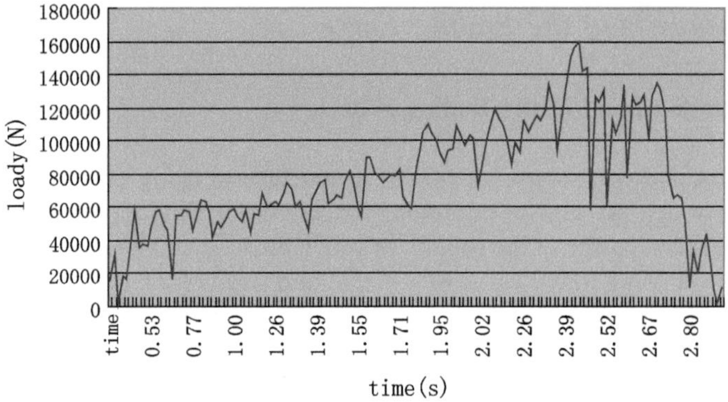

Fig. 8.15 Curve of the radial load varying with time of the roller

hollow threads, the fluctuation of the rolling force is larger and unstable, this is resulted by the whole elastic deformation of the threads.

8.3.1.2 Tangential Load of the Rolling Wheel

Figure 8.16 is the curve of the roller tangential load varying with time; the variation rule coincides with that of the radial rolling force. The tangential force increases gradually in the first and second stages and then rapidly decreases to zero.

8.3.1.3 Axial Load of the Rolling Wheel

Figure 8.17 is the curve of the axial load varying with time of the roller. It is shown that the axial load is changing continuously with the same rule of the solid thread rolling; with the processing of the rolling, it increases gradually, there is no obvious decrease in the finishing stage, and the axial force is larger than the tangential force. Therefore, it should be highly focused in the period of design.

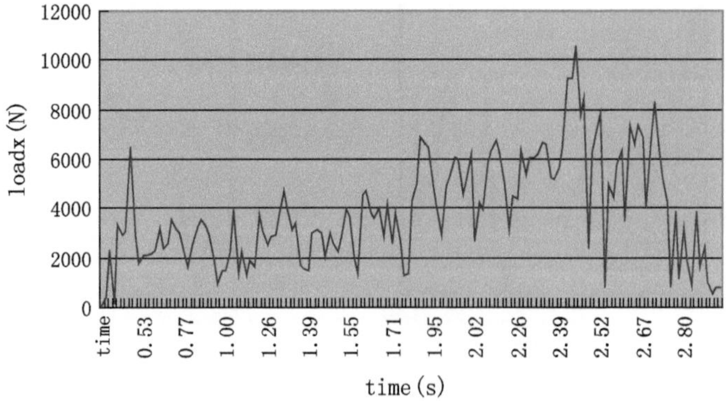

Fig. 8.16 Curve of the tangential load varying with time of the roller

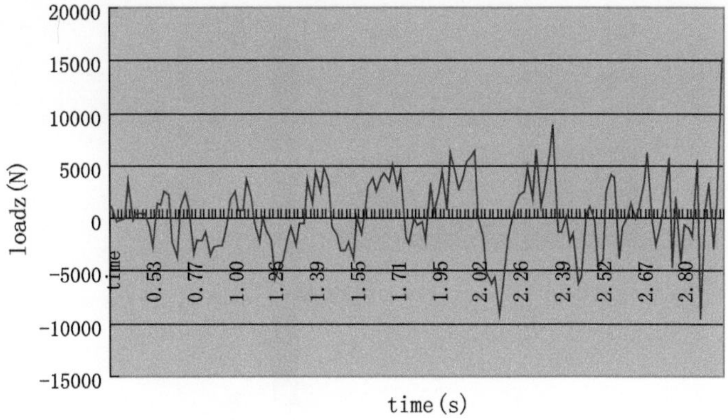

Fig. 8.17 Curve of the axial load varying with time of the roller

8.3.1.4 Torque

Figure 8.18 is the curve of the torque varying with time drawn with the acquisition data. It is shown that the variation rule of the torque in the cold rolling forming of hollow threads is almost the same with that of the tangential force, and coincides with that obtained in the rolling of solid threads.

8.3.2 Velocity Field

Figure 8.19a is the velocity vectorgraph in the pure rolling state of the workpiece and the rolling wheel. It is shown that the direction of every point on the workpiece is along the tangential direction of the workpiece circumference. Because the deformation speed of the workpiece surface is greatly smaller than the rotation

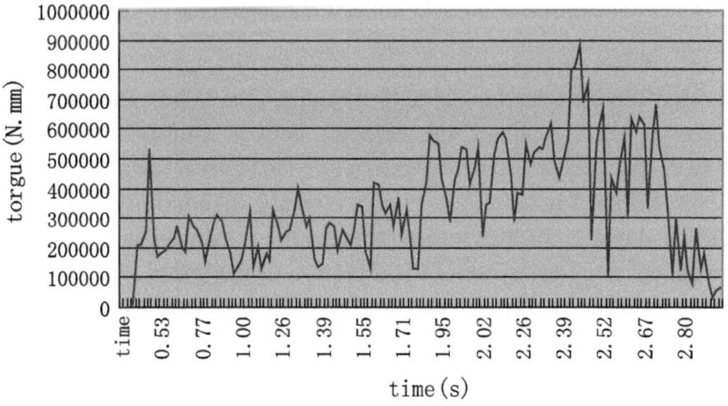

Fig. 8.18 Curve of torque varying with time

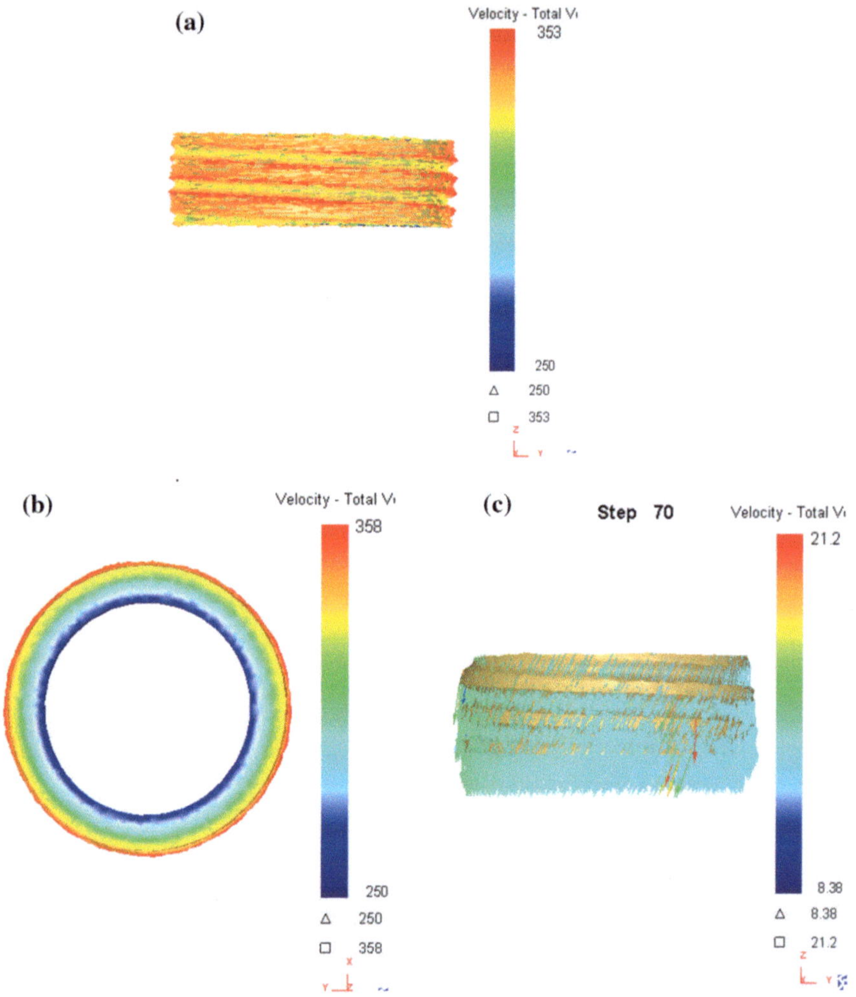

Fig. 8.19 Velocity vectorgraph in the cold rolling forming process of the thread

speed, it is not obvious in the velocity vectorgraph. When there is no swing of workpiece in the rolling wheel, the velocity field is uniform and the velocity is increasing from the inside to the outside while observing from the end surface, as shown in Fig. 8.19b. Figure 8.19c is the velocity in condition there is a sliding between the workpiece and the rolling wheel. The rotation speed of the workpiece is greatly reduced, even to zero, but the direction of the speed is closer to the axis of the workpiece, and the workpiece has moved relatively to the roller along the axial direction.

It is known from the above theoretical analysis that in the cold rolling forming, when the helix angle of the rolling wheel pitch diameter is equal to that of the thread, there is no relative axial movement between the workpiece and the roller; thus, the workpiece is rotating around its axis in condition the sliding does not exist.

8.3.3 Stress State of the Workpiece

8.3.3.1 Variation of the Stress Field

Figure 8.20 is the effective stress of the cold rolling forming thread in every stage of the process. In the figure, the position contacting with the roller is the sectional view along the axis of the workpiece. With the processing of the rolling the tooth height of the thread increases gradually, there is no large variation of the maximum effective stress, which keeps at about 950 MPa, but the range of the maximum effective stress is increasing steadily, and the minimum effective stress remains to be zero, which indicates that there is always a zero stress zone exists in the workpiece.

It can be seen from the simulation that when the pressure applied to the workpiece is not very large, the stress fields of the cold rolling hollow and solid threads are similar, and there is a very large stress at the contacting position with the roller; the farer from the contacting point, the smaller the stress will be, until it reduces to zero. With the increasing normal force, because the bending torque applied to the workpiece is increasing, larger effective stress zone begins to appear on the inner surface of hollow thread.

As shown in Fig. 8.21, at step 120 of the rolling process, that is the stage before the finish of the second rolling stage (stable rolling stage), the effective stress zone in the workpiece is larger, and the position is the inner surface of the workpiece and roller contacting zone. The yield of the hollow thread will firstly occur in the three corresponding zones, which coincides with the analysis of the dangerous points.

8.3.3.2 Stress State of the Characteristic Points in the Threads

With the continuous increasing in the rolling force, the inner surface of the workpiece begins to yield and local plastic deformation begins to generate in the workpiece. Take several characteristic points in the inner surface of the hollow thread at the maximum rolling force at step 123, as shown in Fig. 8.22. Point *A* is a point on the inner surface in the contacting zone between the workpiece and roller, where the effective stress is the maximum. From the principle stress shown in Fig. 8.23, it is shown that it is in a state of three-axial tensile stress. Point *B* around point *A* is in a two-axial tensile stress state, and point *C* is a point between the two rollers with a three-axial compressive stress state. It indicates that the inner surface points of the workpiece and roller contacting zone are subjected to tensile stress while the points between the two roller are in a compressive stress state, which well coincides with the above theoretical analysis.

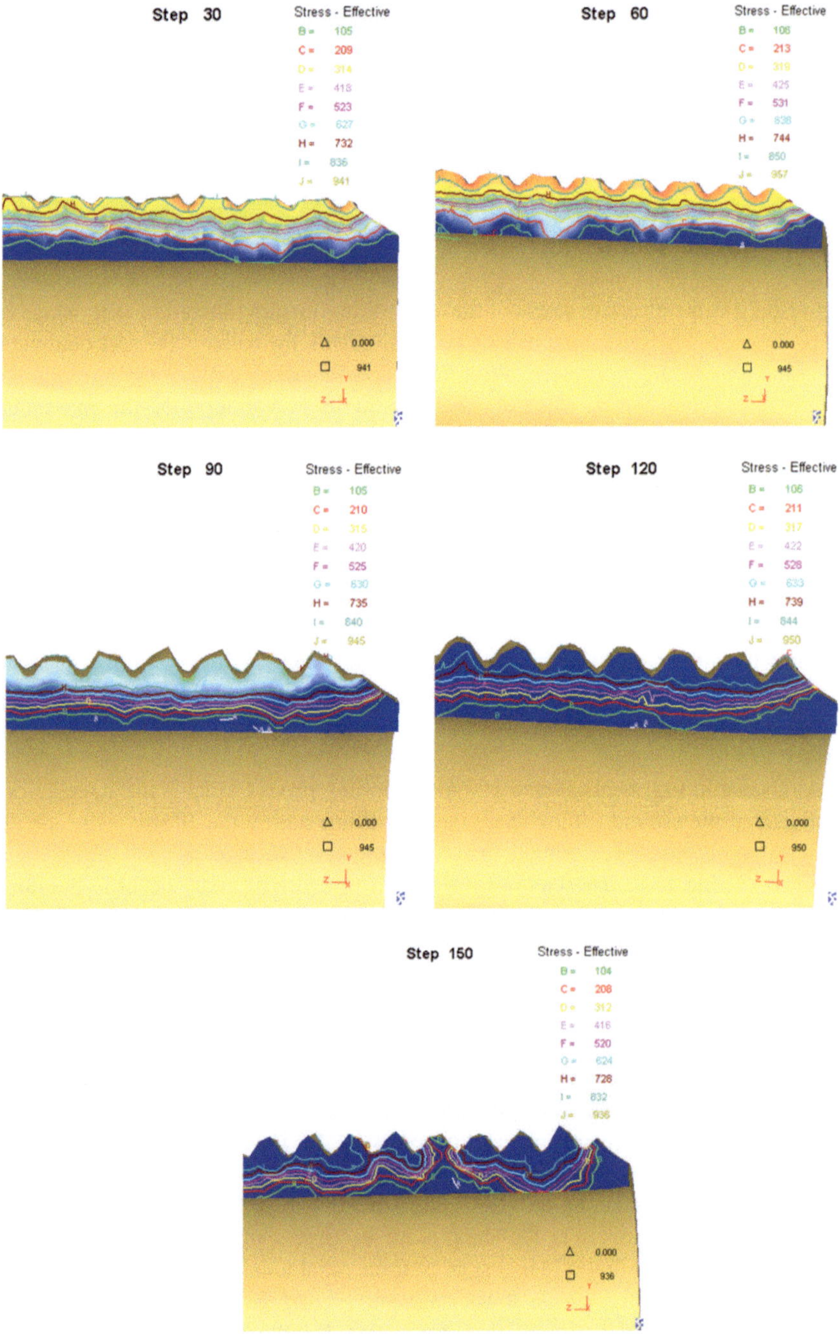

Fig. 8.20 Effective stress in the cold rolling of hollow threads

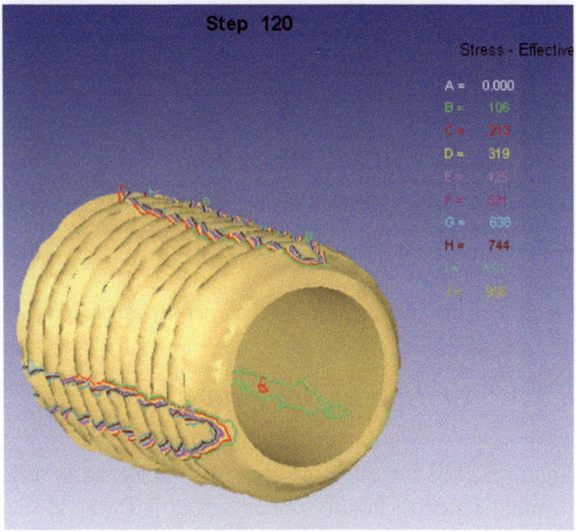

Fig. 8.21 Effective stress of step 120

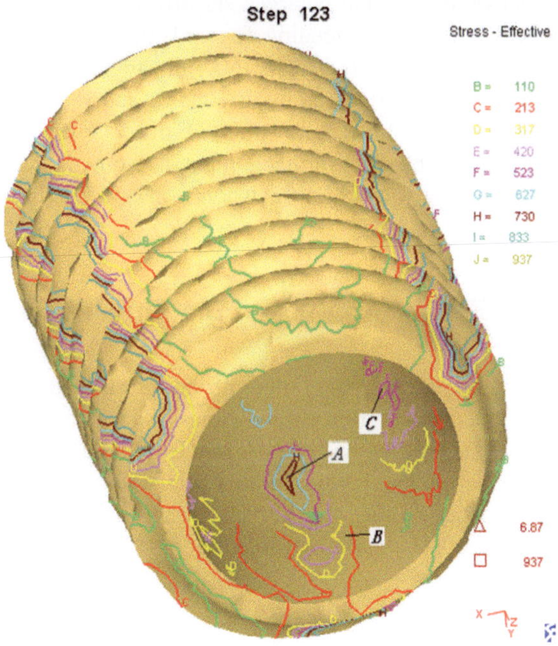

Fig. 8.22 Effective stress of step 123

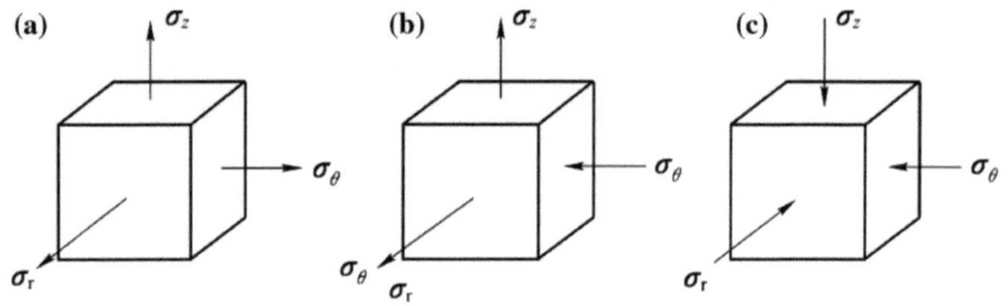

Fig. 8.23 Principle stress of the characteristic points in the inner surface of the hollow thread

8.3.4 Analysis of the Strain State

Figure 8.24 is the effective strain of the thread in different stages of the cold rolling forming process. Positions contacting with the rolling wheel are still the sectional view along the workpiece axis. It can be seen that the maximum effective strain generates in the contacting zone. With the processing of the rolling, the maximum effective strain zone enlarges steadily to the maximum value. When the rolling wheel is no longer feeding forward, the strain then decreases gradually. At the finishing stage, such as step 150, only some irregular points have been deformed, and the value of the strain is smaller and thus obtains the objective of surface finishing. The minimum strain of every stage is increasing gradually; this is because, the bending torque applied on the workpiece increases with the increasing in the rolling force; and the whole workpiece is in a strain state.

8.4 Failure Analysis of the Thread Components

In order to analyze the failure of the thread cold rolling forming process, a thread component with a thin-wall thickness is taken as an example for the simulation. To reduce the calculation time, 10 mm length of the thread is taken in the study, which may result in the flowing of the material along the axis and the incomplete forming of the tooth profile, but the failure analysis will not be affected. Parameters in the simulation are shown in Table 8.3.

8.4.1 Variation of the Effective Stress in the Workpiece

Figure 8.25 is the variation of the effective stress in the workpiece cross section during various stages of the rolling process, which shows the variation of the effective stress in the workpiece during the hollow thread cold rolling process. At the beginning of the rolling, only the effective stress at the contacting position of the

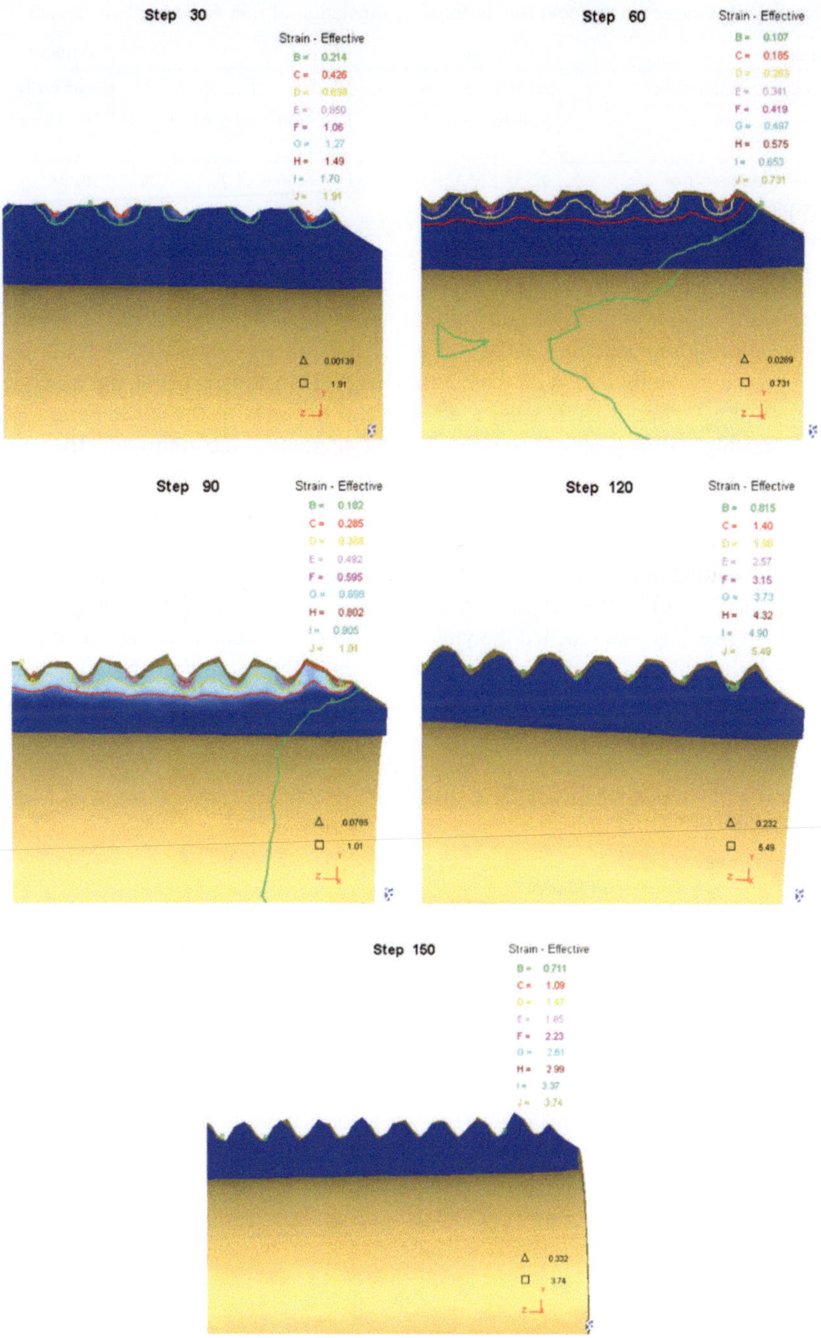

Fig. 8.24 Effective strain of the hollow thread in every stage of the rolling process

Table 8.3 Parameters in the cold rolling forming simulation of thin-walled hollow threads

Items	Parameters	Items	Parameters
Workpiece materials	AISI1020	Model of the roller	Rigid body
Workpiece model	Plastic body	Rotation speed of the roller	5.02655 rad/s
Initial dimension of the workpiece	φ33.483/φ25 mm	Feed rate	3 mm/s
Length of the workpiece	10 mm	Frictional coefficient	0.3

roller and the workpiece is the maximum; at the yield limit of the effective stress, plastic deformation generates at the contacting position of the workpiece to form the tooth profile of the thread. With the continuous of the rolling process, the maximum effective stress zone is increasing gradually and expands to the inner parts of the circle steadily. At step 65 of the stable rolling stage, the maximum effective stress has already reached the inner surface of the circle; points 1, 2, and 3 begin to yield; and only local plastic deformation generates on the circle. The whole component has not been influenced; it is still a circle without whole deformation. At step 72, the stable rolling will be finished; the whole circle presents a larger effective stress station; at the middle part of the inner and outside surfaces of the workpiece, the effective stress has also reached the yield limit (positions mentioned above at points 4, 5, 6 and 4′, 5′, and 6′), and plastic deformation happens in the whole thread. At step 75, the feed rate of the roller is the maximum; at the maximum rolling force, the effective stress of the whole circle is also very large, and the whole component has been yielded. From the simulation results, it can be seen that (step 75 in Fig. 8.26) the shape of the circle is irregular, which has verified the stress strain analysis of the hollow thread and the judgment of the failure [2, 3].

8.4.2 *Repairing Action of the Three Rolling Wheels on the Instability of the Hollow Thread*

Figure 8.26 is the failure process of the hollow threads. It can be seen that at step 70, the workpiece is still keeps a round shape. At the yield limit of the contacting inner surface between the roller and the workpiece, there is no plastic deformation of the whole workpiece. At step 75, the stable rolling stage will soon be finished, the radial displacement of the roller, and the rolling force are the maximum; and the roundness of the hollow component appears, and the whole workpiece comes into the yield state. When the roller is no longer feeding and comes into the finishing stage, the shape of the workpiece is approximately a circle again at step 100. At step 135, the rolling process has been finished, and the workpiece has also been repaired. It can be seen that in the three-wheel cold rolling of hollow threads, gentle instability can be well repaired, which is conform to the theoretical analysis.

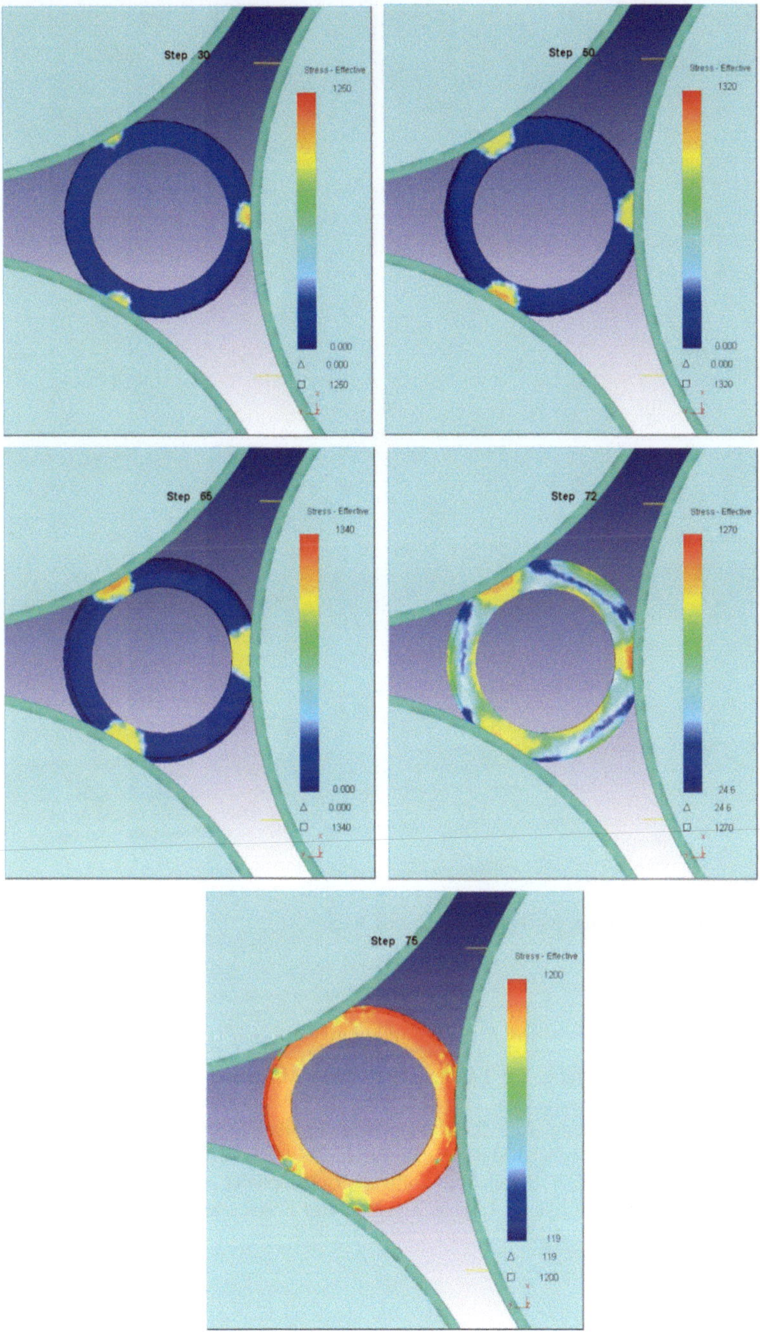

Fig. 8.25 Variation of the effective stress in the workpiece during the hollow thread cold rolling

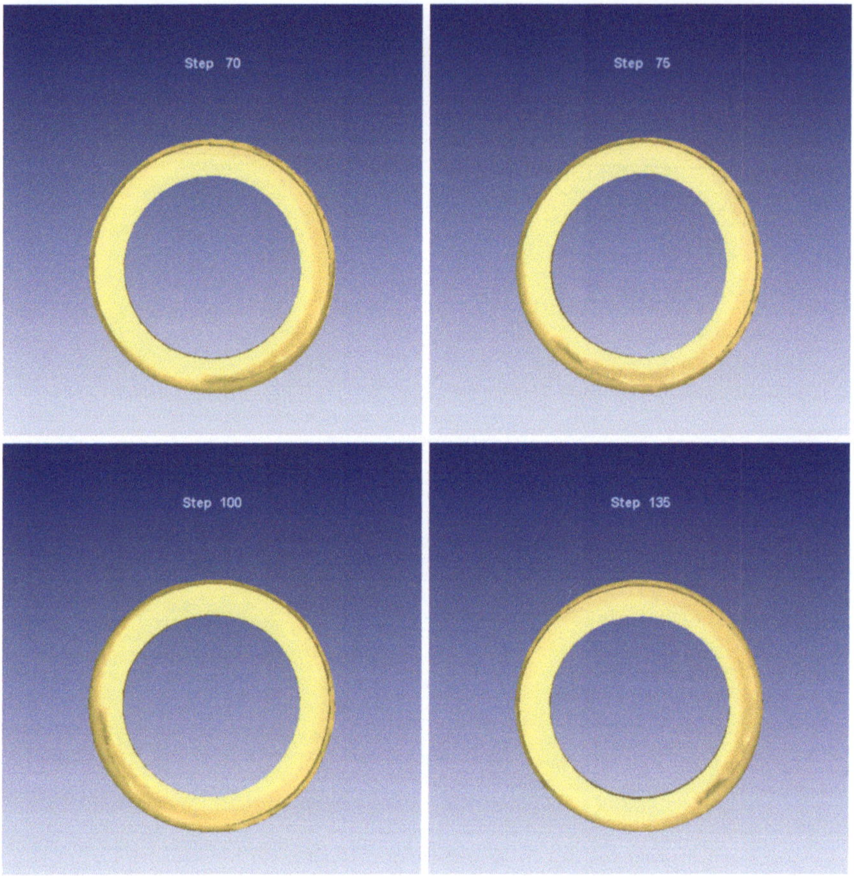

Fig. 8.26 Instability of the workpiece in various stages

References

1. Qi Huiping, Li Yongtang, Fu Jianhua, Liu Zhiqi. Minimum Wall Thickness of Hollow Threaded Parts in Three-die Cold Thread Rolling [J]. Physics Letters B. 2008, 22: 6112–6117.
2. Qi Huiping, Li Yongtang, Liu Zhiqi. Numerical Simulation Study on the Cold Rolling of Hollow Thread [J]. Advanced Materials Research, 2011, 160–162: 849 -854.
3. Liu Hongmei. Numerical Simulation Software Development of the Thread Cold Rolling Forming Process [D]. Taiyuan: Taiyuan University of Science and Technology, 2011.

Chapter 9
Metal Flow and Parameter Optimization of Thread Cold Rolling Forming Process

Abstract Experiments of the cold rolling forming threads with three-wheel rolling machine are carried out; factors influencing the forming process are analyzed. Metal flow rules and parameter optimization of thread cold rolling precision forming are also analyzed and determined.

Solid and hollow thread components were processed with cold rolling precision forming method; the process parameters of the forming process, temperature increasing, microstructure, and microhardness distribution of the forming components have been analyzed; and the minimum wall thickness condition of the hollow thread has also been verified.

9.1 Cold Rolling Forming of Threads with Three-Wheel Rolling Machine

9.1.1 Material of the Billet and the Forming Machine

Cold rolling forming experiments of four groups of hollow threads and a group of solid threads were carried out under different feed rate and wall thickness with the three-wheel rolling machine ZC26-16, shown in Fig. 9.1. The maximum rolling force is 160 kN (the pressure of the system is 13.7 MPa). Diameter range of the rolling thread under the maximum rolling force is within 22–80 mm; the spindle speed is 32, 48, 63, 95, 118, and 178 r/min, respectively; the feed rate is 0.1–4 mm/s; the head number of the rolling wheel is $K = 4$; and the length is $L = 60$ mm, as shown in Fig. 9.2.

The experimental material is 45 steel, HB = 235, the yield strength is not less than 355 MPa, and the tensile strength is not less than 600 MPa. Hollow billets with four groups of different dimensions (mm): $\Phi 44 \times 26.2 - 50$, $\Phi 44 \times 29.9 - 50.2$, $\Phi 44.4 \times 33.1 - 50.1$, and $\Phi 45 \times 36.2 - 50$ and a group of solid billet with a dimension of $\Phi 44 - 50$ have been prepared with cutting method [1].

J. Song et al., *Cold Rolling Precision Forming of Shaft Parts*,
DOI 10.1007/978-3-662-54048-0_9

Fig. 9.1 Three-wheel rolling machine ZC26-16

Fig. 9.2 Rolling wheel used in the experiment

9.1.2 Factors Influencing the Cold Rolling Forming Process of Threads

It has been found that the billet with an inner diameter of 36.2 mm has been seriously deformed in the rolling forming process, the formed shape of the component is greatly irregular, and the rolling process can even not be continued. The other groups of experimental data are shown in Table 9.1.

Table 9.1 shows that the factors influencing the cold rolling forming process of the hollow threads mainly include the following aspects:

(1) Rolling force: Under conditions the other parameters are kept constant, the forming process can be fully carried out with the increase in the rolling force (the external diameter is large), the decrease in the inner diameter is reduced, the rolling time is shortened, and the rolling efficiency is improved.

Table 9.1 Results of the three-wheel thread cold rolling forming experiments

Specifications (external diameter × inner diameter − length)/mm	Pressure/MPa	Rotation speed/(r/min)	Feed rate/(mm/s)	External diameter/mm	Inner diameter/mm	Un-roundness of the inner diameter/mm	Rolling time/s
Φ 44 × 26.2 − 50	8.5	32	2.2	45.8	26		7
Φ 44 × 26.2 − 50	8.5	32	2.2	45.7	25.75		7
Φ44 × 26.2 − 50	10	48	2.2	45.8	25.2		6
Φ44 × 26.2 − 50	10	48	1.1	46	25.2		7
Φ 44 × 29.9 − 50.2	8.5	32	2.2	45.2	29.6	−0.3	7
Φ 44 × 29.9 − 50.2	8.5	32	2.2	45.25	29.6	−0.3	6
Φ 44 × 29.9 − 50.2	10	48	2.2	45.6	29.6	−0.3	7
Φ 44 × 29.9 − 50.2	10	48	1.1	45.9	29.8	−0.4	7.5
Φ 44.4 × 33.1 − 50.1	8.5	32	2.2	44.4	32	−1.1	7
Φ 44.4 × 33.1 − 50.1	8.5	95	2.2	44.7	32.2	−0.9	7
Φ 44.4 × 33.1 − 50.1	10	48	2.2	45	32.6	−0.6	6
Φ 44.4 × 33.1 − 50.1	10	48	1.1	45.2	32.4	0.8	7
Φ 44 × 50	8	48	1.1	45.2			8
Φ 44 × 50	8	48	2.2	45.9			6
Φ 44 × 50	5.5	48	2.2	45.9			5.5
Φ 44 × 50	6	48	2.2	45.6			8

(2) Feed rate: Assuming that sliding is ignored, if the feed rate varies from 2 to 1 m/s, the rolling time will increase along with the decreasing destructive force to the workpiece and the forming is fully conducted, while the reduction amount of the inner diameter is increased.
(3) Wall thickness of the thread: It is the most important factor influencing the rolling of hollow components, and the forming degree of the thread teeth will increase with the increasing wall thickness; when the wall thickness is less than a certain value, the tooth profile will be incompletely formed, and the rolling process can even not be continued.

9.1.3 Influence of the Hollow Thread Wall Thickness on the Forming of Threads

Figure 9.3 shows the tooth profiles of the hollow threads with different wall thicknesses under the same rolling conditions. It is shown that with an inner diameter of 26.2 mm, the tooth profile of the thread is completely formed, which has no obvious difference with that of the solid threads; with an inner diameter of 29.9 mm, the tooth profile is rather flat and incompletely formed; while with an inner diameter of 33.1 mm, the forming teeth is seriously incomplete, and the whole thread component presents a concave shape [2].

With a system pressure of 13.7 MPa, the rolling force is 16 t. Through calculation, with a system pressure of 8.5 MPa, the rolling force is 10 t. Substituting the corresponding parameters in the formula (7.57), it can be obtained as follows:

$$\sigma_{4,5,6} = \frac{My}{S\rho} + \frac{N}{F} = \frac{0.0996PR_0(r_0 - r)}{l\delta(R_0 - r_0)r} + \frac{\sqrt{3}}{3}\frac{P}{l\delta} \tag{9.1}$$

The solved minimum wall thickness suitable for rolling is 7.9 mm, the external diameter is 44 mm, and the inner diameter is 28.2 mm. It can be seen from experiments and Fig. 9.3 that qualified thread component can be obtained with an

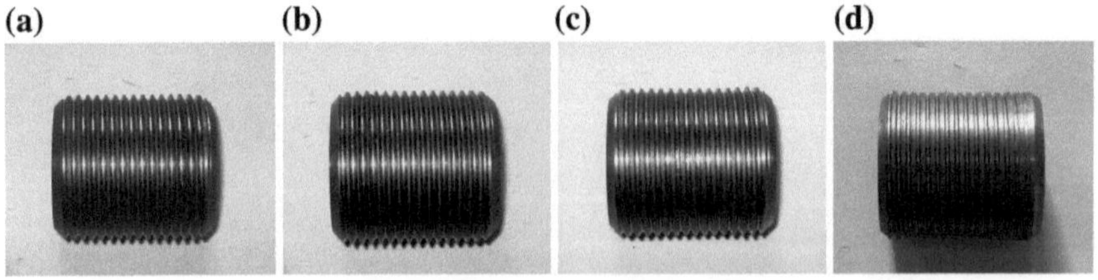

Fig. 9.3 Tooth profile of the hollow threads with different wall thickness under the same rolling conditions. **a** Solid thread, **b** inner diameter of 26.2 mm, **c** inner diameter of 29.9 mm, **d** inner diameter of 33.1 mm

Table 9.2 Stress of the dangerous points under different wall thickness with a pressure of 10 MPa

Specifications (mm)	Stress of points 1, 2, and 3 (MPa)	Stress of points 4, 5, and 6 (MPa)
Φ 44 × 26.2 − 50	539.6203533	253.6519029
Φ 44 × 29.9 − 50.2	863.3644177	426.7384343
Φ 44.4 × 33.1 − 50.1	1421.694784	700.820642
Φ 45 × 36.2 − 50	2431.079385	1219.003162

inner diameter of 26.2 mm; while with an inner diameter of 29.9 mm, the tooth profile of the thread components is incompletely formed and unqualified.

According to the theoretical analysis of Chap. 7, substitute corresponding parameters of 45 steel into formulas (7.56), (7.57), (7.58), and (7.59) and calculate the stress of the dangerous point, respectively, as shown in Table 9.2.

It is shown from the above table that under the inner diameter of 26.2 mm, $\sigma_{1,2,3} > \sigma_S$, while $\sigma_{4,5,6} < \sigma_S$. Although local plastic deformation exists on the workpiece, the whole deformation of the circle has not happened. Therefore, the rolling forming process can be smoothly carried out, and qualified components can be obtained after finishing.

With an inner diameter of 29.9 mm, $\sigma_{4,5,6} > \sigma_S$, plastic deformation occurs on the whole circle, the tooth profile is a bit flat, incompletely formed, and the section presents an ellipse shape. With an inner diameter of 33.1 mm, although the rolling process is finished, the tooth profile is obviously incomplete. And with an inner diameter of 36.2 mm, the rolling process cannot be continued, and the component shape is irregular. The experimental results are well coincided with that of the theoretical calculation.

9.2 Metal Flow Rules in the Cold Rolling Forming Process of Threads

Taking a M22 × 1.5 thread as the experimental sample of the cold rolling forming process, the metal forming rules and the distribution of the microhardness have been analyzed. The rolling wheel tooth type selected is a round addendum whole *A*-type, and the top-and-down tooth depth has been controlled. The dedendum of the workpiece thread is an arc, by which the dynamic strength of the thread can be improved and which has special advantages for the components required for higher thread strength. In the rolling process, the inner diameter of the roller contacts with the external diameter of the workpiece, while the pitch diameter of the thread has been well adjusted, and there will be no error of the external diameter.

The cold rolling forming process is carried out on the numerical control rolling machine Z28KJ-20 manufactured by Qingdao Shengjian Machinery Factory. The rolling wheel is of a round tooth addendum, as shown in Fig. 9.4. The moving guide is automatically lubricated with lubrication oil, and the workpiece is washed

Fig. 9.4 Z28KJ-20 numerical control rolling machine

with cooling oil in the process. The rated power of the motor is 7.5 kW, and the rated rotating speed is 1430 r/min. The rated power of the hydraulic pump motor is 4 kW; the pressure of the hydraulic system is 5 MPa and adjustable. The numerical control system is SINUMERIK 802C base line, and the maximum rolling force is 200 kN.

The basic parameters of the thread are as follows: the external diameter = 22 mm, the pitch = 1.5 mm, the normal pressure angle $\alpha = 60°$, the upper tooth depth = 0.435 mm, and the down tooth depth = 0.487 mm. The cold rolling process parameters are as follows: the rotating speed = 15 r/min and the feed rate = 0.5 mm/r. The head number of the rolling wheel is 9, the pitch diameter is 189.234 mm, the thickness is 60/70 mm, the hole diameter of the shaft is 45 mm, and the large diameter is 190.534 mm. The material of the billet is tempering 45 steel.

The microstructure sample is prepared with wire electrode cutting method; after grinding and polishing, the sample was corroded with 4% nitric acid alcohol solution; the microstructure of the cold rolling formed thread component has been observed with three-dimensional numerical microscopy KEYENCE VHX-600; and the difference in the microstructure machined with cutting methods has also been analyzed.

In the cold rolling process of threads, the component is formed by plastic flow of the metal. Figure 9.5 shows the microstructure and its changing rules on the center, the addendum, the flank, and the dedendum of the rolled threads.

It can be seen that the ferrite and cementite of the billet center are uniformly distributed, and the microstructure is coarse. During the rolling process, the metal is flowing to the surrounding of the tooth under the compressive action of the rolling

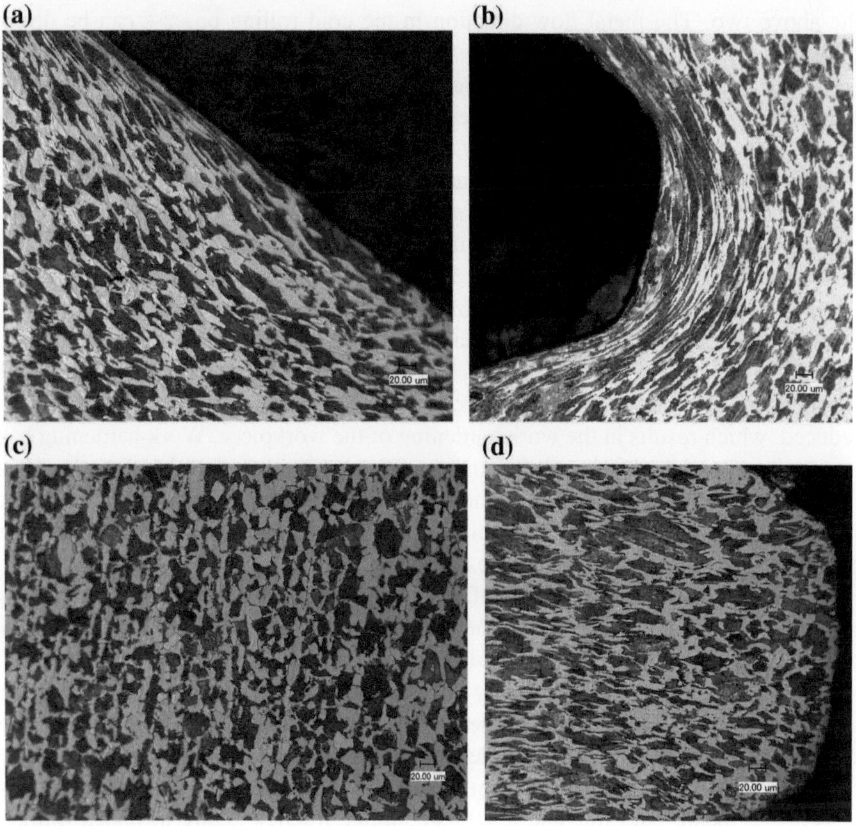

Fig. 9.5 Microstructure of the cold rolling forming thread at different positions. **a** Center of the tooth, **b** addendum, **c** tooth flank, **d** tooth dedendum

wheel. The grain of the billet will then extend along the deforming direction of the metal and distorted. With the increasing in the deforming volume, the grain of the billet has been elongated and presented a kind of fibrous microstructure, and the strip fibrous microstructure is continuously distributed along the tooth profile. During the plastic deformation process, the substructure of the metal grain has also been changed; with the increasing deforming degree, the grain has been stretched, and the sliding width is gradually increasing; and the dislocation density is also improved rapidly. The original grain and the intergranular inclusions have been crushed and refined, and new grains and subgrain microstructure have been formed.

Because the force conditions at the addendum, flank, and dedendum of the component tooth are different in the plastic forming process, the refinement degrees of the grain are also different. The grains at the dedendum are the finest, the grains at the addendum are the coarsest, and the grains at the pitch of the tooth are among the above two. The metal flow direction in the cold rolling process can be determined by the law of the minimum resistant force. Under the action of the rolling force, the minimum metal flow-resistant force in the upper-right zone is the minimum, and the addendum can be formed through free metal flow. At the dedendum of the tooth, the metal flow is not very obvious, which finally results in the bulging on both flanks of the tooth. The metal at the dedendum arc is flowing approximately straight and is compressed to both sides; therefore, the density of the fibrous microstructure is larger than that of the addendum. At the maximum deforming position of the dedendum, the metal is difficult to flow under the perpendicular compressive stress and can only flow downward to form the densest fibrous microstructure. With the further increasing plastic deformation, the resistant force applied on the rolling wheel is also increasing rapidly, the strength and hardness of the metal billet have been greatly improved, and the plasticity and the toughness are reduced, which results in the work hardening of the workpiece. Work hardening has greatly improved the hardness and quality of the cold rolling forming threads. After cold rolling, the metal structure of the thread billet has been refined, and the work hardening degree will be directly influenced by the deformation degree and the density of the fibrous microstructure.

During the cold rolling process of threads, the metal fibrous microstructure has not been cut off. With the development of the plastic forming, the grain has been crushed and refined, and the dislocation density is increased; therefore, the fibrous microstructure of the metal is radially distributed along the surface of the tooth with a streamline style. The closer the dedendum of the tooth, the denser the microstructure is, especially at the dedendum, the metal fiber is bending along the tooth profile densely. Work hardening results in the increasing strength and hardness of the metal. And fine and length streamline microstructure has been formed through metal plastic flow, which also results in the improvement of the cracking toughness and fatigue lifetime of the cold rolling forming threads.

In traditional cutting process of threads, the microstructure of the metal on the surface of the billet has been cut off, while that of the inner part is almost not influenced. From Fig. 9.6, it is shown that the grain is equal-axial distributed from

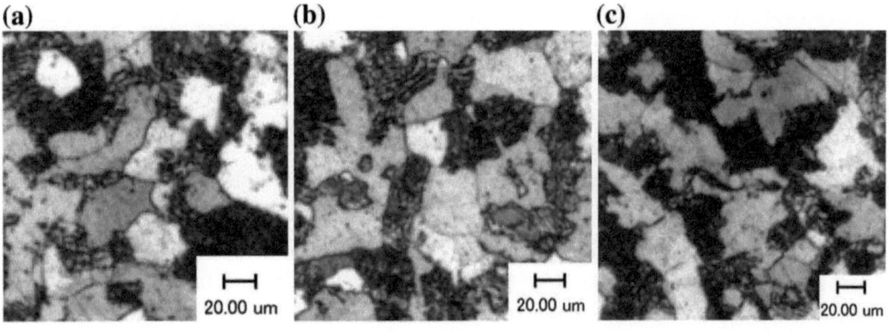

Fig. 9.6 Microstructure at different positions of the tooth on the metal cutting thread. **a** Addendum, **b** flank, **c** dedendum

the addendum to the dedendum of the tooth in metal cutting method, and the grain state has not changed from the surface to the center of the thread. After removing the materials in the groove, the microstructure has not changed; therefore, the hardness and strength of the thread are almost the same after the metal cutting process.

9.3 Hardness Distribution of the Cold Rolling Forming and Cutting Threads

In the cold rolling forming process, the deforming zone and deforming volume on the addendum position are very small and the density of the metal fibrous microstructure is not very large. Therefore, the hardening strength and degree is relatively small. The thickness of the hardening layer is close to 0.6 mm, and the maximum Vickers hardness is about 320 HV. Measuring the hardness along the normal direction of the points on the tooth, the result is shown in Fig. 9.7a. At the position of 0.5 mm, the hardness increases suddenly because the measured point is on the harder pearlite. Compared with the addendum, the fibrous microstructure on the flank of the pitch is rather dense, and the deforming zone and degree is larger. The hardened layer of this position is thicker and reaches 0.8 mm, and the maximum Vickers hardness has exceeded 360 HV. The variation curve of the hardness is shown in Fig. 9.7b. The deforming degree and deforming zone of the dedendum is the maximum, the stretched fibrous microstructure is densest, the hardness is larger than that of the other positions, and depth of the hardening layer is the maximum and exceeds 1.1 mm. The maximum Vickers hardness also exceeds 370 HV. The hardness variation curve at the dedendum is shown in Fig. 9.7c.

During the plastic forming process, the microstructure in the metal has changed and work hardening happened. With the increasing deforming degree, the dislocation density and the deforming resistance increases accordingly, which results in the improvement of the metal strength.

Figure 9.8 shows the hardness distribution along the tooth depth and the measured hardening layer on the normal direction of the outline at the position of the cutting thread addendum, flank pitch, and dedendum. It is shown that the hardness on the cutting tooth profile is obviously lower than that of the cold rolling forming thread, and the hardness distribution of each point on the profile is the same with the hardness of the billet and along the tooth depth. There is no hardening layer on the cutting thread; the hardness at the addendum, flank, and dedendum of the tooth is about 220 HV and without obvious variation along the tooth surface.

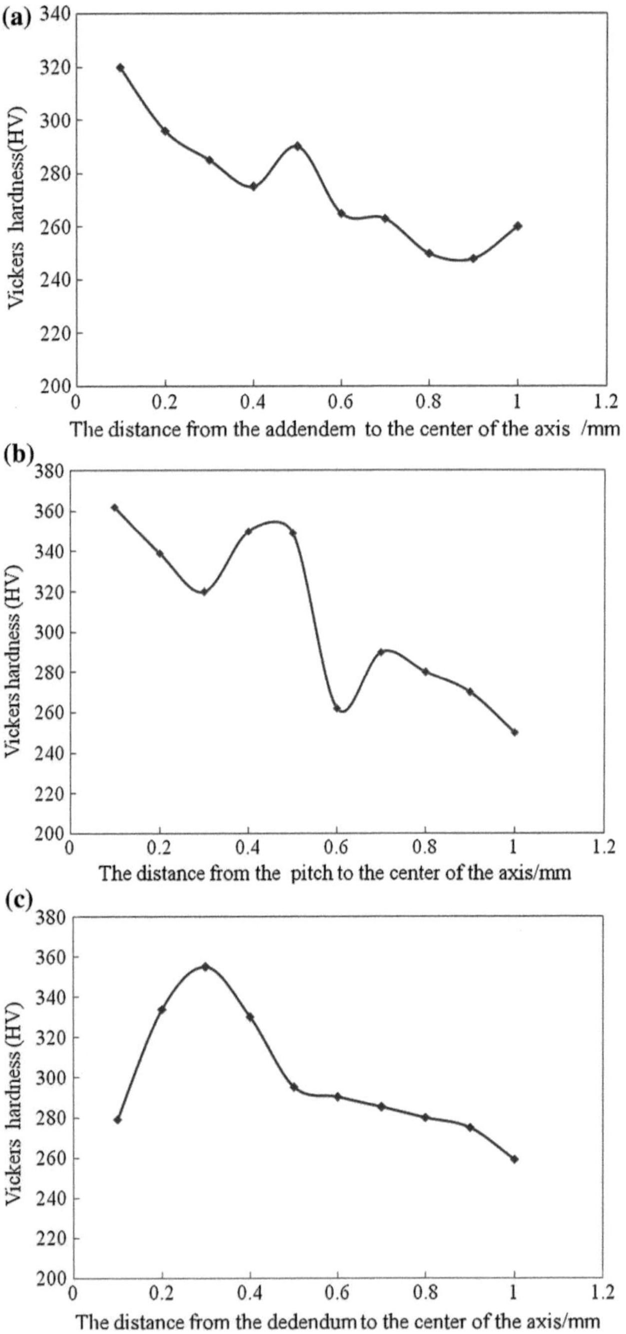

Fig. 9.7 Hardness distributions on the cold rolling forming thread tooth. **a** Hardness distribution of the addendum along the tooth depth, **b** hardness distribution on the pitch along the tooth depth, **c** hardness distribution on the dedendum along the tooth depth

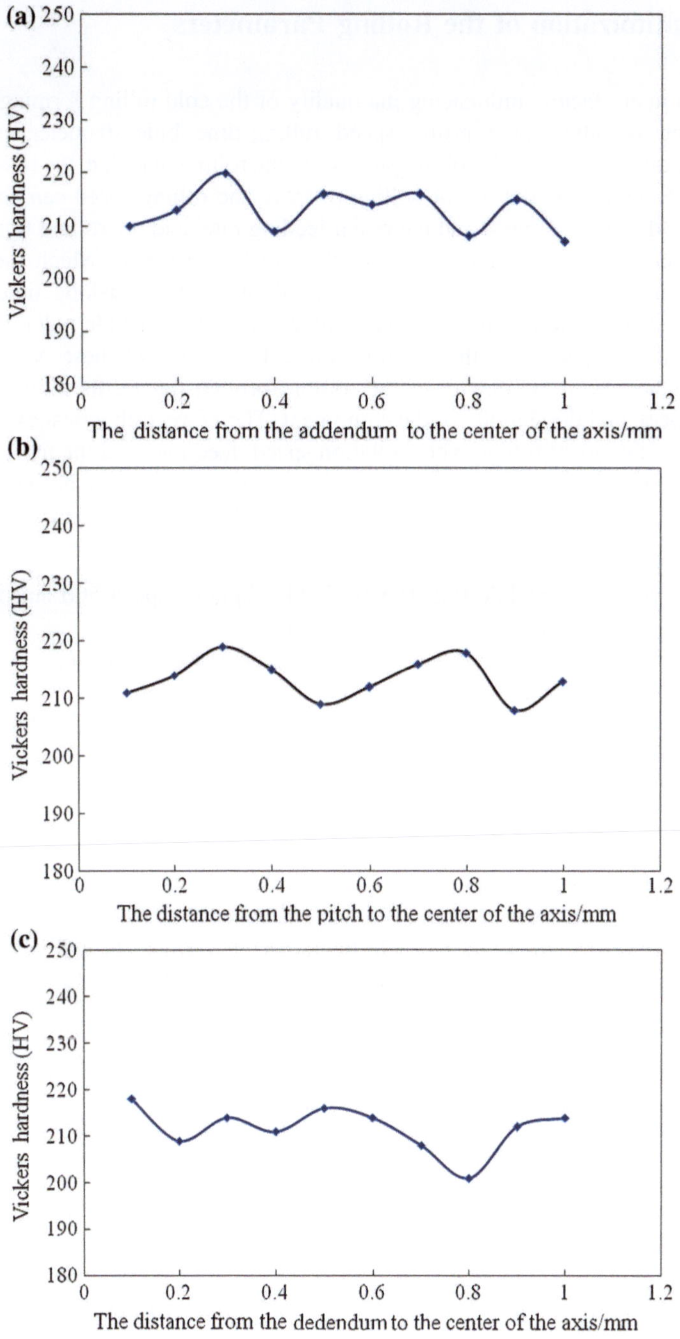

Fig. 9.8 Hardness distributions on the tooth profile of the cutting thread. **a** Hardness distribution along the tooth depth on the addendum, **b** hardness distribution along the tooth depth on the flank pitch, **c** hardness distribution along the tooth depth on the dedendum

9.4 Optimization of the Rolling Parameters

There are many factors influencing the quality of the cold rolling forming components, such as rolling force, rolling speed, rolling time, billet diameter, and lubrication conditions. In the cold rolling process on the rolling machine, the rolling force varies with the conduction of the rolling process; the rolling speed can be decomposed into the spindle speed and the radial feeding rate; and the rolling time can be decomposed into the forming time and the finishing time, in which the rolling forming time is related with the rolling speed, while the finishing time can be controlled freely. The production, controllable parameters include radial feed rate, spindle rotation speed, and the finishing time. The setting of these parameters is directly related with the quality of the rolling forming thread, the lifetime of the rolling wheel, and the stability of the equipment. Therefore, orthogonal experiments on the three factors of rolling wheel rotation speed, feed rate, and the finishing time have been carried out, and the cold rolling forming parameters have been optimized.

The rolling program of the forming rolling machine is shown as follows:

```
NNN;
G1 G22 G90 G94 X51 F300 S500 M03 M11; //spindle speed 500 and feed rate
300, X feeding 51 along the positive direction
G91 X5 F50; //feed rate 50, X feeding 5 along the positive direction
G04 F2; //feeding stopped and stay for 2 s
G91 X-5 F300 M12; //feed rate 300, X feeding 5 along the negative direction
M0; //program finished
GOTOB NNN; //go to next cycle
M2;
```

9.4.1 *Determination of the Orthogonal Factors, Levels, and Index of the Thread Cold Rolling Forming Parameters*

According to the actual condition of the machine, selection of the factors and the experimental levels are as follows:

(1) Spindle speed
The spindle speed in the program is not the rotation speed of the rolling wheel; it is the speed of the stepless speed regulation motor. The effective speed adjusting range of the cold rolling machine is within 0–500 r/min. Therefore, four spindle speeds can be selected: 200, 300, 400, and 500 r/min.

(2) Feed rate
From the aspect of the machine, the selection range of the feed rate is very large, but for cold rolling forming, too large feed rate should not be allowed.

Table 9.3 Table of the factors and levels of the orthogonal experiments

Factors and levels	A: Feed rate (mm/min)	B: Spindle speed (r/min)	C: Finishing time (s)
1	20	200	0
2	30	300	1
3	40	400	2
4	50	500	3

Experiments show that too large feed rate will result in the blocking of the rolling machine and the failure of the rolling process. Referring the results of the theoretical calculation, selected feed rates are, respectively, 20 mm/min, 30 mm/min, 40 mm/min, and 50 mm/min.

(3) Finishing time
Finishing time is the period the moving rolling wheel feeding directionally and keeping rotation in the numerical control program. For 45 steel, the finishing time coefficient should be 1–1.25, the forming time is about 6–10 s, and the theoretical value of the finishing time is about 0–2.5 s. Therefore, the four levels of the finishing time in the orthogonal experiments are 0 s, 1 s, 2 s, and 3 s, respectively.

The factors and levels are shown in Table 9.3.

The external diameter, temperature rising of the thread, and hardness variation are taken as the index of the orthogonal experiment. A 5-factor-4-level orthogonal experiment table L16 (4 × 5) is selected, and three columns are used to arrange the 3-factor-4-level experiments, as shown in Table 9.4.

Table 9.4 3-factor-4-level orthogonal experiment table

Serial number	A: Feed rate (mm/min)	B: Spindle speed (r/min)	C: Finishing time (s)
1	1 (20)	1 (200)	1 (0)
2	1 (20)	2 (300)	2 (1)
3	1 (20)	3 (400)	3 (2)
4	1 (20)	4 (500)	4 (3)
5	2 (30)	1 (200)	2 (1)
6	2 (30)	2 (300)	1 (0)
7	2 (30)	3 (400)	4 (3)
8	2 (30)	4 (500)	3 (2)
9	3 (40)	1 (200)	3 (0)
10	3 (40)	2 (300)	4 (1)
11	3 (40)	3 (400)	1 (0)
12	3 (40)	4 (500)	2 (1)
13	4 (50)	1 (200)	4 (3)
14	4 (50)	2 (300)	3 (2)
15	4 (50)	3 (400)	2 (1)
16	4 (50)	4 (500)	1 (0)

9.4.2 *Hardening Degree Under Different Forming Conditions*

Orthogonal experiment results under different parameters of the cold rolling forming process are shown in Table 9.5.

9.5 Determination of the Optimized Rolling Parameters

From the 16 groups of experiments, it can be seen that the maximum temperature rising is 28 °C, which has no serious influence on the expansion of the billet and the thermal fatigue of the rolling wheel. Therefore, temperature rising is the factor with the minimum influence. The main factor influencing the external diameter is the spindle speed, with the increase in the spindle speed, the external diameter will also increase; therefore, the spindle speed should be as large as possible. The main factor influencing the hardening degree is the feed rate, which has a contrary influencing tendency on the maximum hardness and the thickness of the hardened layer; the other influencing factor is the spindle speed; and the two values increase with the increase in the spindle speed. Furthermore, the hardening degree will influence the mechanical property and the service time of the thread, while the external diameter will influence the shape and dimension precision of the component. Therefore, on conditions, the external diameter is conflict with the hardening degree, and the external diameter should be ensured firstly. The action of the parameters such as the feed rate, and the spindle speed and the finishing time on the indexes should be as follows, successively:

(1) External diameter: the spindle speed → the feed rate → the finishing time;
(2) Temperature rising: the spindle speed → the feed rate → the finishing time; and
(3) Hardening degree: the feed rate → the spindle speed → the finishing time.

From the above analysis, the external diameter is the most important index, the hardening degree, and then the temperature rising. By comprehensive analysis, the optimized parameter combination should be A2B3C3, with a feed rate of 30 mm/min, a spindle speed of 400 r/min, and a finishing time of 2 s; and the demand of precision and efficiency can be satisfied simultaneously.

Cold rolling forming process is then carried out according to the analyzed results, and the workpiece parameters are as follows: The external diameter is 21.96 mm, the temperature rising is 17 °C, the maximum dedendum hardness is 297 HB, the hardened layer thickness of the dedendum is 0.7 mm, the maximum hardness of the flank is 280 HB, the hardened layer thickness of the flank is 0.4 mm, the maximum hardness of the addendum is 265 HV, and the hardened layer thickness of the addendum is 0.5 mm [3].

Table 9.5 Orthogonal experiment results

Serial number	A: Feed rate (mm/min)	B: Spindle speed (r/min)	C: Finishing time (s)	External diameter (mm)	Temperature rising (°C)	Hardening degree (maximum hardness/thickness of the hardened layer) (HB/mm)		
						Dedendum	Flank	Addendum
1	1 (20)	1 (200)	1 (0)	21.58	14.8	280/0.4	267/0.1	252/0.1
2	1 (20)	2 (300)	2 (1)	21.86	19.2	295/0.4	271/0.2	255/0.1
3	1 (20)	3 (400)	3 (2)	21.84	24.5	300/0.5	280/0.2	257/0.2
4	1 (20)	4 (500)	4 (3)	21.82	28	310/0.6	293/0.3	268/0.2
5	2 (30)	1 (200)	2 (1)	21.7	17.2	276/0.5	255/0.1	248/0.2
6	2 (30)	2 (300)	1 (0)	21.66	15	280/0.5	263/0.2	250/0.3
7	2 (30)	3 (400)	4 (3)	21.86	15.2	282/0.5	265/0.3	255/0.5
8	2 (30)	4 (500)	3 (2)	21.92	19	290/0.6	270/0.3	258/0.5
9	3 (40)	1 (200)	3 (0)	21.8	11.9	270/0.5	251/0.2	246/0.3
10	3 (40)	2 (300)	4 (1)	21.86	18.1	275/0.5	255/0.3	249/0.4
11	3 (40)	3 (400)	1 (0)	21.7	18	280/0.6	262/0.3	250/0.5
12	3 (40)	4 (500)	2 (1)	21.9	19.5	282/0.7	265/0.4	255/0.5
13	4 (50)	1 (200)	4 (3)	21.28	10	250/0.6	250/0.3	240/0.3
14	4 (50)	2 (300)	3 (2)	21.9	13.2	268/0.6	261/0.4	247/0.4
15	4 (50)	3 (400)	2 (1)	21.88	18.8	273/0.7	265/0.4	248/0.5
16	4 (50)	4 (500)	1 (0)	21.66	19.2	281/0.7	258/0.4	249/0.5

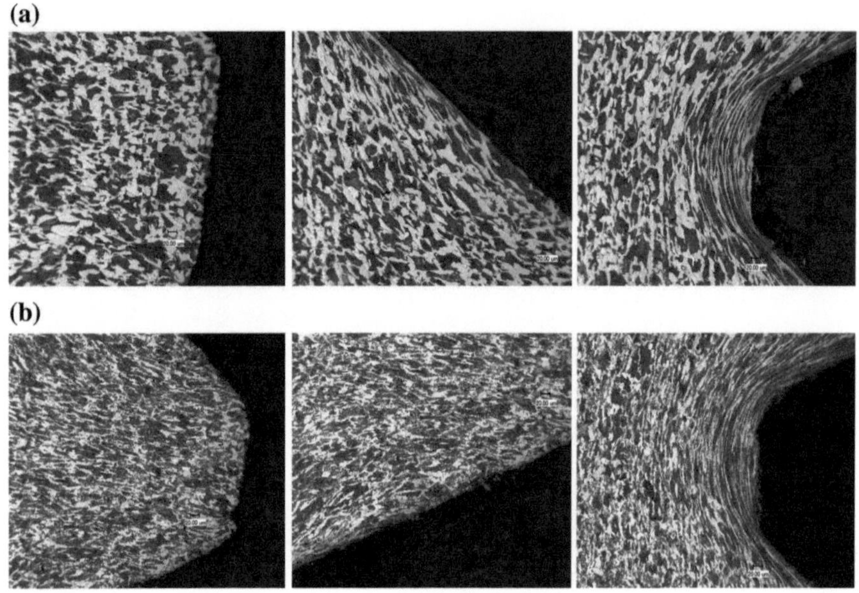

Fig. 9.9 Microstructure of the cold rolling thread addendum, flank, and the dedendum under different process parameters. **a** A feed rate of 20 mm/min, a spindle speed of 200 r/min, and a finishing time of 0 s; **b** a feed rate of 20 mm/min, a spindle speed of 500 r/min, and a finishing time of 0 s

Figure 9.9a and b is, respectively, the microstructure of the thread addendum, flank, and dedendum under the rolling forming parameters of a feed rate of 20 mm/min, a spindle speed of 200 r/min, a finishing time of 0 s, a feed rate of 20 mm/min, a spindle speed of 500 r/min, and a finishing time of 0 s. It can be seen that the metal on the thread tooth surface is distributed in a ribbon pattern and has the same direction with the tooth profile. The metal flow line on the addendum is not very obvious, but the tendency of the metal flowing to the addendum is very obvious. The metal of the dedendum is compacted, the microstructure is seriously distorted, and the metal fibers along the dedendum profile are stretched to form dense metal flow lines. The metal at the flank of the tooth is tending to flow from the root to the top of the tooth, and the metal close to the dedendum is denser. The closer to the top of surface, the coarser the metal is.

At a higher rotation speed, the flow line near the dedendum is smoother, the deforming degree of the flank is larger, and the forming effect of the addendum is better. With the increasing rotation speed, the rolling wheel has a guiding effect to the metal flow companying with the reduction; therefore, the flow lines of the dedendum are smoother and uniform, the forming effect on the addendum is better, and the external diameter is larger. In addition, large rotation speed will promote the flowing of the metal, which will result in a larger deforming degree of the whole tooth, and the work hardening of the tooth will be more obvious.

Under a larger rotation speed, the sliding and the frictional thermal generated by sliding will increase. It is found from experiments that spindle speed has the maximum influence on the temperature rising, and the friction thermal is the main reason for the temperature rising of the thread before and after cold rolling. With the increasing feed rate, the rolling force and the deforming degree of the contacting zone will both increase, which is the main reason of the hardening degree increasing with the increasing feed speed.

References

1. Qi Huiping. Research on the Theory and Process Parameters of Cold Thread Rolling [D]. Taiyuan: Taiyuan University of Science and Technology, 2007.
2. Qi Huiping, Li Yongtang, Fu Jianhua, Liu Zhiqi. Minimum Wall Thickness of Hollow Threaded Parts in Three-die Cold Thread Rolling [J]. Physics Letters B. 2008, 22: 6112–6117.
3. Song Huan. Simulation and Experimental Study on the Cold Rolling Forming of Threads [D]. Taiyuan: Taiyuan University of Science and Technology, 2010.

Chapter 10
Cold Rolling Precision Forming Equipments

Abstract Design of the cold rolling forming machine and dies and measurement of the mechanical parameters of the cold rolling precision forming equipment have been described. Typical cold rolling forming equipment production enterprises, product properties, characteristics, and technique parameters have also been briefly introduced.

10.1 Summary

With the development and improvement of less cutting and precision forming technology, cold rolling forming technology has been increasingly focused by researchers in the field of material processing because of its excellent production forming performance. Well-known shaft cold rolling forming equipment manufactures mainly include Kinefac Corporation in the USA and Bad Duben Corporation in Germany. Kinefac® was founded in 1962 to develop cylindrical die thread rolling machines. In the years since, it has become a world class developer and manufacturer of specialized metal forming, metalworking, and processing equipments. Kinefac Corporation developed the cold and warm thread rolling process and equipments cooperated with many famous aeronautical standard part manufacturers and accumulated rich experiments on thread rolling [1]. Kinefac Corporation drafted out the technical standard of the US aeronautical threads and participated in the establishment of the specifications on the aeronautical and nuclear industrial thread tooth profile. The third-generation aeronautical and nuclear industrial thread rolling and forming equipment Kine-Roller® manufactured by Kinefac Corporation can produce thread, worm, knurl, serration, spline involute, gear, circular groove, and throughfeed components; process such as finishing and rolling cutting can also be conducted with the machine. The numerical control cold rolling forming machine Rollex manufactured by Bad Duben has been focused on the exhibition on the 6th China International Machine Tool Show [2]. The machine is specially designed and developed for cold rolling forming of high-precision involute spline and can be used for surface finishing of the meshed gears after the

J. Song et al., *Cold Rolling Precision Forming of Shaft Parts*,
DOI 10.1007/978-3-662-54048-0_10

milling procedure. The processing procedure can be displayed adopting curves on the screen. The cold rolling forming process parameters can be stored in the computer, and the machine has the functions such as operation suggestion and fault diagnosis. Colonial Corporation in the USA has more than 40 years of experiences on the study of spline cold rolling machine, and the cold rolling machine manufactured by Colonial has reached a level of the world class. In China, Tianjin Automobile Transmission Shaft Factory developed a LDY-700 spline shaft cold rolling forming machine in 1972, but there existed many defects in the aspect of indexing transmission chain and the structures; therefore, the stability of the produced machine was poor and the propagation and application had been limited. At present, many universities are researching on the cold rolling forming machine. Fu zhuang in Yanshan University has developed the structure optimization and design program OPT_XP4 according to the frame structure and the loading characteristics of the cold rolling spline machine, optimized and calculated the frame of the machine, and obtained the common influencing rules of each design variables on the machine frame rigidity. Taking finite element analyzing software MARC, the three-dimensional deformation distribution of the machine has been quantitatively analyzed, and respective suggestions of the improving and optimization scheme have been proposed. Li Jianhua in Huazhong University analyzed the kernel structure of Linux operation system, process scheduling, and module mechanisms firstly, and discussed the structures of the on-time expansion unit RTLinux and RTAI and the realization principle of the multitask scheduling. A general scheme of the cold rolling machine numerical control system was proposed; the realization of the software and hardware structure had been further studied and analyzed. The on-time FIFO and shared memory have been adopted, and various realization mechanisms have been discussed detailed. However, the openness of the numerical control system has not been further improved. Also, taking KDE as the graphics interface developing tools, the calculation speed was slow. Hu Xiangjun in Henan University of Science and Technology studied the dynamics of the spline cold rolling forming and discussed the propagation of the stress waves and the dynamic responding problems, and verification has also been carried out. According to the multibody system dynamic theory, dynamic simulation on the transmission system of the machine in the spline cold rolling forming process has been carried out by adopting ADAMS software. The influence of the stress waves on the spline forming in the forming process has been analyzed, which indicated that the main reason of the front corner collapse on the spline is the plastic deformation caused by the reflection of stress waves on the front end of the spline free surface. The sensitivity of the system is better, but the stability and accuracy are not the best and are expected to improve. Wang Yanyan analyzed the movement rule and motion forms of the cold rolling forming machine, obtained the three primary movements of the cold rolling forming machine. A kind of "zero transmission" design scheme that can substitute the indexing transmission chain by AC servomotor has been proposed, and the model of the transmission system of the zero transmission numerical control cold rolling machines has been set up. The synchronous control method of the parallel zero transmission cold rolling machines and the control strategy applying the module reference fuzzy adaptive control algorithm into the synchronous control system of the cold rolling machine have been proposed. The simulated results basically coincided with the experimental results [3].

Li Yongtang, Zhang Dawei, Liu Zhiqi et al. in Taiyuan University of Science and Technology have studied the cold rolling precision forming of spline systematically [4]. Through theoretical calculation and simulation, the contacting area between the dies and the workpiece, the unit average pressure during the deforming process, and the process parameters such as the rolling force and the moment have been obtained. Modeling and simulation of the hydraulic system have been carried out, and the dynamical characteristics and the selection of the mechanical parameters of the cold rolling precision forming machine have also been studied.

Although the principle of the cold rolling precision forming technology was developed early in the 1970s, and researches in the field have also achieved better progresses, in China, the cold rolling forming production is mainly depended on empirical data and try-and-error method, and components with higher precision requirements must be produced with imported equipments from abroad. Machine manufacture in present is lack of theoretical guide, and forming precision of the components is difficult to improve. As a kind of new component forming technology, there is special requirement for the forming equipment. The study for the spline cold rolling precision forming machine will provide a guarantee for the application of this new technology.

Spline is a kind of straight-tooth part, defects such as incomplete tooth is easy to occur in the forming process, and instant detaching and recontacting between the workpiece and the rolling wheel is existing in the process. According to the theoretical analysis, it is known that because of the instant variation of the deforming zone and the violent fluctuation of the deforming force in the cold rolling forming process, the force condition of the machine transmission system is very serious. In the design and manufacture of the cold rolling forming machine, influences of the forming force on the transmission system clearance and the elastic deformation should be considered, which is very important for the forming precision and stability of the process. Also, the measurement of the actual dynamic load of the rolling process is of great importance for the design of the transmission system and control system of the machine, and the verifying of the theoretical results.

In this chapter, Z28KJ-20 numerical control rolling machine of Qingdao Shengjian Machinery Factory will be taken as an example to study the determination of the parameters such as the rolling wheel speed, the feed rate, and the design of the cold rolling precision forming machine and dies. And cold rolling precision forming of spline has been studied for the measuring and testing of the mechanical parameters. Simultaneously, characteristics, application scope, and technique parameters of the cold rolling precision forming machines manufactured by famous manufactures of the world have also been briefly introduced.

10.2 Design of the Cold Rolling Forming Machine and the Rolling Wheel

10.2.1 Structure and Parameters of the Equipment

The structure of the numerical control spline cold rolling machine developed by Qingdao Shengjian Machinery Factory is shown in Fig. 10.1. It mainly includes the frame, the transmission mechanism, the hydraulic system, the electric control system, the lubrication and cooling system, and the fixture system. Spindle system is the core part of the forming; the two spindles are installed on the left and right sliders of the feeding mechanism, the forming dies (rolling wheels or rollers) are installed on the spindle, and the synchronous rotation of the two rollers is guaranteed by the transmission system.

The frame structure of the rolling machine is a U-shaped style, which not only increases the rigidity of the whole machine, but also solves the problem of the shock absorption. The two spindles are connected with the same servomotor through the constant velocity joint, and the spindles are driven by the high-precision integral transmission system to ensure the synchronization of the two spindles. The two sliders are driven by the two sets of closed-loop proportional servohydraulic systems of the feeding mechanism to realize the accurate location of the two sliders. There are three work stations of the workpiece, and the work stations are controlled by the promoting of the central bracket. After the power source and control power supply of the machine have been connected, self-check of the system will be carried

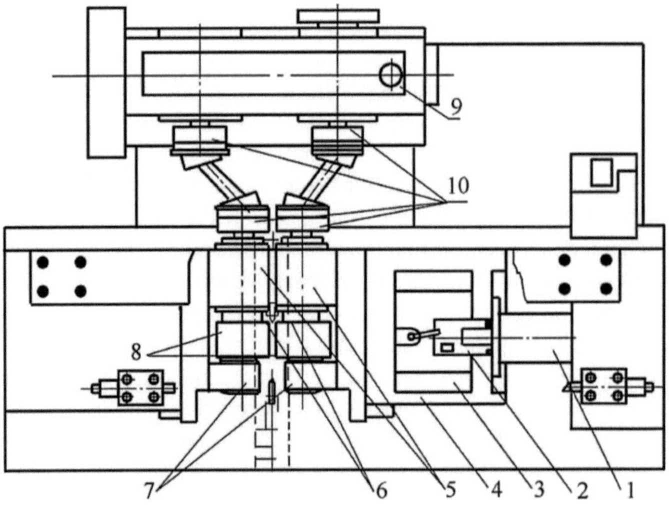

Fig. 10.1 Structure of the numerical control spline rolling machine. 1—hydraulic cylinder; 2—proportional servovalve 3—guide; 4—slipway; 5—rear bracket of the spindle; 6—spindle; 7—front bracket of the spindle; 8—rolling wheel; 9—transmission box; 10—constant velocity joint

out and the hydraulic system will start. Then, one of the sliders will return to the reference point, the bracket will return back to the right position, and the rolling of the workpiece can be started after setting of the system parameters.

The machine parameters of the numerical control rolling machine are as follows:

Maximum rolling force: 200 kN;
Maximum rolling diameters of the workpiece: 80 mm;
Rotation speed of the spindle (frequency control): 0–80 r/min;
Feeding stroke (numerical control): 100 mm;
Feed rate (numerical control): 0–10 mm/s;
Power of the principle motor (variable-frequency motor): 4 kW; and
Mass: 3300 kg.

10.2.2 Design of the Synchronous Spindle Damping Absorption

The main transmission system of the cold rolling machine is shown in Fig. 10.2. The two spindles are driven by the same motor, and the requirement of the spindle rotation synchronization is higher. In the practical cold rolling forming of spline, unstable always appears in the process, noise of the machine is larger, and serious

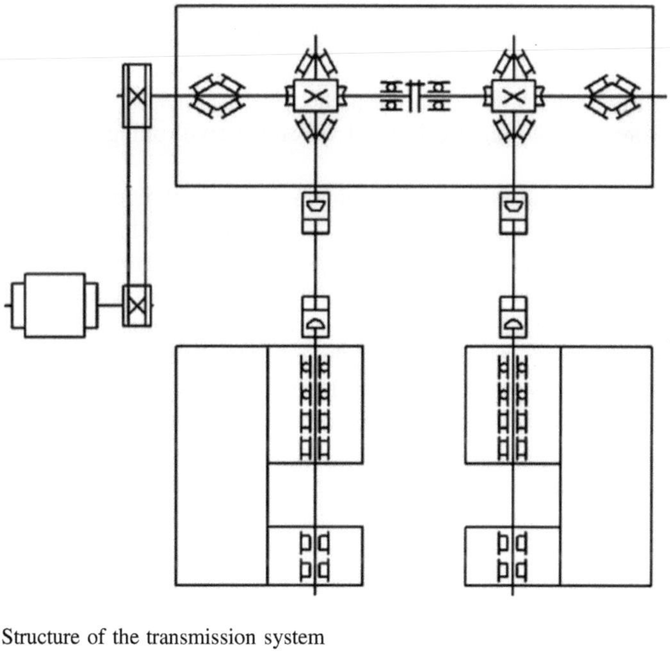

Fig. 10.2 Structure of the transmission system

defects such as plastic instability and tooth disorder will occur in the rolling process.

From the forming process of the spline teeth described in Chap. 2, it is obvious that in the cold rolling precision forming of spline, the rolling wheel and the workpiece are engaged without backlash. During the rolling process, the contacting point is changing instantly; contacting and disengagement of the multipoints may appear at various times, and variation of the forming torque applied on the spindle is violent. By calculating of the contacting zone a single tooth on the workpiece rotating through, it can be obtained that during the rolling of a tooth, the torque will experience twice severe variation from zero to the maximum value then to zero. Because of the elastic deformation of the transmission system and the clearance between the transmission components, in the instant variation of the rolling moment, larger synchronous errors of the transmission system will occur, which will result in the incompatible deformation and defection of the workpiece, the noise of the machine will increase, and the precision of the component will be decreased.

In order to reduce the impact of torque fluctuation on the machine in the forming process, a new type of synchronous spindle damping absorption equipment has been designed. A constant frictional torque is applied previously on the spindle, and while rotating, the force must be overcome by the spindle. Because the direction of the frictional torque is always converse with the rotation direction, in the cold rolling precision forming process, the rotation direction of the spindle is invariable, and because of the existing of the frictional torque, the influence of the component clearance in the transmission system on the synchronous precision can be eliminated. When fluctuations of the torque resulted by the resistant force of the deformation take place, or the value or direction of the torque has been changed, the vibration of the machine spindle will decrease under the damping action of the frictional torque, which will keep the higher repeat circumferential position accuracy; therefore, the forming precision of the spline cold rolling forming process is ensured.

The structure of the synchronous damping absorption equipment is shown in Fig. 10.3. A frictional ring is installed in the shell, and round pin 10 is used for the location of the frictional ring in shell 6. The spindle is connected with the spindle sleeve 2 by a key 7, and the sleeve 2 is running through the central hole of shell 6 and the frictional ring 3; the external diameter of the spindle sleeve keeps with a uniform clearance with the shell 6 and the inner diameter of the frictional ring 3. The frictional ring 3 is pressed on the right end of the spindle sleeve 2 with piston ring 8 by the hydraulic oil entered from hole 4 to act as a damping. The magnitude of the frictional torque can be adjusted through the adjusting of the hydraulic system. Under the action of the hydraulic force, the frictional ring is always pressed on sleeve 2, and the wear of the end surface can be automatically compensated. When the spindle stops and adjustment of the workpiece teeth positions is needed, the valve 12 can be loosed for air exhaust, the frictional ring 3 and the spindle sleeve 2 are disconnected, and then, the spindle can rotate freely and the adjustment can be carried out.

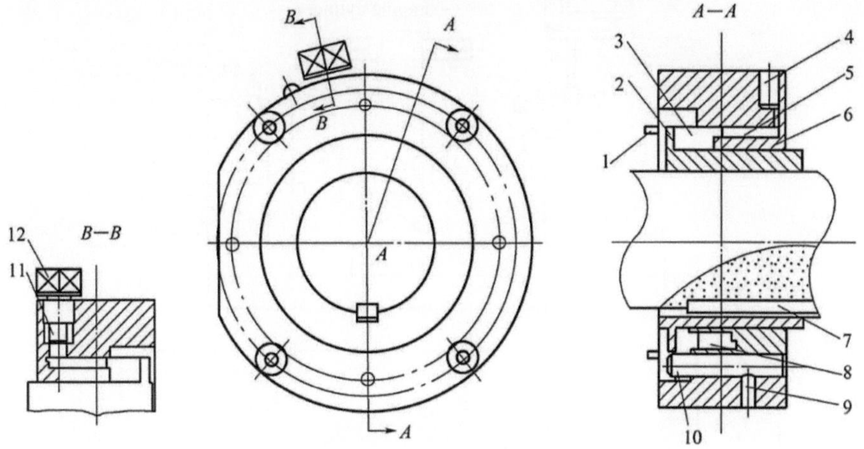

Fig. 10.3 Synchronous damping absorption structure of the spindle. 1—screw; 2—spindle sleeve; 3—frictional circle, 4—entrance of the hydraulic oil; 5—oil cavity; 6—shell; 7—key; 8—piston ring; 9—holding screw; 10—pin; 11—seal gasket; 12—exhaust valve

After the application of the synchronous spindle damping absorption equipment, the force state of the main transmission system in the cold rolling forming machine has been improved and the measured variation of the dynamic rolling torque is stable. The noise in the forming process has been reduced, and the forming precision of the component has also been improved.

10.2.3 Modeling and Simulation of the Hydraulic System

10.2.3.1 Structural Principle of the Hydraulic System

In the cold rolling forming of the spline, the spindle of the machine should be positioned accurately to gain a good dynamical response; in conditions, the forming load has been changed. High-response electrohydraulic control system has been adopted in the feed rate system of the spindle. The feeding motion of the spindle is realized under the driven power slider, which is driven by the hydraulic cylinder. The displacement of the slider is measured in real time by the grating ruler. Through the treatment of the numerical control system and the controller of the proportional servovalve, a closed-loop valve-control-cylinder control system has been set up. The hydraulic system can also provide constant pressure for the damping absorption system simultaneously. The principle of the designed hydraulic system is shown in Fig. 10.4 [3].

Fluid supply of the hydraulic power source by the variable displacement pump has been adopted. The output displacement of the pump can be automatically adjusted according to the required flow volume of the system. Thus, the energy and

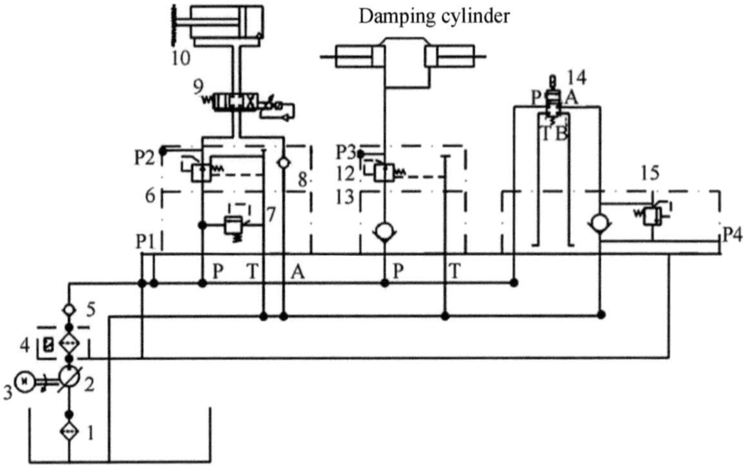

Fig. 10.4 Principle schematic diagram of the hydraulic system. 1—oil filter; 2—piston pump; 3—motor; 4—pressure piping oil filter; 5—straight check valve; 6—pressure release valve; 7—relief valve; 8—straight check valve; 9—proportional servovalve; 10—hydraulic cylinder; 11—multiple pressure gauge switch; 12—superposition pressure reducing valve; 13—superposition check valve; 14—electromagnetic directional valve; 15—back pressure valve

heat generation of the system can be reduced. The maximum work pressure of the system is set up by relief valve 15.

In the cold rolling forming of spline, the variation of the forming force is rather large, which will result in the violent fluctuation of the pressure. Because a constant working pressure of the damping circuit is required, in order to prevent the interference between the slider driving circuit and the damping circuit, a pressure release valve has been installed in each oil feed loop. The working pressure of the damping circuit is set up by the pressure release valve 12, and the working pressure of the servovalve entrance in the feed and driving circuit is set up by the pressure release valve 6.

The valve-control-cylinder position servocontrol system of the slider driving part consists of the servovalve 9 and the hydraulic cylinder 10. The position signal is measured by the grating ruler and then fed back to the electric control system, and the high-precision control of the spindle position is realized under the control of the numerical control system.

10.2.3.2 Setting up of Simulation Model on the Hydraulic System

In the cold rolling forming process of spline, the two rolling wheels are required to feed at an invariable speed, or the central distance should be a constant. Because the changing range of the metal deformation resistant is larger in the forming process, the motion speed and position precision of the hydraulic system driving the slider of the rolling machine are difficult to ensure in the working process, and the requirement for the dynamic characteristic of the hydraulic system is higher.

In this book, the hydraulic system of the numerical control rolling machine Z28KJ-20 manufactured by Qingdao Shengjian Machinery Factory of China has been taken as the research object; the simulation model of the hydraulic system in the spline cold rolling machine has been set up with the simulation software AMEsim. The main working parameters of the system are identified by experiments, and dynamic simulation of the hydraulic system working process is carried out to discuss the influence of outside interfere on the response of the system and predict the dynamic variation rules of the system.

(1) Selection and establishment of the components
The hydraulic system is driven by the slider of the spline cold rolling machine, the piston of the hydraulic cylinder is fixed, and the body of the cylinder is moveable. Proper components have been selected from the hydraulic component design library (HCD) of AMEsim to establish the hydraulic cylinder as shown in Fig. 10.5.
(2) In the environment of AMEsim, the simulation model of the hydraulic system shown in Fig. 10.6 is set up with the Sketch pattern by calling the hydraulic library, the mechanical library, and signal library provided by the system.
(3) The model selection of the main components and the setting up of the parameters are as follows:
Under the pattern of submodel, appropriate model of each graph in the simulation model has been selected, and the parameters are set up under the pattern of parameter:

    (1) Hydraulic cylinder: The inner diameter of the cylinder is 160 mm, and the stroke is 100 mm. Because it is a single-rod hydraulic cylinder, the piston diameter of the left chamber is set up to 0 mm and that of the right chamber is 80 mm.
    (2) The type of the proportional servovalve 9: atos 0811 404 043.

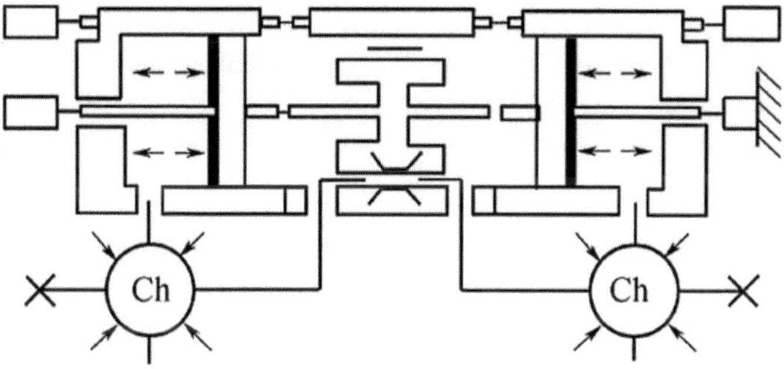

Fig. 10.5 Hydraulic cylinder

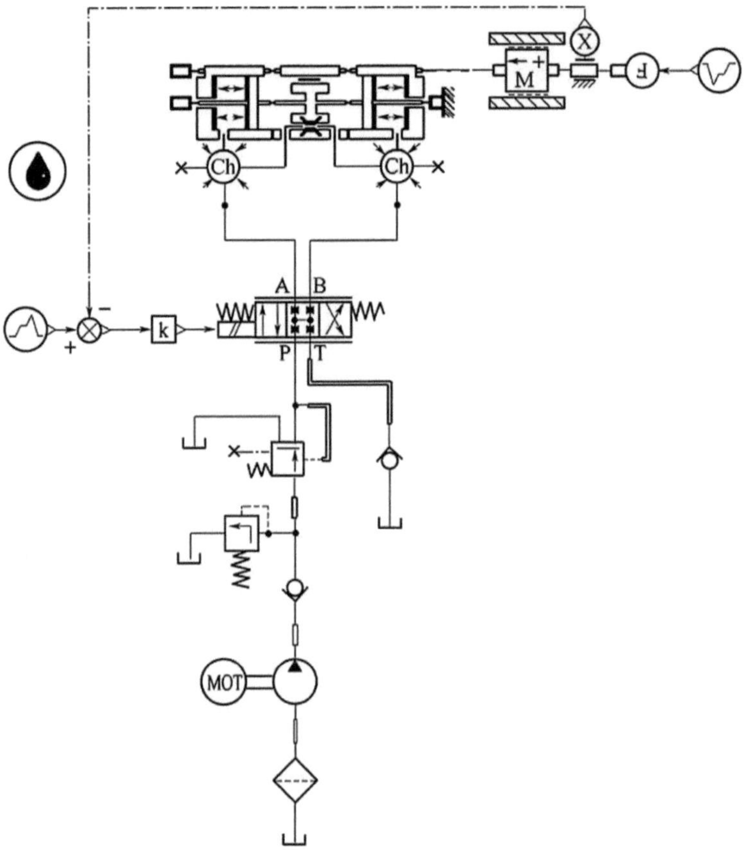

Fig. 10.6 Simulation model of the hydraulic system of the spline cold rolling forming machine

(3) Hydraulic pump 2 is a constant pressure fixed displacement pump, the mechanical efficiency is considered to be a constant, and the displacement of the hydraulic pump is 15 L/min.
(4) Type of the relief valve 7: MRF-03P-K2-30, which is set to 10 MPa according to the maximum pressure of the system.
(5) Type of the pressure release valve 6: MPR-03P-K2-30.
(6) Type of straight check valve 5 and 8: CL-NPT0404-50-10.
(7) Motor 3 is with a constant speed of 1500 r/min and a power of 4 kW.

10.2.3.3 Dynamic Simulation of the Hydraulic System

The density of the oil used in the simulation is 850 kg/m^3, the dynamic viscosity is 5.1×10^2 Pa·s, and the temperature is 40 °C.

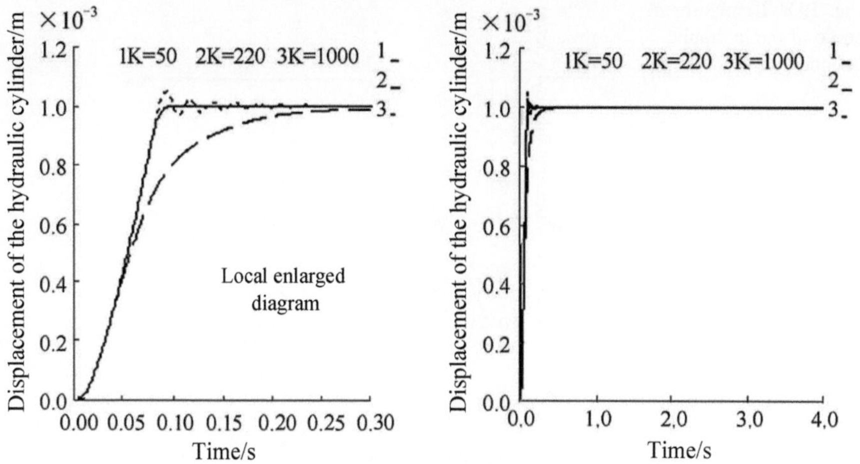

Fig. 10.7 Displacement curve of the hydraulic cylinder

(1) Dynamic characteristic simulation without load

The response of the system to an input step signal is studied. According to the classical control theory, it is known that the position error of the system is in an inverse proportion to the open circle gain; under conditions, other parameters keep invariable. The proportional gain of the system is an important parameter. By setting up a reasonable proportional gain, the response speed of the system can be improved, and the deviation between the displacement feedback and the expected value can be decreased. The batch processing function of the software AMEsim is utilized to analyze the influence of the proportional gain on the deviation of the position and the selection of the appropriate gain value. The proportional gain of the system is selected as 50, 220, and 1000, the simulation time is 4 s, and the acquisition period is 0.001 s. The displacement variation curve of the hydraulic cylinder obtained by the simulation is shown in Fig. 10.7.

Figure 10.7 shows that the response speed increases with the increasing of K. While K > 220, overshoot of the system will occur, which is not expected in the rolling forming of spline; otherwise, overcut of the spline will happen. While K = 1000, overshoot and oscillation of the system will both occur. Therefore, at K = 220, there is no overshoot, the response of the system is the fastest, and the adjust time $t_s = 0.1$ s.

(2) Simulation of the hydraulic system in the cold rolling forming of spline

In the cold rolling forming process of spline, the motion of the slider can be divided into three stages: fast approaching to the workpiece; keeping the central distance; and returning rapidly. Through identifying by experiments, 0–1 s is the fast approaching stage, and the stroke is about 5.3 mm. During 1–3 s, the hydraulic

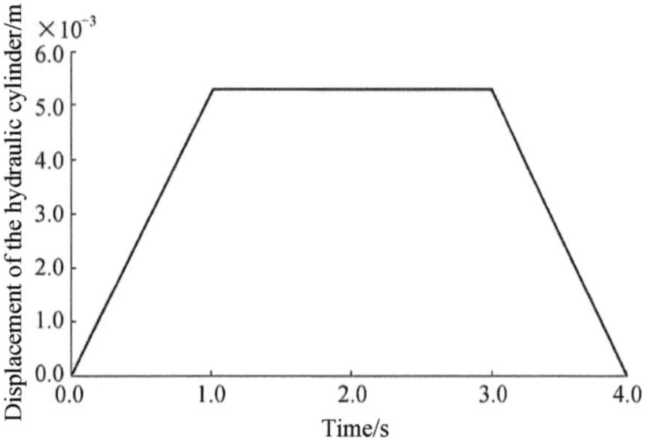

Fig. 10.8 Displacement curve of the hydraulic cylinder

cylinder keeps at the initial position and the finish rolling is conducted. After 3 s, the hydraulic cylinder will return to the original position.

According the above stages of the spline cold rolling, add a step load of 20 kN at 0.6 s, and the dynamic response of the hydraulic system is observed. The simulation time is 4 s, and the acquisition period is 0.001 s. The variation curve of the displacement, the speed, and the pressure of the hydraulic cylinder obtained by the simulation is shown in Figs. 10.8, 10.9, and 10.10.

(1) From Fig. 10.8, it is shown that the displacement transition of the system is smooth and without shocking. At 0–1 s, the hydraulic cylinder feeds quickly and reaches the specified location at 1.001 s, which has delayed 0.001 s than the set value. Because the precision forming time is 2 s, the delayed time 0.001 s will not affect the processing accuracy of the component and the demand of the system can be satisfied.
(2) From Fig. 10.9, it is shown that after the starting of the system, the open speed of the valve is very fast and the hydraulic cylinder reaches the specified speed only after 0.05 s with a little overshoot, and the speed will be stabled in 0.1 s. In this stage, the roller approaches the workpiece rapidly, the demand for the stability of the speed is not very higher, overshoot is allowable, and the speed

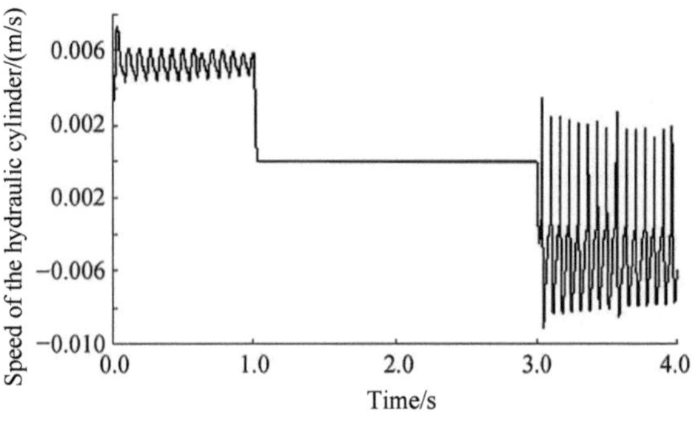

Fig. 10.9 Speed curve of the hydraulic cylinder

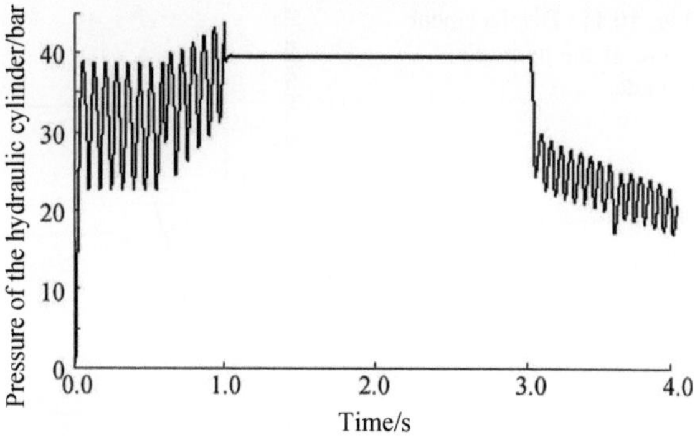

Fig. 10.10 Pressure curve of the hydraulic cylinder

of the system is higher enough. After the contacting between the roller and the workpiece, force is loaded, the opening of the valve is very small, and the motion speed of the hydraulic cylinder is lower and stable, which can satisfy the requirement of the system.

(3) The pressure is depended on the load. From Fig. 10.10, it is shown that at 0–0.6 s, the working pressure of the hydraulic cylinder is rather lower and pressure fluctuation exists. When loading at 0.6 s, the pressure increases abruptly. During 1–3 s of the spline rolling, the pressure is not changed to satisfy the requirement of the system. At the returning of the hydraulic cylinder at 3 s, the pressure of the system will decrease suddenly once more.

(3) Influence of the rolling force

In the finish rolling process of spline, because of the periodical variation in the contacting area, every time a tooth has been rolled, there is an equivalent cycle of pulse loading. It is obtained from experiments that the amplitude of the resistant deforming force is within 0–20 kN. With a rotation speed 15 r/min of the machine spindle, the corresponding time of a tooth rolling is 20 ms. Therefore, the load can be simplified as a square signal with a period of 20 ms. At 0–0.3 s, the system is adjusted stably; and at 0.3–2 s, rolling forming of the spline will start. The simulation time is 2 s, and the acquisition period is 0.001 s; the variation curves of the main performance parameters after the simulation are shown in Figs. 10.11 and 10.12.

Figure 10.12 shows that, at 0.355 s, the position of the hydraulic cylinder is invariable in the rolling of spline. With abrupt increase in the load, there is a gentle deviation of the displacement. When the hydraulic cylinder has reached the specified location and measured by the grating ruler, a signal will be given by the numerical control system to close the proportional valve. There exists a period from the proportional valve receiving the signal to the closing of the valve core, and the length of the period is depended on the transfer frequency of the valve core. During this period, there still some oil enters the hydraulic cylinder through the proportional valve and the hydraulic cylinder is continuously moving.

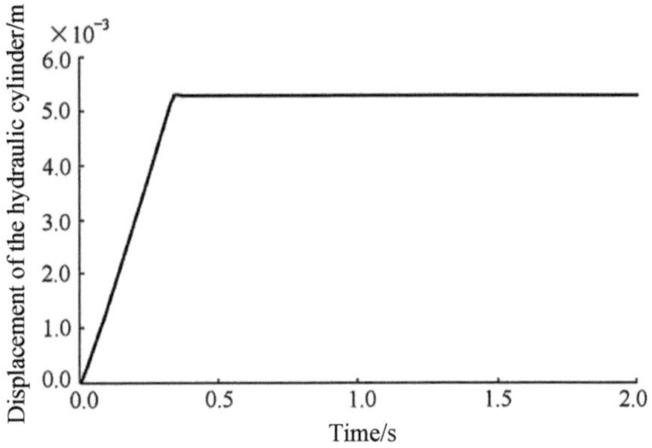

Fig. 10.11 Displacement curve of the hydraulic cylinder

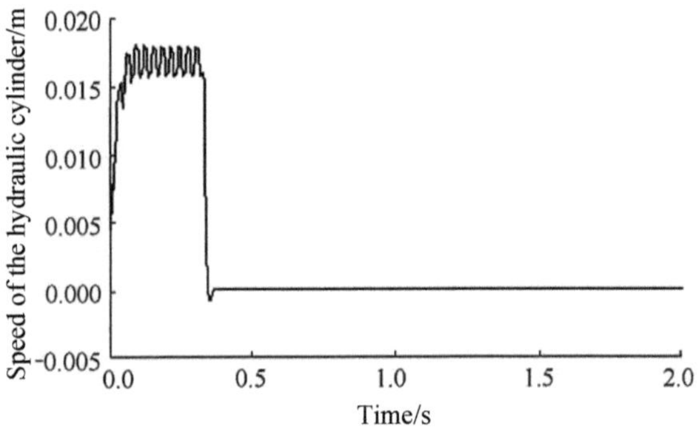

Fig. 10.12 Speed curve of the hydraulic cylinder

In addition, after complete closing of the proportional valve, because of the inertia of the hydraulic cylinder and the compressibility of the liquid, the hydraulic cylinder will still move for a small displacement before stop. Therefore, the actual displacement of the hydraulic cylinder is a little larger than the specified one. Fig. 10.9 shows that in the stage of finish rolling, the speed of the hydraulic cylinder increases abruptly with the sudden increase in the load and it keeps fluctuating until to 0.05 s, the system reaches a stable state. In the period of the rolling process at 0.355 s, the speed of the hydraulic cylinder is zero, and the rolling condition can be satisfied. With the variation of the load, the speed of the hydraulic system also fluctuates in a certain degree for a period of time and then tends to be stable. The main reason is because there is a time delay and adjustment period on the conversion of the signal. During the adjusting period, the amplitude of the fluctuation decreases gradually and tends to be stable.

To sum up, the dynamic performance of the hydraulic system is better, with a powerful interfere resistant capability and higher control accuracy of the control system and a good stability, which can satisfy the demands of the spline forming. In the simulation of the hydraulic system of spline cold rolling machine based on

AMEsim, visualized modeling of the hydraulic system and open analyzing environment have been adopted, and the static simulation of the hydraulic system can thus be realized with a better result.

10.2.4 Design of the Rolling Wheel

The die of the cold rolling forming is the rolling wheel, and design of the parameter, material, and manufacturing precision of the rolling wheel have great influences on the forming of the spline shaft. In the cold rolling precision forming of spline shaft, the rolling wheel is bearing a great pressure and bending moment and the tooth surface is under the action of the frictional force caused by sliding; therefore, the material of the rolling wheel should have an appropriate hardness, higher toughness, and wear resistance. The material of the rolling wheel is always Cr12MnV or W6Mo5Cr4V2. To ensure the stability of the dimension and increase the wear resistance, deep cryogenic treatment of the rolling wheel should be carried out after quenching. Also, in order to increase the frictional coefficient in the tooth indexing stage, proper rotation and biting can be guaranteed, and a layer of WC or TiC coating with a thickness of $\Delta = 0.035$–0.050 mm should be plated on the surface of the rolling wheel tooth. Here, the design regular of the rolling tooth parameters is mainly discussed.

According to the principle of spline cold rolling precision forming, in the cold rolling forming process, the axis of the workpiece and the two rollers is in a same plane. The three axes are parallel; the relative motion between the workpiece and the two rollers is the same with the meshing motion of the plane gears. In design, the main parameters of the rollers are determined according to the corresponding principles of the involute gear engagement. The main parameters of the rolling wheel should be designed according to the parameters of the forming spline shaft tooth, such as the modulus m, the tooth number Z, the pressure angle α, the pitch diameter d_b, the addendum diameter d_a, the root diameter d_f, and the tooth thickness of the addendum circle S_a.

(1) Modulus and tooth number

In the cold rolling precision forming of spline shaft, the motion between the roller and the workpiece is similar to gear transmission, and the spline tooth is formed with generating method. According to the gear meshing conditions, the modulus of the workpiece should be equal with the roller, and the pressure angle of the workpiece and the roller should be equal, that is:

$$m_{\text{workpiece}} = m_{\text{roller}},\ \alpha_{\text{workpiece}} = \alpha_{\text{roller}} \tag{10.1}$$

The number of the roller tooth is restricted by many conditions, in which the rotation and biting condition should be satisfied firstly. In addition, the maximum addendum circle diameter is limited by the structure of the machine, and the roller tooth number Z_2 satisfies:

$$Z_2 \leqslant [D/m - 1] \tag{10.2}$$

where D is the maximum addendum circle diameter of the roller.

According to the rotation biting condition formula (2.7) and (10.2), the category of the roller tooth number Z_2 can be determined. The more the roller tooth number, the smaller the wear of each tooth is in the forming process; therefore, the roller with more tooth number has a longer lifetime than the roller with fewer teeth. In the design of the rolling wheel, the tooth number is always the integrate time of 10. Under conditions allowed by the machine structure, the tooth number of the roller should be as close as possible to [D/m − 1]. The meshing overlapping coefficient in the forming process can be improved under the maximum value of the rolling wheel, which is favorable for decreasing the accumulate error of the tooth pitch, as analyzed in Chap. 6.

(2) Modification of the roller tooth profile

According to the analysis of the tooth forming mechanism in Chap. 2, in the cold rolling precision forming process, the metal flow direction on the active tooth surface of the workpiece is different with that of the driven side, the value and direction applied by the roller tooth are different, and it will results in the incline of the tooth profile and increase in the pitch circle pressure angle. Through the design of the roller tooth profile parameters, and modification of the roller tooth profile, the precision of the forming component tooth profile can be improved. According to the data provided by Qingdao Shengjian Machinery Factory, it is obtained from the simulation that by decreasing the pressure angle of the roller pitch circle appropriately, the tooth profile of the workpiece can be correspondingly modified. For the three kinds of the spline pressure angle in GB/T3478.1-1995, the modification value α of the roller pitch circle pressure angle is as follows:

$$\begin{cases} \alpha = 30^\circ, & \alpha = 30' \sim 45' \\ \alpha = 37.5^\circ, & \alpha = 45' \sim 1^\circ \\ \alpha = 45^\circ, & \alpha = 1^\circ \sim 1^\circ 15' \end{cases} \tag{10.3}$$

where α is the pressure angle of the formed spline pitch circle.

After modification of the pressure angle on the pitch circle, the tooth thickness of the roller pitch circle can be determined by the following formula:

$$s_2 = d_2(s/d + \operatorname{inv}\alpha - \operatorname{inv}\alpha') \tag{10.4}$$

where

s is the tooth thickness of spline pitch circle;
d is the pitch circle diameter of the spline; and
α' is the modified pressure angle of the roller.

The cold rolling forming spline root must be round to improve the strength of the spline dedendum. The dedendum of the spline is extruded by the addendum of the rolling wheel; therefore, in order to extend the lifetime of the roller, the addendum of the roller should be designed as an integral arc or with a fillet, but the end point of the roller tooth involute must be guaranteed.

The pressure angle of the rolling wheel addendum arc is as follows:

$$\alpha_{a2} = \arccos\left(mZ\cos\alpha \,/\, d_a\right) \tag{10.5}$$

The tooth thickness of the rolling wheel addendum circle is as follows:

$$s_{a2} = d_2\left(s_2 \,/\, d_2 + \operatorname{inv}\alpha' - \operatorname{inv}\alpha_{a2}\right) \tag{10.6}$$

where

s_2 is the tooth thickness of the roller pitch arc;
d_2 is the diameter of the roller pitch circle; and
α' is the pressure angle of the roller pitch circle.

10.3 Measuring System of the Mechanical Parameters

In order to study the force conditions of the cold rolling machine and verify the theoretical calculation results of the spline cold rolling precision forming force, a measurement system of the cold rolling mechanical parameters has been set up. The resistant deformation force in the rolling process is transferred to the machine through the rolling wheel, and the information of the deforming force can be obtained from the force of the machine spindle system. The torque of the spindle and the radial feeding force of the roller have been collected for the analysis of the deforming force in the cold rolling process.

10.3.1 Measurement Method of the Spindle Torque

The spindle of the cold rolling machine is rotating under the driven by two cardan shafts. The cardan shaft is considered to transfer pure torque only, and the torque of the spindle is equal to that of the cardan shafts. Limited by the space of the cold rolling forming machine, torque sensor cannot be installed in the transmission system; therefore, a machine body sensor is designed to measure the torque of the spindle in the forming process.

Under the premise ensuring the power transfer, the cardan shaft has been modified into an elastic shaft, and specialized strain gauges has been pasted on the elastic shaft to constitute a strain bridge. By providing working power for the strain bridge, electric signal of the shaft torque can be measured, which composes of the basic torque sensor; and the torque information of the spindle can be obtained with electric measuring method. In the working state of the spline cold rolling machine, the cardan shaft is in a rotation condition and feeding transversely; if bridge voltage is input into the strain bridge rotating with the cardan shaft, output strain signal of the elastic shaft can be measured, and conductive slip ring is often used to transfer these signals between the rotation and the static part. Friction will results in the wear and heat generation of the slip ring, and the rotation speed of the shaft and the lifetime of the slip ring will be limited. In addition, unreliable contacting will lead to the fluctuating of the signal and the increasing of the measuring error. The output strain signals will even be failure to gain because of the large noise.

Referring to the latest research achievements [5, 6], the non-contact transfer of the strain signals from the rotation component to the static part has been realized, and the measuring principle is shown in Fig. 10.13. The secondary coil of the energy circle rotary transformer, the printed circuit board of the shaft, and the primary coil of the signal circle are fixed on the elastic shaft, where the rotation of the shaft, the exciting circuit, the primary coil of the energy circle rotary transformer, and the secondary coil of the signal circle fixed on the outer shell of the sensor are stationary. The strain signal caused by the transfer of the torque on the cardan shaft is transformed to the static secondary coil from the rotation shaft; then through signal processing, circuit filtering, and reshaping the sensor shell, it has transferred to output frequency signal in proportion to the torque applied to the elastic shaft. In this kind of signal transfer method, there is no friction between the rotational component and the static parts, the defect of the conductive slip ring is overcome, and the signal is stable and reliable.

Figure 10.14 is the modified numerical control cold rolling machine Z28K-20, and the cardan shaft has been modified into a specialized torque sensor and then installed into the machine to form part of the transmission system. After calibrating the signal output by the torque sensor on the frame of the machine, the dynamic torque in the cold rolling forming process can be measured with the data acquisition system.

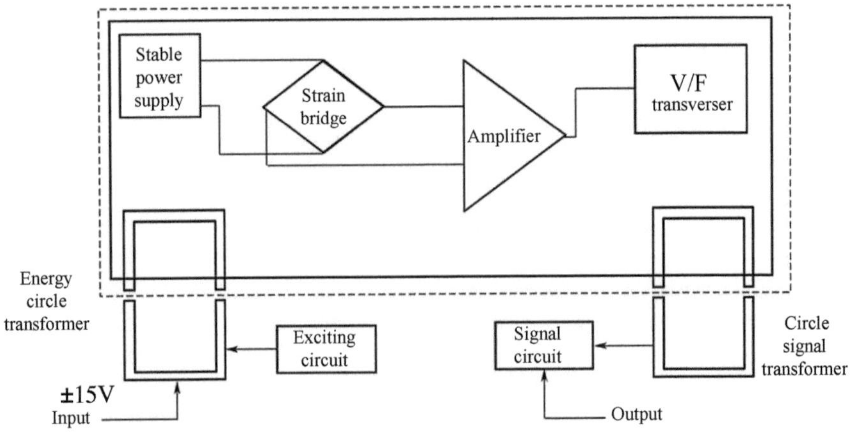

Fig. 10.13 Measurement principle of the torque

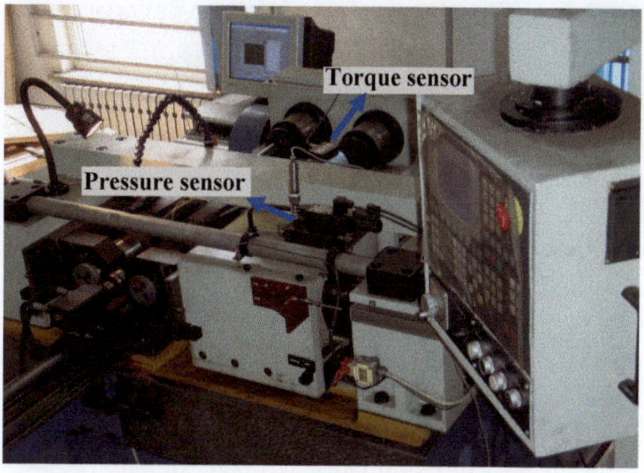

Fig. 10.14 Numerical control cold rolling machine installed with mechanical measuring system

10.3.2 Measuring of the Radial Feeding Force of the Slider

In the spline cold rolling machine, the rolling wheel is pressed into the billet through the radial feeding of the slider, and the rolling force in the forming process is provided by the radial feeding of the slider. The motion of the slider is driven by the hydraulic cylinder; therefore, by measuring the pressure at the entrance and outlet of the hydraulic cylinder, the driving force of the hydraulic cylinder can be calculated and the indirect measuring of the rolling force can thus be realized. Fix up a pressure sensor at the entrance and outlet of the hydraulic cylinder, respectively (shown in Fig. 10.14), the pressure of the entrance and the outlet in the cold rolling process can be measured, and through Eq. (10.7), the radial pressure of the slider can be obtained, which is also the rolling force:

$$F = p_1A_1 - p_2A_2 \tag{10.7}$$

where

F is the rolling force;
p_1 is the pressure at the entrance of the hydraulic cylinder;
p_2 is the pressure at the outlet of the hydraulic cylinder;
A_1 is the effective area of the oil inlet side of the hydraulic cylinder; and
A_2 is the effective area of the oil return side of the hydraulic cylinder.

In Eq. (10.7), the frictional force on the slider guider and the hydraulic piston has been neglected; compared with the rolling force in the forming process, the proportion of the frictional force is relatively small; therefore, the measuring precision cannot be affected.

10.3.3 The Data Acquisition System

The hardware of the data acquisition system consists of the sensor, the signal conditioning circuit, the data acquisition card, and the computer, and the diagram of the measuring system is shown in Fig. 10.15. The sensor, signal conditioning circuit, and the data acquisition card are the core elements of the data acquisition system. The sensors are calibrated before using, through the signal conditioning circuit; the signals are converted into the unified simulation signals within the category that can be received by the data acquisition card. Because the output of the pressure sensor is voltage signal, the output of the torque sensor is a pulse signal; therefore, the pulse signal of the torque sensor can be transferred into a voltage signal by the pulse-voltage transformer and the signals output by the sensors can be adjusted into the simulation voltage signals of 0–5 V by the signal conditioning circuit. Data acquisition program has been written, and the measuring signals of each sensor are transferred into numerical signals by the data acquisition card and collected into the computer; then, the variation in the rolling force and the torque in the forming process can be displayed dynamically.

The data acquisition card selected in the system design is a PCI.8333 analog input and output card produced by Beijing Zhongtai Yanchuang Technology Co. Ltd. The data acquisition software is programmed based on the virtual instrument platform LabVIEW. A common computer is used for the system controller, and the functions of the traditional instruments such as oscilloscope, the spectrum analyzer, the signal generator, and the logic analyzer have been integrated together. Therefore, data acquisition, storage, processing, and analyzing the force and torque in the forming process can be realized integratedly. Using the data acquisition card of NI Corporation, the DAQ library provided by LabVIEW can be directly called for port operation, the programming is very convenient, and the external functions should be called in LabVIEW to driven the PCI.8333 data acquisition card. In the mechanical parameter acquisition system design, the dynamic link library file pciSkplx.Dll of PCI.8333 is called by the CLF functional modular of LabVIEW to realize the open (OpenDevicePlx) and close (CloseDevicePlx) of the equipment and the start of AD. The back panel of the data acquisition program under the platform of the LabVIEW is shown in Fig. 10.16. By adjusting the pressure and torque signals transferred by the signal conditioning circuit of the data acquisition system, the mechanical parameters in the cold rolling forming process can be dynamically

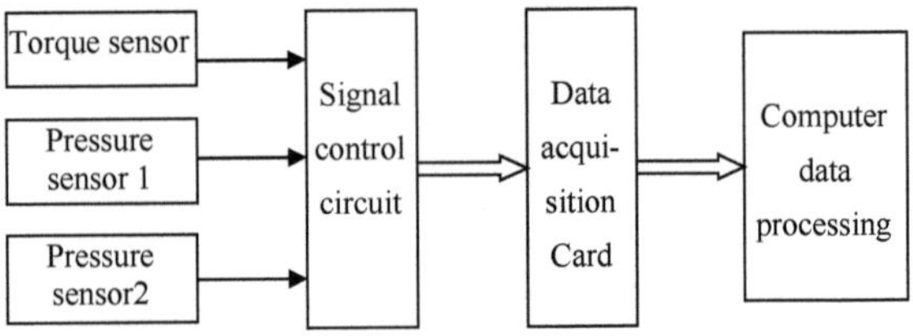

Fig. 10.15 Diagram of the measuring system

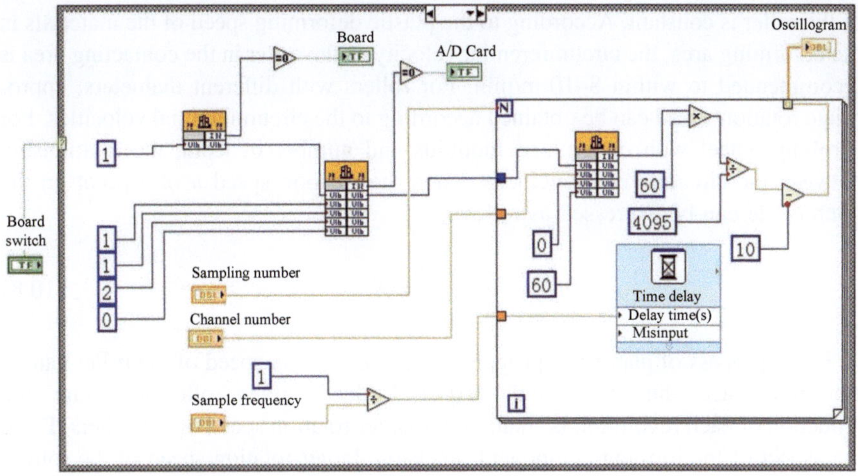

Fig. 10.16 LabVIEW data acquisition program

displayed in the computer and the results can be stored at the same time for later analysis and research.

10.4 Determination of the Process Parameters

Appropriate selection of the process parameters is the guarantee of high-precision component forming. The process parameters of spline cold rolling forming include the following: rotation speed of the rolling wheel, feed rate of the rolling wheel, and the rolling time.

10.4.1 Rotation Speed of the Rolling Wheel

Rotation of the rolling wheel is the main motion of the cold rolling forming machine, and the rotation speed of the rolling wheel is one of the most important process parameters in the spline cold rolling forming process. The deforming speed of the workpiece material will be determined by the circumferential velocity of the roller, and the selection of the parameters is related with the physical mechanical properties of the material. Inappropriate rotation speed of the rolling wheel will result in the wrinkle or scaled microgrooves on the surface of the forming components.

The size and shape of the rolling wheel are often described by parameters such as the modulus and the number of teeth. In the forming process, the rotation speed of the roller is constant. According to the plastic deforming speed of the materials in the deforming area, the circumferential velocity of the roller in the contacting area is recommended to within 8–10 m/min. For rollers with different diameters, appropriate rotation speed can be obtained according to the circumferential velocities. For a rolling wheel with determined modulus and number of teeth, the relationship between the circumferential velocity v and the rotation speed n of a point on the pitch circle can be expressed as follows:

$$v = \frac{2\pi n}{60} m Z_2 \tag{10.8}$$

In the process of parameter programming, the rotation speed of the roller can be determined according to formula (10.8). Under a given roller feed rate, the reduction of each revolution is small under larger rotation speed of the rollers. From the aspect of the forming component precision, larger rotation speed of the rolling wheel should be selected to improve the rolling production. But at too large rotation speed, because the deforming speed is higher, rapid plastic flow of the material is liable to result in defects of the forming components and will increase the noise of the forming process, especially in the initial tooth indexing and pressing period. The tooth profile of the workpiece is not complete, the forming force has a large impact on the machine, and too large rotation speed of the roller will result in the vibration of the machine and the instability of the forming process.

10.4.2 Feed Rate of the Rolling Wheel

In the theoretical study of spline cold rolling forming process, the deforming amount is described with the parameter of per revolution reduction of the workpiece. The reduction is related with the quality of the cold rolled shaft, the mechanical property of the material, and the capability of the rolling equipment. Inappropriate reduction will results in the generation of the accumulated pitch error and roundness error of the spline, even the stuffiness of the machine and failure of the forming process. In the working parameters of the cold rolling forming machine, the reduction per revolution is controlled by the two parameters of the rotation speed and the radial feed rate of the rolling wheel. The relationship among the per revolution reduction Δs, the tooth number Z of the spline, the tooth number Z_2 of the rolling wheel, the rotation speed n of the spindle, and the feed rate v can be calculated with the following formula:

$$\Delta s = \frac{v}{n} \cdot \frac{Z}{Z_2} \tag{10.9}$$

For shaft parts of common low-carbon steel, the recommended reduction is $\Delta s = 0.05\,\text{mm/r} \sim 0.15\,\text{mm/r}$. The radial feed rate of the rolling wheel can be determined by formula (10.9). In actual process parameter programming, the feed rate of the roller should be a little smaller with a higher material hardness and the vice versa. With a higher feed rate, the production efficiency is higher. But with a too higher feed rate, the force applied on the roller will increase, the lifetime of the rolling wheel will be reduced, and the accumulated pitch error and tooth profile error will also be generated.

In the forming parameters, the total rolling time is also an important parameter related with the forming quality. In the parameter programming, the contacting rolling time between the roller and the workpiece should be determined. By the control of the finishing time, lower tooth precision caused by too short rolling time can be prevented, and the fatigue damage of the workpiece and the rolling wheel will be caused by too longer rolling time and excessive rolling. The total rolling time t from the contacting between the roller and the workpiece to the finish of the rolling process can be divided into two parts: time t_1 from the beginning of the roller pressed into the workpiece to the minimum center distance and time t_2 of the finishing stage.

t_1 is determined by the feeding time and the working strokc of thc rolling wheel:

$$t_1 = \frac{d_z - d_f}{v} \tag{10.10}$$

where

d_z is the diameter of the billet and
d_f is the dedendum diameter of the spline.

The finishing time should not be too longer, and 3–5 revolutions of the workpiece are appropriate; therefore, t_2 can be obtained according the following formula:

$$t_2 = (3 \sim 5)\frac{z}{z_2 \cdot n} \tag{10.11}$$

Therefore, the total rolling time t is as follows:

$$t = t_1 + t_2 = \frac{Z}{Z_2 \cdot n}\left[\frac{d_z - d_f}{\Delta s} + (3 \sim 5)\right] \tag{10.12}$$

10.4.3 Control of the Spindle Positions

The spindle position control of the spline cold rolling machine is realized by the numerical control programming, and the working position of the spindle should be determined in the forming numerical control program. The start and end point of the precision forming should be properly set up, especially the end point of the rolling wheel, which has the maximum influence on the forming of spline. If the end point is near the front, the center distance is larger; the tooth cogging of the workpiece is shallow, the rise per tooth is insufficient, and the tooth profile is not fully formed. If the point is near the rear, disorder of the teeth, stuffiness of the machine, and even damage of the rolling wheel will be generated.

In the cold rolling precision forming of spline, the axis center of the workpiece is invariable, the two rolling wheels are feeding radially. Taking the axis of the workpiece for the reference position, according to the engagement transmission principle, the position of the axis center at the finish of the feeding and the stage of finishing can be calculated. The original zero point of the cold rolling forming numerical control machine is not on the center axis of the workpiece, it is on the position of the maximum distance between the two spindles. In addition, considering the processing efficient, the roller should feed rapidly far from the workpiece and turn to normal feeding after approaching the workpiece. In batch production, the position of the end point should be calculated firstly in the control of the spindle position, and the start point of the stroke should be set up at the position of the end point. Before the start point, it is the rapid feeding stroke, and the working stroke is a little larger than the actual rolling contacting stroke. A length ΔL can be taken for the load and unload of the billet. The feeding stroke can thus be determined as follows:

$$L = (d_z - d_f) + \Delta L \tag{10.13}$$

where ΔL is a constant, which is commonly 2–3 mm.

10.5 Mechanical Parameters in the Cold Rolling Precision Forming Process

10.5.1 Dynamic Load of the Forming Process

In the cold rolling forming process, dynamic load is the important basis for the design of the equipment, and the variation rule of the dynamic load in the forming process will provide basic data for the design of the machine structure and the transmission system. A kind of integrated test system for the measuring of the cold rolling forming parameters has been developed, and synchronous on line measurement of the mechanical parameters such as the rolling force and the rolling torque can be realized. The mechanical parameter testing system is installed on the numerical control cold rolling machine Z28K-20; then, the cold rolling forming experiments have been carried out, and the synchronous dynamic variation rule of the rolling force and the measured spindle torque is shown in Fig. 10.17 [7].

From the analysis of the spline cold rolling forming process in Chap. 2, it can be seen that there are four stages in the cold rolling forming process. According to the acquisition data of mechanical parameters obtained by time, the relative position of between the roller and the workpiece can be calculated according to the feed rate of

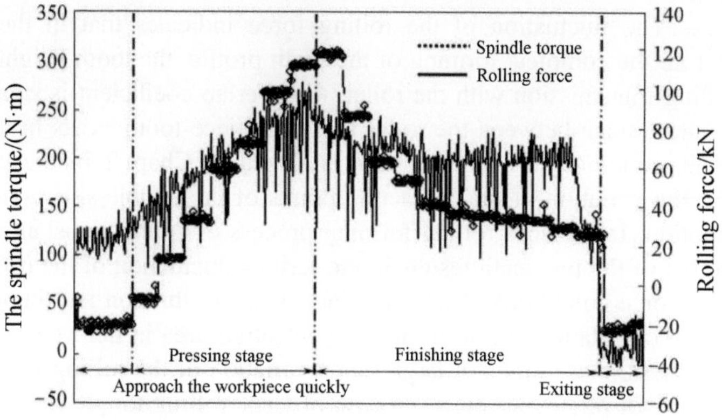

Fig. 10.17 Dynamical variation of the rolling force and the torque in the cold rolling process

the machine, and after data processing, the dynamic torque of the machine spindle and the characteristics of the rolling force at different time periods of the forming process can be obtained. Before the contacting of the roller and the workpiece, the rolling force and the torque of the spindle is basically constant, the measured value is various kinds of frictional resistant force of the machine movement, which is not related with the deforming resistance. In the pressing forming stage, the rollers begin to contact with the workpiece and continuously feed radially under the driven of the slider. From the contacting to the minimum center distance of the roller, the rolling force and the spindle torque increase gradually with the increasing slider feed rate, and it reaches the maximum value at the finishing of the feeding. During the finishing stage, the slider and the roller no longer feed any more, the central distance keeps invariable, and with the improvement of the tooth profile, the rolling force and the spindle torque decrease gradually. It can be found from the dynamic variation of the measured rolling force that great fluctuation of the rolling force exists in the forming process, while the rolling force is relatively stable.

The dynamic load characteristic of the forming process indicates the following: Because of the continuous feeding of the rolling wheel, the compressive degree of the materials on the workpiece increases gradually; the contacting area between the roller and the workpiece tooth, the rolling force, and the spindle torque also increase accordingly. Each tooth of the workpiece will be rolled for several times, because work hardening of the workpiece materials exists in the cold rolling forming process, the yield stress of the material will improve after rolling, it is presented in the deforming resistance, and the rolling force and spindle torque measured will thus increase correspondingly. In the finishing stage, the involute tooth profile has been formed, the plastic deformation increases gradually, the motion of the workpiece and the roller is equivalent to the meshing movement of gears, and the deforming resistance has transferred into the contacting elastic force of gear transmission. With the perfect of the workpiece tooth, the rolling force and the spindle torque decrease from the maximum value to the contacting elastic force and the frictional force of the machine. The fluctuation of the rolling force indicates that in the forming process, before the complete forming of the tooth profile, the tooth height is small, while meshing transmission with the roller, the overlap coefficient is small than 1, and non-contact stage between the roller and workpiece tooth exists in the rolling process. This condition has already been analyzed in Chap. 2 in the theoretical analysis for the variation of the contacting points of the double-side non-backlash meshing motion. In the cold rolling forming process of spline, constant detaching and contacting of the two teeth results in the serious fluctuation of the rolling force and rigorous force conditions of the machine, such as vibration and impacting. In the theoretical calculation according to the contacting area in deforming zone and the average stress, these kinds of large scope variation of the rolling force and the rolling torque will often take place. In case of large rolling torque fluctuation, the highly synchronous motion of the two spindles cannot be ensure and the rolling process can even not be carried out. In the experimental cold rolling machine, damping absorption mechanism has been designed to reduce the vibration of the spindle on conditions of serious variation of the deforming force (see Sect. 10.2.2). Therefore, there is no great fluctuation in the measured dynamic rolling torque curve at the contacting and detaching period of the spline and roller teeth, which has provided a guarantee for the forming of high-precision spline shafts.

10.5.2 Influence of the Process Parameters on the Maximum Forming Force

It can be found from the dynamic variation of the load in the spline cold rolling forming process that the forming force is the maximum when feed rate of the roller is the maximum. The maximum force is one of the main parameters for the design of the rolling wheel, and the design and type selection of the machine. The maximum forming force is related with the material of the billet, the geometrical parameters of the workpiece, and the forming process parameters. In the forming process, the selection of the appropriate process parameters should be mainly considered to reduce the force and deformation of the machine and improve the precision of the forming components. According to the experimental conditions, two groups of experiments have been carried out taking rotation speed of the spindle and the feed rate as variables. The influences of the roller rotation speed and the feed rate on the maximum forming force have been studied in the spline cold rolling precision forming process. Parameters of the forming spline are as follows: the modulus $m = 0.75$, the pressure angle $\alpha = 45°$, the number of tooth $Z = 34$, and the length of the spline $L = 35$ mm; material of the billet is 45 steel; and the experimental machine is Z28K-20 numerical control cold rolling machine.

Table 10.1 Experimental parameters

Serial numbers	1	2	3	4	5
Rotation speed of the motor/(r/min)	450	525	600	675	750
Rotation speed of the roller/(r/min)	12	14	16	18	20
Feed rate of the roller	$v = 50\,\text{mm/min}$				

10.5.2.1 Influence of the Roller Rotation Speed on the Maximum Rolling Force

The rotation speed of the rolling wheel is set up to be invariable as follows: $v = 50\,\text{mm/min}$. Cold rolling forming experiments of the spline shaft under different spindle rotation speeds set up in Table 10.1 have been carried out. The mechanical parameters during the rolling process have been measured. The maximum spindle torque and variation of the maximum rolling force with the rotation speed of the roller are shown in Fig. 10.18.

Experiments show that the rotation speed of the roller has a little influence on the maximum torque and maximum radial force in the cold rolling forming process. The fluctuation of the maximum torque and the maximum radial force is within 10%. In the scope 14–16 r/min of the spindle rotation speed, the maximum torque and the radial force are obviously smaller than that of the other rotation speeds. And within the range, the maximum deformation resistance (the torque and radial force) of the spindle will both increase obviously no matter increasing or decreasing of the spindle rotation speed. Therefore, under a given feed rate, there is a best rotation speed of the roller, with a better match of the two parameters: The compatibility of deformation is better and the force applied to the machine in the rolling process is the minimum. In the designed experiments, the optimized parameters are as follows: Feed rate of the roller is 50 mm/min, and the rotation speed of the roller is 14–16 r/min.

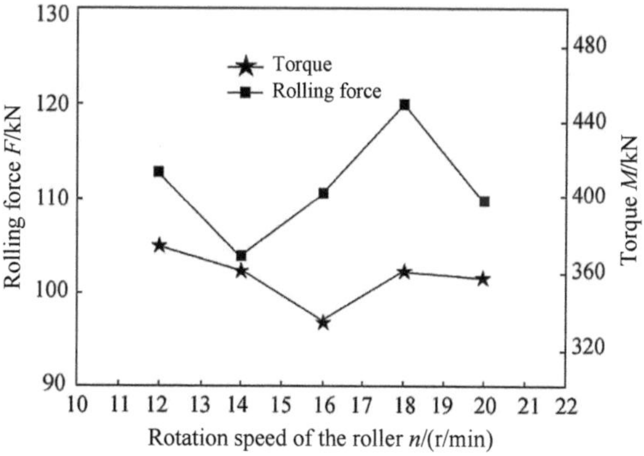

Fig. 10.18 Influence of the roller rotation speed on the maximum torque and rolling force

Table 10.2 Experimental parameters

Serial numbers	1	2	3	4	5
Feed rate (mm/min)	30	40	50	60	70
Rotation speed of the motor (r/min)	$n_0 = 600$				

10.5.2.2 Influence of the Roller Feed Rate on the Maximum Rolling Force

Assuming that the rotation speed of the roller is invariable, the rotation speed of the motor is as follows: $n = 600\,\mathrm{r/min}$. The spline shaft is rolling with the feed rates set up in Table 10.2. The mechanical parameters of the forming process are measured, and the curve of the maximum spindle torque and the maximum rolling force that varies with the feed rate is shown in Fig. 10.19.

In the cold rolling precision forming of spline, with the increase in the feed rate of the rolling wheel, the reduction per revolution, the deformation speed, the maximum rolling force, and torque will increase accordingly. Also, with the increase in the feed rate, the increasing amplitude of the maximum torque is larger than that of the maximum rolling force. It indicates that in the forming process, because of the increase in the feed rate, the tangential force causing the deformation of the billet has a higher proportion in the deforming resistance, which results in the increasing spindle torque, and the variation in the maximum radial force is not obvious.

The rotation speed and the feed rate of the rolling wheel are two important process parameters of the spline cold rolling precision forming process, which should be well matched to guarantee the quality of the forming components. Feed rate is the main influencing factor to the maximum rolling force and the maximum rolling torque of the cold rolling forming process, and the influence of the roller rotation speed is smaller than that of the feed rate. Therefore, from the aspect of the maximum rolling force, larger spindle rotation speed and smaller feed rate should be selected in the cold rolling forming process.

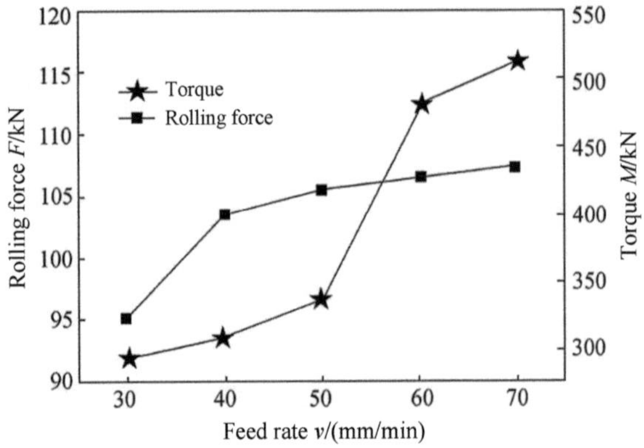

Fig. 10.19 Influence of the roller feed rate on the maximum torque and rolling force

10.5.3 Comparison Between the Theoretical Calculation and Experimental Results

In the experiments of the rolling force and rolling torque measuring in the laboratory, the rolling wheel should be designed specialized every time the spline parameters have been changed. Therefore, it is not efficient to arrange experiments for different modulus and different tooth numbers of the spline shafts. In Chap. 3, the calculation method and program of the rolling force and the rolling torque have been proposed, and in order to verify the feasibility of the theoretical methods, experiments shown in Table 10.3 have been designed and carried out. And theoretical and experimental results of the rolling force and rolling torque under the three groups of different process parameters have been compared to verify the theoretical calculation method.

The experimental material 45 steel has been treated with hardening and tempering method. In the calculation and programming, the parameters of the material are obtained from the Machine Design Handbook. The work hardening is considered as the isotropic hardening model, and the modulus and pressure angle of the spline and the tooth number of the rolling wheel are all standard values.

In the calculation of the rolling force and the rolling torque, the peak value of the rolling force and torque is taken each time of the single tooth contacting. It can be seen that the trends of the rolling force and rolling torque calculated in the whole rolling process coincide with that in the actual rolling process. In engineering application, the maximum rolling force and maximum rolling torque are the most important parameters which should be focused. Under the three groups of process parameters, the comparison of the maximum rolling force and torque between the theoretical calculation and measuring results in the spline cold rolling process are shown in Fig. 10.20.

The theoretical calculated maximum rolling force is larger than that of the measured value within an error of 15%, which is because larger value of the uncertain parameters has been selected in the theoretical calculation method and the error has been further amplified in the iterative calculation. To improve the calculation accuracy, the modified coefficient of the area and the frictional coefficient can be further modified by experiments.

Table 10.3 Process parameters

Parameters of the formed spline	Modulus $m = 0.75$, pressure angle $\alpha = 45°$, tooth number $Z = 34$, length of the spline $L = 35$ mm
Tooth number of the rolling wheel	$Z_2 = 230$
Process parameter 1	Feed rate $v = 50$ mm/min, rotation speed of the roller $n = 16$ r/min
Process parameter 2	Feed rate $v = 60$ mm/min, rotation speed of the roller $n = 12$ r/min
Process parameter 3	Feed rate $v = 40$ mm/min, rotation speed of the roller $n = 18$ r/min

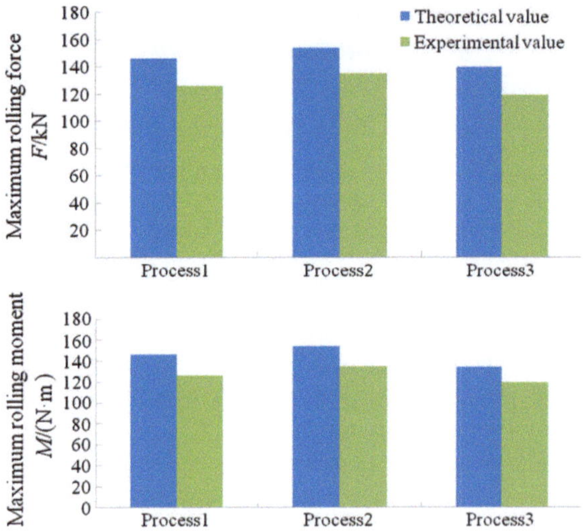

Fig. 10.20 Comparison of the theoretical and experimental results of the maximum forming force

10.6 Brief Introduction of the Cold Rolling Precision Forming Equipments

10.6.1 Cold Rolling Forming Equipments of Kinefac Corporation(Kinefac) in the USA [1]

Kinefac® was founded in 1962 to develop cylindrical die thread rolling machines. It is first in the innovative application of cylindrical die rolling, extrusion, radial forming, and center drive turning for the production of shaft-like parts. In the years since, it has become a world-class developer and manufacturer of specialized metal forming, metal working, and processing equipment. Today the Kinefac Division products are sold worldwide.

10.6.1.1 Kine-Roller® Cylindrical Die Rolling Machines

Kinefac is at the leading edge of rolling machine and die technology, using advanced engineering skills for the design and development of equipment for the high-speed production of precision parts meeting rigorous requirements of statistical control.

Applications of Kine-Roller® machine are as follows:

(1) Thread and worm rolling;
(2) Knurl and serration rolling;

(3) Spline and involute rolling;
(4) Flanging and annular forming;
(5) Roll straightening and finishing; and
(6) Throughfeed part forming.

10.6.1.2 PowerBox Two Cylindrical Die Machines

PowerBox with its load-carrying box structure provides maximum stability of part diameter and straightness for the rolling of large diameter, high hardness, and precision threads while maintaining a fixed centerline allowing for simplified tooling design. This high-performance infeed and throughfeed thread rolling machine is typically used to produce precision thread forms on cylinder head bolts, rock drill shafts, gas compressor rods, pipe flange studs, turbine casing bolts, piston rods, engine head studs, blower shafts, jack screws, cylinder tie rods, propeller hub studs, tower anchor bolts, control rods, lead screws for gas or coal power plants, steam and wind turbines, nuclear reactor vessels, power transmission lines, and other similar products that require a precision thread roller to meet their needs.

(1) Features

PowerBox two cylindrical die machines have a fixed rolling center line.

(1) CNC die actuation and spindle drive with automatic match are available options.
(2) PowerBox machines are of a building block design. Therefore, spindle size, die drive horsepower, and speed and hydraulic die actuation can be adjusted for the performance optimization for specific applications.
(3) Standard work support tooling, special application tooling, and Kine-Mat automated part handling are available where required.

Figure 10.21 is the structural features of the transmission axis unit of 2 die horizontal axis Kine-Roller® cylindrical die rolling machines.

(2) Types and technique parameters

Parts of the types of PowerBox two cylindrical die machines are shown in Fig. 10.22, and the technique parameters are listed in Table 10.4.

For example, the MC-300 PowerBox provides maximum stability of part diameter and straightness for the rolling of large diameter, high hardness, and precision threads while maintaining a fixed centerline allowing for simplified tooling design. This high-performance infeed and throughfeed two die cylindrical thread rolling machine is typically used to produce precision thread forms on pressure vessel studs, cylinder head bolts, rock drill shafts, gas compressor rods,

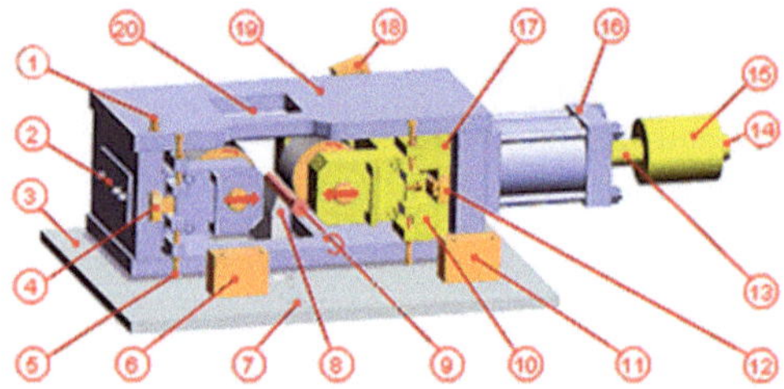

(1) taper adjustment (2) left skew pivot (3) machine base plate (4) skew readout vernier (5) skew adjustment screw (6) powerbox support rollers (7) work support mount (8) work support blade adj. (9) fixed rolling center line (10) cylinder rod extension (11) optional mechanical stop (12) in-line hydraulic cylinder (13) automatic centralizer (14) force enveloping "powerbox" (15) rolling area access

Fig. 10.21 Structural features of the transmission axis unit of 2-die horizontal axis Kine-Roller® cylindrical die rolling machines [1]

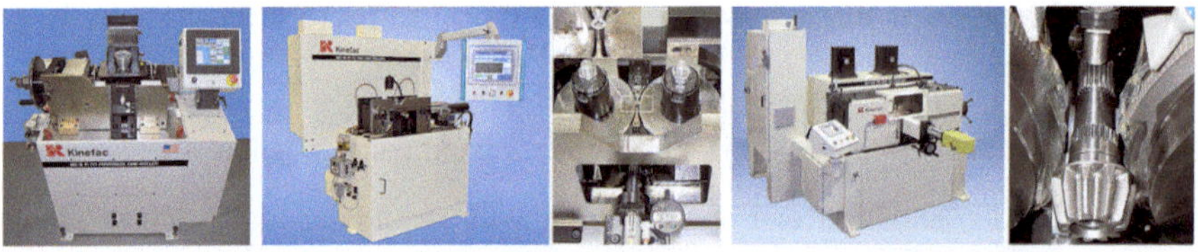

(a) MC-8 (b) MC-15 (c) MC-35

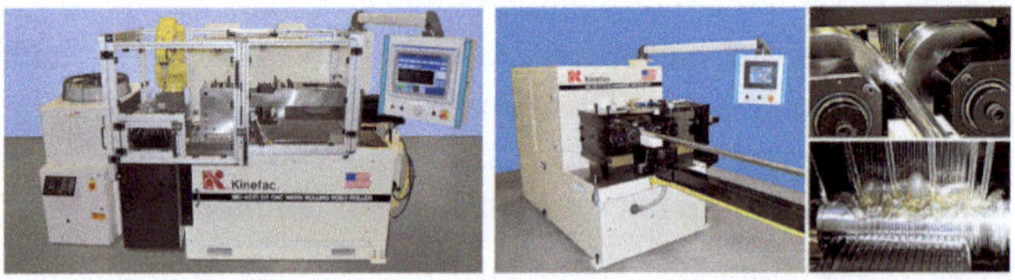

(a) MC-40 (b) MC-60

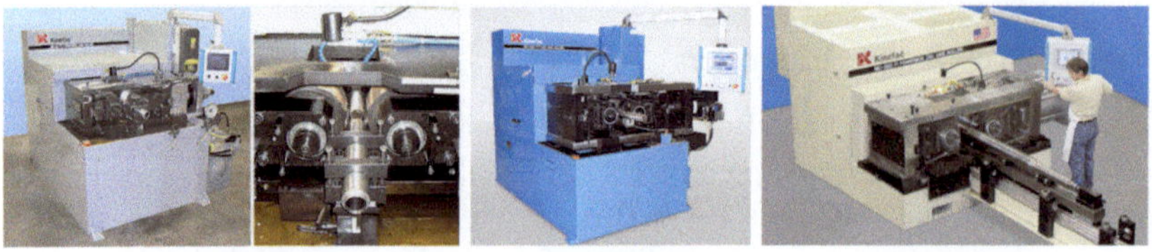

(c) MC-80 (d) MC-200 (e) MC-300

Fig. 10.22 Part of the types of PowerBox two cylindrical die machines [1]

Table 10.4 Technique parameters of the PowerBox two cylindrical die machines [1]

Types	Unit	MC-8	MC-15	MC-35	MC-40	MC-60	MC-80	MC-150	MC-2000	MC-300
Maximum radial die load	kN	90	222	1335	356	534	712	1335	1779	2936
Center distance range	mm	114–159	127–234	304–457	152–317	152–304	171–314	228–457	203–393	254–520
Standard spindle diameter	mm	54	54	76	75	80	100	114	127	152
Spindle skew range	°	–	–	–	–	±8	±8	±5	±5	±2.5
Standard die gap	mm	51	114	50	120	117	165	266	266	292
Standard die drive	kW	2.23	9	25	19	22	26	22	34	34
Die diameter range	mm	100–160	100–171	203–406	140–254	152–254	165–254	203–304	216–304	254–381
Maximum rolled O. D.	mm	25.4	72	50	101	101	101	101	152	177

pipe flange studs, turbine casing bolts, piston rods, engine head studs, blower shafts, jack screws, cylinder tie rods, propeller hub studs, tower anchor bolts, control rods, lead screws for gas or coal power plants, steam and wind turbines, nuclear reactor vessels, power transmission lines, and other similar products that require a precision thread roller to meet their needs. The MC-300 Kine-Roller may be equipped with the Kine-Trol NC rolling control system that provides the operator with complete control of the rolling process and also can be easily adapted with automatic loading to maximize production, or can be manually loaded for smaller production requirements. A Kine-Spin centrifuge system is available as an option to clean and reclaim used coolant. The technique parameters of PowerBox machine are listed in Table 10.4.

10.6.1.3 Double-Arm Two Cylindrical Die Machines

(1) Features
   (1) Double-arm two cylindrical die rolling machines use a unique double pivot arm design and a sliding block device to provide uniform die penetration to the rolling axis.
   (2) Double-arm two cylindrical die rolling machines are of a building block design available with other spindle diameters and die drive horsepower and speed.
   (3) It is available with precision spindles in which the dies are rigidly mounted to the spindle and ground to the spindle centerline. In addition, it may be specially calibrated to provide 0.0001 diameter adjustments.

The structural features of the transmission axis unit of the double-arm two cylindrical die rolling machines are shown in Fig. 10.23.

(2) Types and technique parameters of the double-arm two cylindrical die machines

Types and specifications of the double-arm two cylindrical die machines are shown in Fig. 10.24, and the technique parameters are listed in Table 10.5.

For example, the MC-10 Kine-Roller® is designed for both throughfeed and infeed rolling. This high-performance machine is ideally suited for thread rolling, motor shaft serration rolling, ball screw form rolling, contour thread rolling, and other unique forms typically found in aerospace fasteners, automotive components, medical devices, off-road vehicles, appliances, hand and power tools wind power, and other industries requiring precision roll forming. The MC-10 Kine-Roller® can be adapted with automatic loading to maximize production, or manually loaded for smaller production requirements. A Kine-Spin centrifuge system is available as an option to clean and reclaim used coolant.

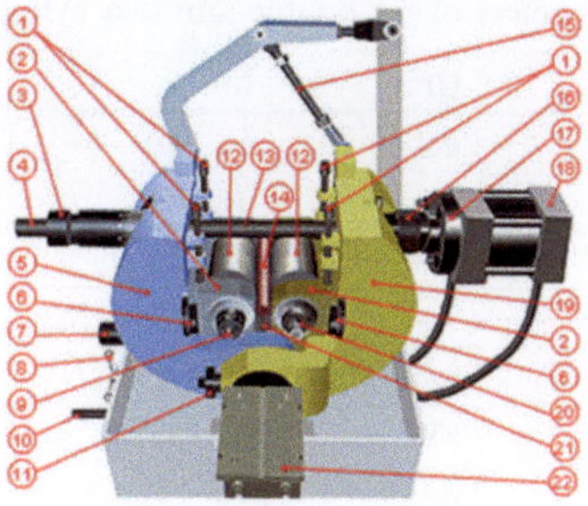

(1) skew adjustment screws (2) spindle bearing housing (3) size adjustment nut (4) size dial lock-up (5) left arm (6) skew vernier (7) rolling height adjustment (8) work support lock (9) axial matching dial (10) rear work support adjust (11) taper adjustment (12) die (13) draw bar (14) work (15) centrality adjustment (16) die open stop nut (17) stroke stop block (18) hydraulic cylinder (19) right arm (20) spindle assembly (21) rolling support (22) front work support

Fig. 10.23 Structural features of the transmission axis unit of the double-arm two cylindrical die machines [1]

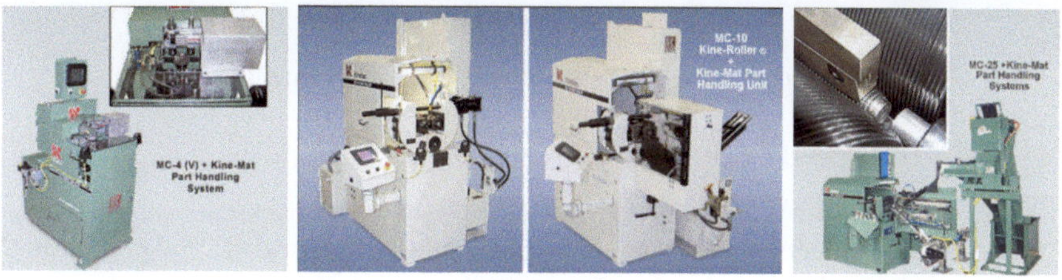

(a) MC-4 (b) MC-10 (c) MC-25

Fig. 10.24 Types of the double-arm two cylindrical die machines [1]

10.6.1.4 Three Cylindrical Die Rolling Machines

(1) Types and specifications

The types and specification of the three cylindrical die rolling machines are shown in Fig. 10.25.

The MC-6 Kine-Roller® is a high-performance three cylindrical die rolling machine available with a horizontal or vertical rolling axis. Due to its ability to distribute loads more evenly than a conventional two die rolling machine, the MC-6 is primarily designed to roll forms such as splines or threads on tubular parts as found on pneumatic or hydraulic fittings, fasteners, automotive components,

Table 10.5 Technique parameters of the double-arm two cylindrical die machines [1]

Types	Unit	MC-4	MC-10	MC-25
Maximum radial die load	kN	6	240	240
Center distance range	mm	76–127	101–171	111–222
Standard spindle diameter	mm	38	50	63.5
Spindle skew range	°	–	±12	±12
Standard die gap	mm	66	143	114
Standard die drive	kW	2	15	18
Die diameter range	mm	76–111	95–127	114–152
Maximum rolled O. D.	mm	40	88	101

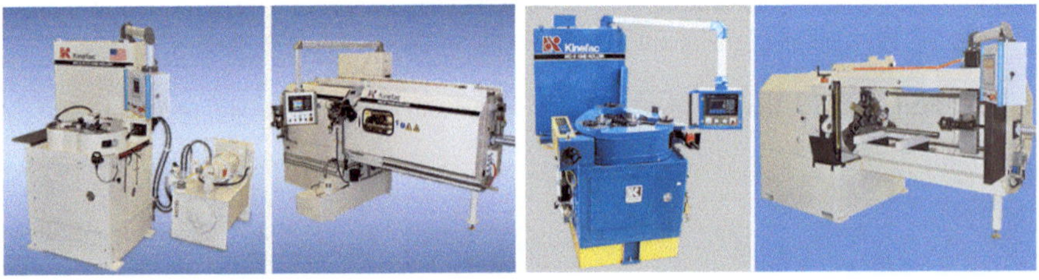

(a) MC-6 (b) MC-9

Fig. 10.25 Types of the three cylindrical die rolling machines [1]

off-road vehicles, appliances, and hand and power tools wind power although itswell suited to roll these or other forms as on solid parts as well.

The MC-9 Kine-Roller® is a high-performance three cylindrical die rolling machine available with a horizontal or vertical rolling axis. Due to its ability to distribute loads more evenly than a conventional two die rolling machine, the MC-9 is primarily designed to roll forms such as splines or threads on tubular parts as found on pneumatic or hydraulic fittings, fasteners, automotive components, off-road vehicles, appliances, and hand and power tools. The MC-9 Kine-Roller® can be adapted with automatic loading to maximize production, or manually loaded for smaller production requirements. A Kine-Spin centrifuge system is available as an option to clean and reclaim used coolant.

(2) Main technique parameters

Technique parameters of the three cylindrical die rolling machines are shown in Table 10.6.

Table 10.6 Technique parameters of the three cylindrical die rolling machines [1]

Types	Unit	MC-6	MC-9
Maximum radial die load	kN	222	444
Center distance range	mm	47–76	63–85
Standard spindle diameter	mm	38	63
Spindle skew range	°	–	–
Standard die gap	mm	142	101
Standard die drive	kW	11	15
Die diameter range (mm)	mm	85–110	107–152
Maximum rolled O. D.	kW/A	76	88

10.6.1.5 Rolling Dies and Typical Components

Kinefac can provide the user with its unique die performance optimization program. The types of dies mainly include threads, worms, knurls, ball screws, splines, and serrations, as shown in Fig. 10.26. Figure 10.27 is the typical cold rolling forming components of Kinefac.

10.6.2 Cold Rolling Equipments of Profiroll Corporation in Germany [2]

The Profiroll business concept is strongly oriented on customer's sophisticated technical requests for profiling rotationally symmetrical parts using the most modern cold forming technology, leading finally in a development of an efficient manufacturing with profile rolling. The customer will be served with machines, handling and conveying systems, rolling dies as well as reliable processes from one

Fig. 10.26 Rolling dies of Kinefac [1]

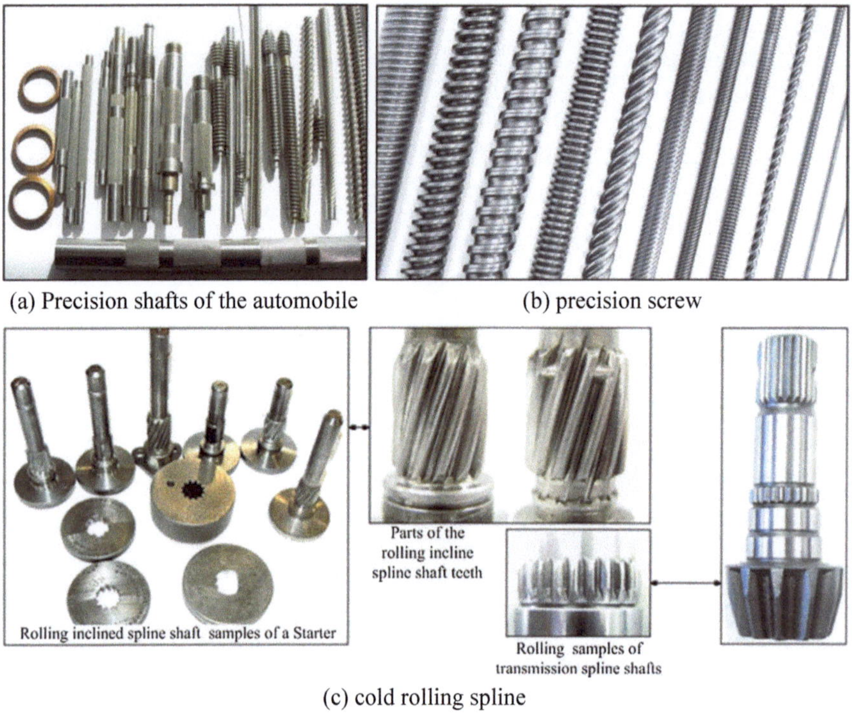

Fig. 10.27 Typical cold rolling forming components of Kinefac [1]

source. From 1999, Profiroll begun to develop spline cold rolling machine Rollex, and in 2001, 840D CNC numerical control system has been installed on the machine.

10.6.2.1 Rollex Serial Spline Cold Rolling Machines

The Rollex® is more than an innovation offered by Profiroll in the field of spline rolling. This flexible machine type enables a rolling of splines onto most different types of shafts. The CNC-control combined with symmetrical circular dies allows a manufacturing of splines with highest quality demands. Likewise, the Rollex® is able to coldform profiles, threads and to perform finishing rolling due to its reliable profile rolling process.

One of the recent innovations of Profiroll engineers is the rolling of splines with circular dies based on Rollex® technology. The biggest advantage using circular dies—as with thread rolling—is the infinite rolling motion giving an unlimited coldform length. Results are a higher quality and much lower changeover times. Actually with the rolling on hollow shafts, Profiroll took over technology leadership serving capable industrial processes.

(1) Circular die process

   (1) Circular dies for kinematic rolling with symmetrical geometry;
   (2) This correlates with a rolling rack of infinite length;
   (3) A workpiece-specific calibration phase achieving least tolerances is the result;
   (4) Lowest changeover times due to compact circular dies; and
   (5) Up to 5 sets of dies mountable in one tooling layout.

(2) Process description

   Workpieces are to be positioned inside the work area, clamped between centers. The rolling slide infeed starts, whereas the circular dies on the slides penetrate the workpiece and coldform the requested spline profile. After forming all profiles, the workpiece is getting retracted from work area to be replaced by operator.

(3) Features

   (1) Special drive chain for spline rolling;
   (2) Controlled workpiece acceleration;
   (3) Multiaxes CNC;
   (4) 2-rolling slides;
   (5) Graphical user interface;
   (6) Process visualization;
   (7) Data management;
   (8) User guidance;
   (9) Machine sizes for different applications; and
   (10) HP (high precision) version with servomechanical drives.

(4) Types and specifications of Rollex serial spline rolling machines

   Types and specifications of Rollex serial spline rolling machines are shown in Fig. 10.28.

(5) Main technique parameters

   Technique parameters of Rollex spline rolling machines are listed in Table 10.7.

(6) Rollex dies

   Rollex dies and spline rolling dies with constant outer diameter have been developed to coldform splines with CNC-machines (e.g., Rollex Series)—spline rolling dies with configurations depending on customer demand.

   Advantage of Rollex dies compared to incremental rolling dies is the unlimited number of cold forming steps. While splines made by incremental rolling dies or racks have to be formed within one rotation (or stroke), Rollex dies allow a flexible feeding process that is only controlled and limited by NC-axes of the machine. This leads to new possibilities to form high modules, form splines on hollow shape parts, and reach new levels of quality.

   Additionally, Rollex dies can be used for reversing processes. A process includes the change of rotation direction one or more times to calibrate both

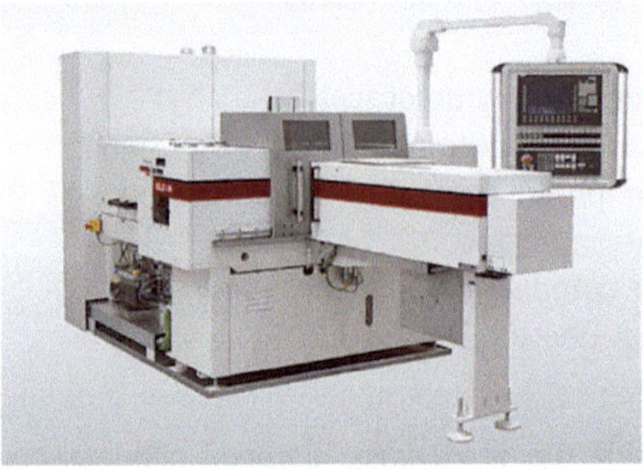

(a) Rollex S HP spline rolling machines

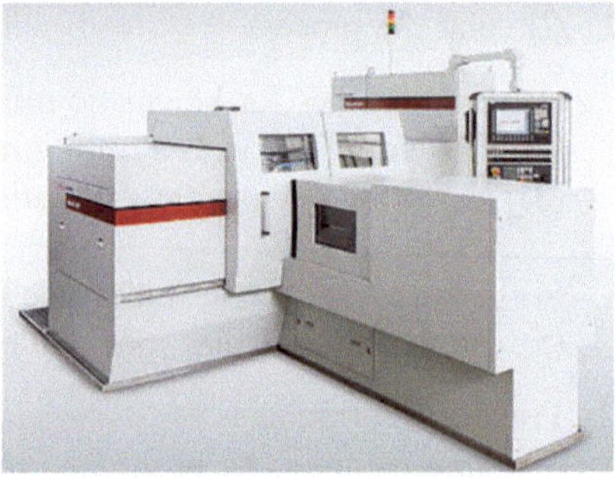

(b) Rollex HP spline rolling machines

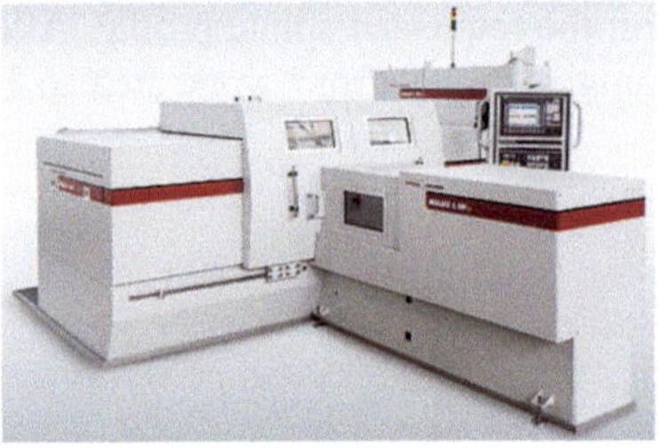

(c) Rollex L HP spline rolling machines

(d) Rollex XL HP spline rolling machines

Fig. 10.28 Rollex serial of spline rolling machines [2]

Table 10.7 Technique parameters of Rollex spline rolling machines [2]

Types	Unit	Rollex S HP	Rollex HP	Rollex L HP	Rollex XL HP
Workpiece diameter max.	mm	40	70	100	150
Spindle diameter	mm	120	120	120	120
Die diameter max.	mm	250	300	330	330
Die mounting length	mm	80	180	180	180
Speed (stepless)	RPM	0–50	0–40	0–40	0–30
Connecting load	kW/A	23/63	41/100	41/100	48/160
Width	mm	2900	3500	4200	4400
Length	mm	3600	4350	4600	5000
Height	mm	2200	2200	2200	2500
Weight	kg	3000	7500	11000	13700

Fig. 10.29 Dies of Rollex [2]

tooth flanks in order to optimize quality or to form special elements such as assembly chamfers, shown in Fig. 10.29.

(1) Many regrinding steps and therefore lower tooling cost per workpiece;
(2) Possibility to insert special shape elements such as blocked teeth, assembly chamfers, or flexibility due to infinite rolling length; and
(3) Fast clamping system for convenient changeover within short time.

(7) Typical components
Typical components manufactured by Rollex are shown in Fig. 10.30.

10.6.2.2 Thread and Profile Rolling Machines

The Profiroll thread rolling machines are outstanding in terms of workpiece quality requirement, possibility of connection to an automated handling system and output.

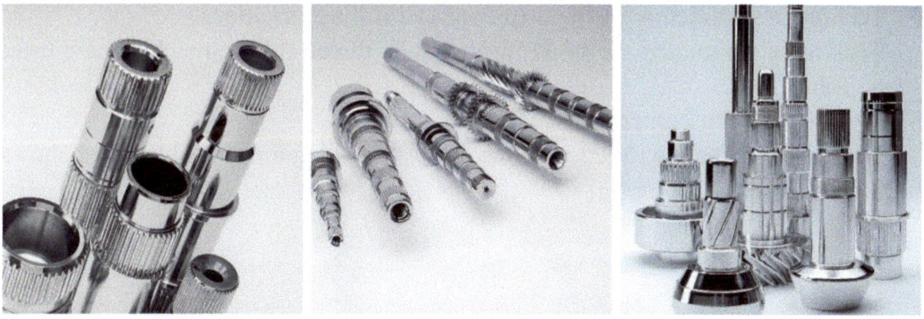

Fig. 10.30 Typical components rolling by Rollex serial machine [2]

The innovative are 2-slide thread rolling machines for high-quality demands. The precision drives and guideways together with a Profiroll CNC based on Siemens 840D allow rolling of challenging external profiles and especially parts with low wall thickness.

(1) Characteristics of all Profiroll machines are as follows:

   (1) A statically and dynamically very stiff iron cast C-frame;
   (2) Best accessibility to open work area for insertion and extraction of parts;
   (3) Best accessibility during changeover;
   (4) Rolling force from 50 kN to 1000 kN;
   (5) Infeed and throughfeed rolling; and
   (6) A drive and control system created for operator needs.

(2) 3-rolls thread rolling machines
The 3-rolls thread rolling machines, developed for rolling of hollow parts; satisfy with a high rolling quality resulting from centered positioning of parts and symmetrical 3-side load due to the usage of 3 rolling dies. The main characteristic of the unique Profiroll design is the central position of all 3 rolling slides in a preloaded double frame. This ensures highest rigidity. By using the 3-rolls principle, the needed force is brought via 3 contact zones into the workpiece material instead of 2. Thus, even rolling of hollow parts is possible without compromise in quality or damaging of workpiece.

(3) Features of the 3-rolls thread rolling machines

   (1) 3-slide system with symmetrical support of dies for centered positioning of workpiece;
   (2) Adjustment for your quality requirement;
   (3) Close double frame that leads to minimized bending;
   (4) Multiaxes CNC/AC control;
   (5) Graphical user interface;
   (6) Process visualization; and
   (7) Process data management.

(4) Types and specifications of the 3-rolls thread rolling machines
Types and specifications of the 3-rolls thread rolling machines are shown in Fig. 10.31.

(5) Technique parameters of the 3-rolls thread rolling machines
Technique parameters of the Profiroll 3-rolls thread rolling machines are listed in Table 10.8.

(6) Typical forming components
Typical components rolled with the 3-rolls thread rolling machines are shown in Fig. 10.32.

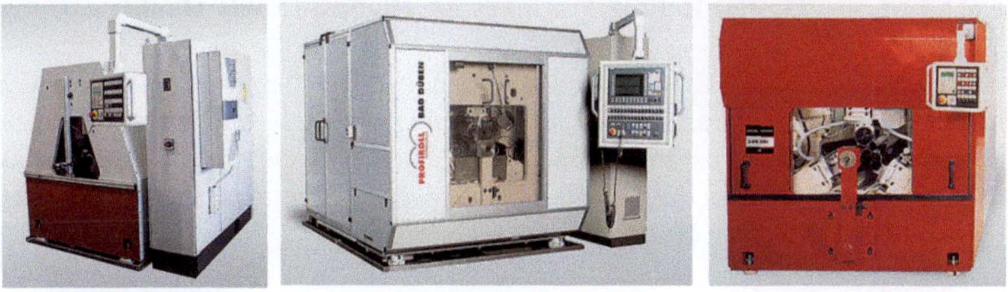

(a) 3-PR 15e/AC (b) 3-PR 40 CNC/AC (c) 3-PR 60e

Fig. 10.31 Types and specifications of the 3-rolls thread rolling machines [2]

Table 10.8 Technique parameters of the 3-rolls thread rolling machines [2]

Types	Unit	3-PR 15e/AC	3-PR 40 CNC/AC	3-PR 60e
Rolling force max.	kN	150	400	600
Workpiece diameter min.	mm	15	25	25
Die mounting length	mm	100	160	160
Spindle diameter	mm	54	80	80
Speed(stepless)	mm	10–150	10–150	30–120
Swivel angle	°	±10	±10	±10
Rolling die diameter max.	mm	160	235	235
Connecting load	kW/A	22/100	63/250	70/200
Width	mm	2100	2100	2500
Length	mm	2000	3000	3000
Height	mm	2000	2300	2300
Weight	kg	5000	12000	12600

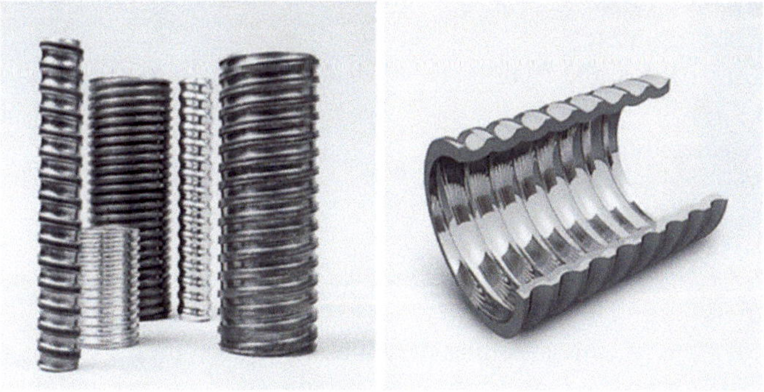

Fig. 10.32 Typical forming components of the 3-rolls thread rolling machines [2]

10.6.2.3 CNC Thread Rolling Machines for Highest Profile Quality

A machine line of CNC controlled one and two slide rolling machines with extended features, e.g., free programmable rolling parameters, online user guidance, and service programs. Due to the easily adjustable CNC, the reproducibility of all quality characteristics at cold formed parts is given. The innovative allows rolling high-precision parts even in small series economically.

(1) Features

(1) Machine body as c-frame for improved rigidity;
(2) One or two slide machines;
(3) Perfect accessibility due to open work area;
(4) Infeed and throughfeed rolling;
(5) To be connected easily to an automation;
(6) Multiaxes-CNC/AC;
(7) Graphical user interface;
(8) Process visualization;
(9) Process data management;
(10) User guidance;
(11) Central lubrication; and
(12) Optional: Quality monitoring of stroke, force, torque, and axial movement.

(2) Applications

(1) Metric and UN thread;
(2) Lead screws;
(3) Whitworth thread;
(4) Trapezoidal thread;
(5) Round thread;
(6) Unsymmetrical thread;
(7) Masonry drills;
(8) Worm profiles;
(9) Knurls RAA and RGE;
(10) Splines;
(11) Finish rolling;
(12) Grooves; and
(13) Special profiles.

(3) Types of CNC thread rolling machines for highest profile quality and technique parameters
Types of the CNC thread rolling machines for highest profile quality are shown in Fig. 10.33, and the main technique parameters are listed in Table 10.9.

(4) Typical components
Typical components with the highest profile quality rolled with the CNC machine are shown in Fig. 10.34.

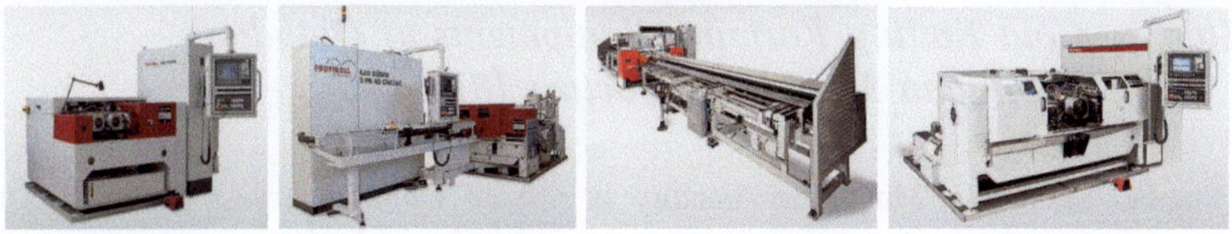

(a) PR 30 CNC/AC (b) 2-PR 40 CNC/AC (c) 2-PR 60 CNC/AC (d) 2-PR 100 CNC/AC

Fig. 10.33 Types of CNC thread rolling machines for highest profile quality [2]

Table 10.9 Technique parameters of CNC thread rolling machines for highest profile quality [2]

Types	Unit	PR 30 CNC/AC	2-PR 40 CNC/AC	2-PR 60 CNC/AC	2-PR 100 CNC/AC
Rolling force max.	kN	300	400	600	1000
Workpiece diameter	mm	5–80	5–100	60–200	20–200
Die mounting length max.	mm	200	200	300	300
Spindle diameter	mm	80	80	100	130
Rolling die diameter max.	mm	235	235	300	300
Speed	mm	0–150	10–150	0–100	0–80
Swivel angle	°	±10	±10	±10	±10
Connecting load	kW/A	27/100	39/160	50/160	100/315
Width	mm	2360	2700	3600	3600
Length	mm	2440	2250	3000	3300
Height	mm	2150	8500	2200	2000
Weight	kg	5500	12000	12600	13000

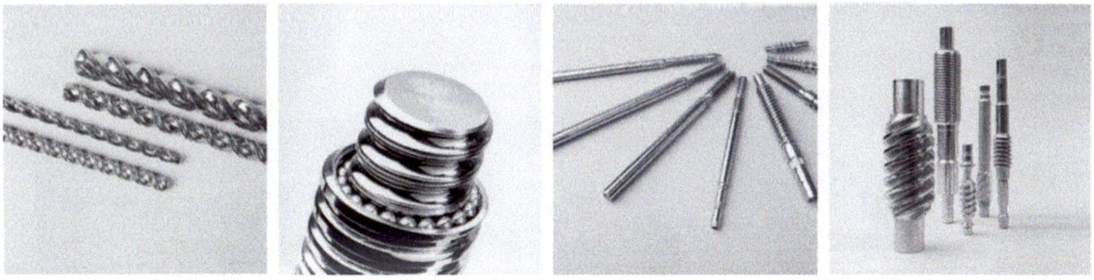

Fig. 10.34 Typical components of CNC thread rolling machines for highest profile quality [2]

10.6.3 Cold Rolling Forming Equipments of Qingdao Shengjian Machinery Factory [8]

Qingdao Shengjian Machinery Factory is one of China's largest producers of shaping machines and screw rolling machines, with nearly 60-year machine tool experience, one provincial cold rolling R&D center, and complete production technique and testing methods. They have developed more than 30 types of screw rolling machine, spline cold rolling machine, and spline cold tapping machine.

10.6.3.1 Spline Cold Rolling Forming Machine

The serial of spline cold rolling machines has the characteristics of better rigidity, higher precision, stable property, higher reliability, and compact structures and is convenient for operation and maintenance. The machine is controlled by PLC, and manual, semiautomatic, and automatic operation working patterns can be realized.

(1) Types and specifications

Types and specifications of the spline cold rolling forming machine of Qingdao Shengjian Machinery Factory are shown in Fig. 10.35.

(2) Application fields

This serial of machines is mainly used in the processing of helical tooth, fine tooth, oil groove, threads, and so on. It is widely used in the fields such as mechanical, chemical, automobile, motorcycle, electric motor, aerospace, railway, shipbuilding, and agricultural machine.

(3) Main technique parameters

The main technique parameters of the spline cold rolling machine of Qingdao Shengjian Machinery Factory are listed in Table 10.10.

(a) Z28J-16 (b) Z28J-25 (c) Z28J-35

Fig. 10.35 Spline cold rolling machine of Qingdao Shengjian Machinery Factory [8]

Table 10.10 Technique parameters of spline cold rolling machine of Qingdao Shengjian Machinery Factory [8]

Types / Items	Z28J-16	Z28J-25	Z28J-35
Maximum rolling force/kN	160	250	350
Center distance/mm	120–240	180–280	180–360
Hydraulic stroke/mm	20	2 × 8	2 × 10
Spindle diameter/mm	54	75	85(75)
Key width/mm	12	20	22(20)
Total power of the motor/kW	8	12	15
Outer size of the machine/cm	142 × 137 × 135	187 × 148 × 125	211 × 192 × 127
Weight/kg	2200	4470	5000

10.6.3.2 Numerical Control Rolling Machines

(1) Types and specifications

The types and specifications of the Z28K series NC rolling machines of Qingdao Shengjian Machinery Factory are shown in Fig. 10.36.

(2) Application fields

Z28K serial numerical control rolling machines is independently developed high-performance numerical control rolling machines. They are mainly used in the rolling of various precision external threads (such as common threads and trapezoidal threads), small modulus involute spline and worms, oil groove, roll finishing, knurling, bar straightening, and cold rolling forming of some rotational components, especially suitable for rolling of large volume components multispecies.

Z28K serial numerical control rolling machine has already applied in the automobile field of China. The quality and competitiveness of the products have been greatly improved.

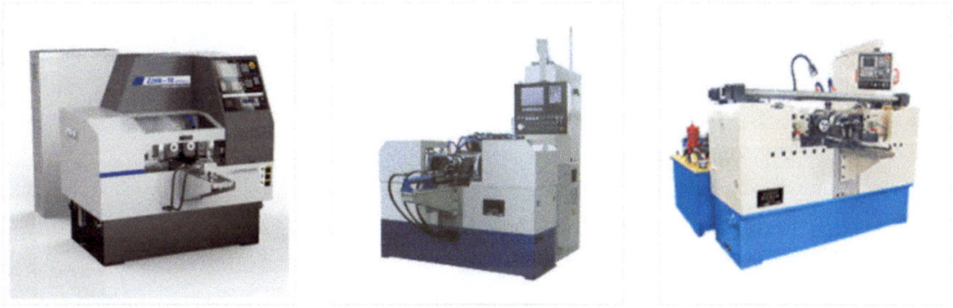

(a) Z28K-16 (b) Z28K-25 (c) Z28K1-35

Fig. 10.36 Numerical control cold rolling machine [8]

(3) Property characteristics

The frame of the Z28K serial numerical control rolling machine is U style; not only can the whole rigidity of the machine be increased, but also the problem of vibration can be solved. Two sets of whole closed-loop large power AC servosystem have been adopted for the driving of the spindle. Driven by the high-precision integrated transmission system, the synchronous of the two spindles has been ensured. The precision localization of the two sliders is guaranteed by the driving of the whole closed-loop proportional servohydraulic system.

Three sets of rolling wheels can be installed on the spindles of Z28K numerical control rolling machine to realize the multistation rolling of the same workpiece. Four kinds of rolling methods, radial rolling, axial rolling, repeatedly rolling, and reciprocating rolling, and three kinds of working patterns and adjusting, manual and automation, can be realized. Through the human–computer interaction interface, the parameters of the machine and the workpiece can be inputted and adjusted easily. The dynamical processing can be observed, and simulating processing can also be realized. Through the circle detection of the signals such as the current, pressure, temperature, and distance by the control system, early warning, fault prompt, and emergency stop can be realized, which provides a reliable guarantee for the safety of the operators.

In production, the relative angle of the two spindles and the center distance between the two sliders can be conveniently adjusted with the electric handwheel with a higher precision.

And only 1–3 components are demanded for the determination of the spline processing parameters, which is advantageous for processing of multivariety and small batch splines. The forming parameters can be stored in the computer, in conditions the specification of the spline has been changed, the roller should be

Table 10.11 Technique parameters of the NC rolling machines of Qingdao Shengjian Machinery Factory [8]

Types Parameters	Z28K-35 Numerical control rolling machine	Z28K-25 Numerical control rolling machine
Rolling force (numerical control)/kN	50–350	50–250
Spindle center distance (numerical control)/mm	180–320	160–310
Spindle diameter/mm	85	75
Rotation speed of the spindle (numerical control)/(r/min)	0–60	0–110
Maximum modulus of the spline that can be rolled/mm	1.5	1.25
Roller diameter/mm	Φ190–260	Φ160–230
Total power of the motor/kW	22	19
Outer size/mm	2010 × 1810 × 2420	1850 × 2540 × 2100
Weight/kg	5100	4000 s

Fig. 10.37 Thread and spline components manufactured by Qingdao Shengjian Machinery Factory [8]

changed, the processing of another specification of spline can be realized only by loading of required files, and the operation is very simple.

(4) Main parameters of the machine and typical components

The main technique parameters of the NC rolling machines of Qingdao Shengjian Machinery Factory are listed in Table 10.11. And the typical components rolled with the NC rolling machines are shown in Fig. 10.37.

References

1. http://www.kinefac.com [OL].
2. http://www.profiroll.com [OL].
3. Liu Zhiqi. Theoretical and Experimental Study on the Cold Rolling Precision Forming of Spline Shafts [D]. Lanzhou: Lanzhou University of Technology, 2012.
4. Li Yongtang, Zhang Dawei, Fu Jianhua, Song Jianli. Unit Average Pressure in the External Spline Cold Rolling Process [J]. China Mechanical Engineering, 2007, 24: 2977–2980.
5. Li Zhigang. The New Equipment and Technology on Cross Wedge Rolling Precision Performing of Railway Axles [D]. Changchun: Jilin University, 2010.
6. Ying Fuqiang. Forming Mechanism, Simulation and Experimental Research on Cross Wedge Rolling of Involute Gear-shaft [D]. Shanghai: Shanghai University, 2007.
7. Liu Zhiqi, Song Jianli, Li Yongtang, Li Xudong. Analysis and Experimental Study on the Precision Cold Rolling Process of Involute Spline [J]. Journal of Mechanical Engineering, 2011, 47 (14): 32–38.
8. http://www.qdqs.com.cn [OL].

MIX
Papier aus verantwortungsvollen Quellen
Paper from responsible sources
FSC® C105338

If you have any concerns about our products,
you can contact us on
ProductSafety@springernature.com

In case Publisher is established outside the EU,
the EU authorized representative is:
Springer Nature Customer Service Center GmbH
Europaplatz 3, 69115 Heidelberg, Germany

Printed by Libri Plureos GmbH
in Hamburg, Germany